EAI/Springer Innovations in Communication and Computing

Series editor
Imrich Chlamtac, European Alliance for Innovation, Ghent, Belgium

Editor's Note

The impact of information technologies is creating a new world yet not fully understood. The extent and speed of economic, life style and social changes already perceived in everyday life is hard to estimate without understanding the technological driving forces behind it. This series presents contributed volumes featuring the latest research and development in the various information engineering technologies that play a key role in this process.

The range of topics, focusing primarily on communications and computing engineering include, but are not limited to, wireless networks; mobile communication; design and learning; gaming; interaction; e-health and pervasive healthcare; energy management; smart grids; internet of things; cognitive radio networks; computation; cloud computing; ubiquitous connectivity, and in mode general smart living, smart cities, Internet of Things and more. The series publishes a combination of expanded papers selected from hosted and sponsored European Alliance for Innovation (EAI) conferences that present cutting edge, global research as well as provide new perspectives on traditional related engineering fields. This content, complemented with open calls for contribution of book titles and individual chapters, together maintain Springer's and EAI's high standards of academic excellence. The audience for the books consists of researchers, industry professionals, advanced level students as well as practitioners in related fields of activity include information and communication specialists, security experts, economists, urban planners, doctors, and in general representatives in all those walks of life affected ad contributing to the information revolution.

Indexing: This series is indexed in Scopus, Ei Compendex, and zbMATH.

About EAI

EAI is a grassroots member organization initiated through cooperation between businesses, public, private and government organizations to address the global challenges of Europe's future competitiveness and link the European Research community with its counterparts around the globe. EAI reaches out to hundreds of thousands of individual subscribers on all continents and collaborates with an institutional member base including Fortune 500 companies, government organizations, and educational institutions, provide a free research and innovation platform.

Through its open free membership model EAI promotes a new research and innovation culture based on collaboration, connectivity and recognition of excellence by community.

More information about this series at http://www.springer.com/series/15427

Umar Zakir Abdul Hamid • Fadi Al-Turjman
Editors

Towards Connected and Autonomous Vehicle Highways

Technical, Security and Social Challenges

Springer

EAI
RESEARCH MEETS INNOVATION

Editors
Umar Zakir Abdul Hamid
Team Lead (Autonomous Vehicle Planning & Control Algorithm Development)
Sensible 4 Oy
Espoo, Finland

Fadi Al-Turjman
Department of Artificial Intelligence Engineering
Research Center for AI and IoT
Near East University
Mersin 10, Turkey

ISSN 2522-8595 ISSN 2522-8609 (electronic)
EAI/Springer Innovations in Communication and Computing
ISBN 978-3-030-66041-3 ISBN 978-3-030-66042-0 (eBook)
https://doi.org/10.1007/978-3-030-66042-0

This Springer imprint is published by the registered company Springer Nature Switzerland AG
The registered company address is: Gewerbestrasse 11, 6330 Cham, Switzerland

'Circulate knowledge and teach the ignorant, for knowledge does not vanish except when it is kept secretly (to oneself).'

Preface and Acknowledgement

We are living in the era of the Fourth Industrial Revolution, where a lot of new technologies emerge. Most of these new inventions are identified as having the potential of disrupting not only the fields that they are based in but also the way human beings live. One of the most notable examples is the autonomous vehicle (AV). Studies have identified that autonomous vehicles are part of the bigger picture of future mobility, where the discussions revolve not only on how to make the 'vehicle move autonomously'. Reports have been made where the future of mobility is said to be composed of Autonomous, Connected, Electric and Shared (ACES) Vehicles. As such, to allow AV to be realised as a disruptive invention, the discussions should be not only on the technical aspects but also on how to bring the new technology to the masses. In the past decade, when the publicity circulating autonomous vehicles has been attracting a lot of discussions, not a few timelines have been created to bring the connected autonomous vehicles (CAV) in the open-road as early as 2020 as a product. Over time, this has been realised as bearing more challenges compared to what has been thought. It is also identified that the main challenges not only lie in technological but also social and security aspects, among many others.

This book is written based on this notion and our combined experiences in automated driving and smart city. The editors consist of practitioners and researchers who have worked extensively in the field of autonomous vehicles and smart city as well as connectivity. The book tries to provide comprehensive multi-angle discussions on the topics of fully connected and automated vehicle highway implementation in a single resource. It covers the current progress of the works towards autonomous vehicle highway development, which encompasses discussions on the technical, social, policy as well as security aspects of topics on connected and autonomous vehicles (CAV). This in turn will be beneficial to a vast number of readers who are interested in the topics of CAV, automated highway and smart city, among many others.

The 12 chapters of this book are divided into four main sections. In the first section, an introductory chapter is written by the editors of the book to provide a very brief overview on the connected and autonomous vehicles and the importance of this book for public discussions. In the second section, 'Connected and Autonomous Vehicle (CAV) as Part of Future Mobility', discussions on the implementation of CAV as part of future mobility solutions are given. In the third section, 'Policy, Cyber Security and Data Management of CAV', the authors discuss formulating a safe and secure implementation of CAV for end-users. This includes discussions of cybersecurity for connected cars, policymaking and data management for CAV. In the final section, 'CAV Developments and Experiments', state-of-the-art experiments done in the field are reported, where the challenge of testing, weather and connectivity towards CAV performance are discussed, among many others.

Summarising, for the full potential of CAV to provide an alternative and solution to the mobility sector which complements the current transportation system, it is a necessity to discuss the topic from a multi-angle point of view. Thus, it is hoped that this book attracts the attention on the importance of this topic not only of engineering practitioners of CAV, but also other fields such as legal, ethical and policymaking, among many others. The editors would like to thank all of the reviewers and authors involved in turning this work into reality.

Espoo, Finland Umar Zakir Abdul Hamid

Mersin, Turkey Fadi Al-Turjman

List of Reviewers

We would like to express our gratitude and appreciation to the following reviewers of the book:

- Vishrut Jain, Technische Universiteit Delft, Netherlands
- Mauro Bellone, Chalmers University of Technology, Sweden
- Lin Zhang, Tongji University, China
- Selim Solmaz, Virtual Vehicle Research Center, Graz, Austria
- Marlies Mischinger, Virtual Vehicle Research Center, Graz, Austria
- Anitha R., Sri Venkateswara College of Engineering, India
- Max Turner, Ethernovia, Netherlands
- Fakhrul Razi Ahmad Zakuan, Moovita, Malaysia
- Pratham Oza, Virginia Tech, USA

Contents

About the Editors

Umar Zakir Abdul Hamid A PhD holder, Umar Zakir Abdul Hamid is among the pioneering researchers in the autonomous vehicle (AV) field in Malaysia and was a member of Intelligent Drive Team of Vehicle System Engineering Research Group (Universiti Teknologi Malaysia). The team leads Advanced Driver Assistance Systems (ADAS) and AV-related research in Malaysia, where previous collaborators include Proton Holdings Berhad and Smart Mobility Research Center (Tokyo). Umar Zakir served as an AV Scientist with Moovita (Singapore & Malaysia) from 2017 to 2018, before embarking on a journey as Senior AV Research Engineer in Espoo, Finland. He is now the Team Lead Engineer for the Autonomous Vehicle Planning & Control Algorithm Development of Sensible 4. He is also one of the recipients of the Finnish Engineering Award 2020 for his contributions to the development of all-weather autonomous driving solution with Sensible 4.

Team Lead Engineer (Autonomous Vehicle Planning & Control Algorithm Development), Sensible 4 Oy, Espoo, Finalnd.

Fadi Al-Turjman received his PhD in computer science from Queen's University, Canada, in 2011. He is a full professor and a research center director at Near East University, Nicosia, Cyprus. Prof. Al-Turjman is the head of Artificial Intelligence Engineering Dept., and a leading authority in the areas of smart/intelligent IoT systems, wireless and mobile networks' architectures, protocols, deployments, and performance evaluation in Artificial Intelligence of Things (AIoT). His publication history spans over 350 SCI/E publications, in addition to numerous keynotes and plenary talks at flagship venues. He has authored and edited more than 40 books about cognition, security, and wireless sensor networks' deployments in smart IoT environments, which have been published by well-reputed publishers such as Taylor and Francis, Elsevier, IET, and Springer. He has received several recognitions and best papers' awards at top international conferences. He also received the prestigious *Best Research Paper Award* from Elsevier Computer Communications Journal for the period 2015–2018, in addition to the *Top Researcher Award* for 2018 at Antalya Bilim University, Turkey. Prof. Al-Turjman has led a number of international symposia and workshops in flagship communication society conferences. Currently, he serves as book series editor and the lead guest/associate editor for several top tier journals, including the *IEEE Communications Surveys and Tutorials* (IF 23.9) and the Elsevier Sustainable Cities and Society (IF 5.7), in addition to organizing international conferences and symposiums on the most up to date research topics in AI and IoT.

Part I
Introduction

Introductory Chapter: A Brief Overview of Autonomous, Connected, Electric and Shared (ACES) Vehicles as the Future of Mobility

Umar Zakir Abdul Hamid and Fadi Al-Turjman

1 Introduction

Etymologically, transportation refers to the process to move people or goods to a different place [1]. Throughout the years, the way of human movement from one place to another has evolved, and it has dramatically changed in the last 100 years [2, 3]. Impact of road vehicles technology has spurred across many sectors and have benefitted not only the road use-cases but also other applications, which include freight and agriculture sectors [4]. As transportation is an ever-evolving field, it is not a surprise when it is now an essential component of the Fourth Industrial Revolution discussion [5]. Advancement on the 'future mobility' is assumed to possess the potential to disrupt how human beings are travelling and dwelling in the near future.

2 Future Mobility

According to the herein-cited study [6], 'Future Mobility' refers to the method of transportations that will be used by humans in the next few years. In the same report which is prepared by McKinsey & Company, the disruption of mobility sectors

U. Z. A. Hamid (✉)
Team Lead (Autonomous Vehicle Planning & Control Algorithm Development), Sensible 4 Oy, Espoo, Finland
e-mail: umar.hamid@sensible4.fi

F. Al-Turjman
Department of Artificial Intelligence Engineering Research Center for AI and IoT, Near East University, Mersin 10, Turkey

U. Z. A. Hamid, F. Al-Turjman (eds.), *Towards Connected and Autonomous Vehicle Highways*, EAI/Springer Innovations in Communication and Computing,
https://doi.org/10.1007/978-3-030-66042-0_1

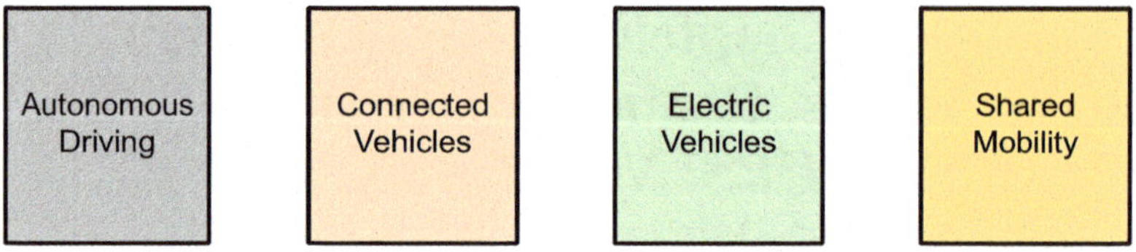

Fig. 1 Collectively, ACES is said to be the future of mobility in the following years [6]

is expected to be spearheaded by four elements, i.e. Autonomous, Connected and Electric vehicles, as well as Shared Mobility, which can be abbreviated as ACES (Fig. 1).

To amplify the notion, the idea of ACES as the disruptive factor of the mobility sector in the future has consistently been highlighted in the KPMG's Autonomous Vehicles Readiness Index in three consecutive years (2018–2020) [7–9]. Subsequently, the highlights stimulate not only the discussions of ACES developments but also the studies of the potential decline of the personal vehicle possession, which eventually will potentially allow the replacement of the areas used for building highways for other purposes and thus reducing the pollution caused by the vehicles [10].

In the next section, details of each ACES component are briefly described in their subsections. This is established to aid the perusal of the readers who might be new to this topic. As the scope of the book is on the wide implementation of CAV highway and its challenges, the terms CAV and ACES might be used intermittently.

2.1 Autonomous Vehicles

Autonomous vehicles, depending on the level of vehicle automation, will reduce or eliminate the participation of human drivers in the driving process [11]. One of the main factors that motivate the research on AV is the potential advantages that it will cause to reduce human error and eventually decrease the number of road fatalities [12]. As of 2020, the AV field has consistently grown in the number of research and development outputs. To propel the field, a lot of higher education institutions have introduced specialized Master of Science courses for Autonomous Vehicles in order to nurture the talents, which include Cranfield University and University of Warwick [13, 14].

However, contrary to the 'overly ambitious reports' that occurred around several years earlier (where some of the reports even anticipated AV to be production-scale deployment-ready in as early as 2020), practitioners and academia have softened and re-calibrated their expectations towards more scalable expectations [15].

This has been done upon the realization that the wide implementation of AV is not a standalone discussion, but instead, it is an interdisciplinary effort. However, despite the extension of the timeline, progress is still being made. This is shown with the acquisition of several startups by big technology corporations, which include

Zoox purchase by Amazon in mid-2020 [16]. However, main challenges such as AV implementations in mixed-traffic scenarios are still the challenges which need to be addressed.

2.2 *Connected Vehicles*

The term AV usually comprises 'decentralized autonomous platforms' which is not connected. However, to be fully implemented in real traffic, the AVs need to be interacting with each other. Thus, the connectivity between vehicles is an important topic that needs to be addressed to enable connected and automated vehicles ecosystems. A connected car concept revolves around the ability of the vehicle to be connected to other platforms.

In general, vehicle-to-everything technology (commonly known as vehicle-to-x or V2X) enables the said communication by allowing the host vehicle to be connected to other platforms with certain protocol. As the topic is rapidly evolving, the current discussions also involve Internet-of-Vehicle, 5G internet and the latest, 6G connection [17–19]. In the context of ACES, a secure and robust connectivity between vehicles will enhance the automated driving experience and enable vehicle platooning as well as emergency remote intervention. Vehicle connectivity has seen increasing interests from major companies, with Ericsson, Nokia and Telia are among the examples doing the research on the said topics for the use case of CAV [20–22]. However, challenges remain in the topics of cybersecurity, data management as well as network latency, among many others [23].

2.3 *Electric Vehicles*

The pollution caused by conventional vehicle engine technology such as internal combustion engine (ICE) motivated the study and development of electric vehicles (EV) [24]. Generally, the most common types of electric vehicles include battery electric vehicles (BEV), fuel cell electric vehicles (FCEV) and hybrid electric vehicles (HEV) [25]. With the awareness of global warming that has increased over the years, automakers have realized the potential of EV as one of the solutions for future mobility [26]. Besides, with the electrification of road vehicles, it will also leverage and expedite the CAV implementations.

Among notable recent developments in the automotive industry with regard to EV is when the multinational technology company Amazon invested in Rivian [27]. Besides, several companies such as Tesla and Canoo are among the well-known companies in the EV sectors [28]. However, despite the wide reputation, the coverage of charging stations for EV is still not wide. To be able to be scaled for wide mass production, among identified issues that the automakers and practitioners need to address for the EV topics are driving range, competitive price (to allow scalable productions) and lack of charging stations infrastructures, among many others [29]. Summarizing, EV is a crucial part of CAV as it will aid in providing better computational devices and validation platforms for the CAV.

2.4 Shared Mobility

Shared mobility is the use of a shared transportation platform as a means to move by different people. Among examples of shared mobility are car sharing and ride sharing, which introduced companies such as Uber, Grab and Gojek [30, 31]. However, among challenges for shared mobility are data management, passenger security and safety, user acceptance, crime prevention policy as well as vandalism. For example, despite having a relative size of success in certain European countries, bike-sharing faces vandalism challenges in some parts in Southeast Asia [32–34]. Thus, in addition to the said challenges, education and policy-making should also be widely enforced to improve the mass deployment of shared mobility. These require collaboration with interdisciplinary bodies such as government, academician and traffic enforcement.

Furthermore, in relation to the ACES concept, it is apparent the importance of shared mobility in pushing the mass-deployment of CAV. Sensible 4, a Finnish-based autonomous vehicle software company is among the companies which widely embraced the shared mobility concept into their CAV implementation, where they have publicly worked with several companies in the said topic [35, 36].

3 Challenges for Wide-Scale CAV Implementations

Based on the discussions, it is clear that a wide-scale ACES implementation will bring a lot of benefits to humanity. However, challenges remain. To recap, the hurdles that need to be solved are safety, cybersecurity, infrastructure, user acceptance, technological aspects and network issues, among many others. This, in return, requires collaboration between various organizations from different fields and not only automotive practitioners.

4 Conclusions and Expectations

In this chapter, the editors of this book try to provide some introductory ideas on ACES to the readers. This is to guide the new readers before reading the subsequent chapters. As has been mentioned, for the wide-scale implementation of connected and autonomous vehicles which will eventually replace the conventional road vehicles, a lot of challenges need to be addressed.

This book is written not to provide solutions to all of the known problems of this topic, but instead, it is aimed to serve as a catalyst that will hopefully trigger more interdisciplinary multi-angle discussions that will eventually enable mass deployment of ACES, subsequently disrupting the transportation sectors.

Acknowledgements The editors would like to thank all of the authors and reviewers that have contributed for the book publication and Eliška Vlčková from the European Alliance for Innovation (EAI) for the tremendous assistance throughout the editorial process.

References

1. Transportation | Definition of Transportation by Merriam-Webster, Transportation | Definition of Transportation by Merriam-Webster (2020). [Online] Available at: https://www.merriam-webster.com/dictionary/transportation. Accessed 13 July 2020
2. Y. EL Mokaddem, F. Jawab, *Research Advances in Intelligent Transportations Systems*
3. D. Sperling, Electric vehicles: Approaching the tipping point, in *Three Revolutions*, (Island Press, Washington, DC, 2018), pp. 21–54
4. M. Jaller, C. Otero-Palencia, A. Pahwa, Automation, electrification, and shared mobility in urban freight: Opportunities and challenges. Transp. Res. Proced. **46**, 13–20 (2020)
5. K. Schwab, *The Fourth Industrial Revolution* (2017)
6. McKinsey & Company, The future of mobility 2020. (McKinsey, 2020), [Online] Available at: https://www.mckinsey.com/industries/automotive-and-assembly/our-insights/the-future-of-mobility-is-at-our-doorstep. Accessed 13 July 2020
7. R. Threlfall, *Autonomous Vehicles Readiness Index* (Klynveld Peat Marwick Goerdeler (KPMG) International, 2018)
8. KPMG, Autonomous Vehicles Readiness Index – KPMG Global (2019), [Online] Available at: https://home.kpmg/xx/en/home/insights/2019/02/2019-autonomous-vehicles-readiness-index.html
9. R. Threlfall, Autonomous Vehicles Readiness Index (AVRI) – KPMG Global (2020), [Online] KPMG. Available at: https://home.kpmg/xx/en/home/insights/2020/06/autonomous-vehicles-readiness-index.html#download. Accessed 13 July 2020.
10. U.Z.A. Hamid, K. Pushkin, H. Zamzuri, D. Gueraiche, M.A.A. Rahman, Current collision mitigation technologies for advanced driver assistance systems–a survey. Perintis eJournal **6**(2) (2016)
11. Sae.org, SAE J3016 automated-driving graphic (2019), [Online] Available at: https://www.sae.org/news/2019/01/sae-updates-j3016-automated-driving-graphic
12. X. He, Y. Liu, C. Lv, X. Ji, Y. Liu, Emergency steering control of autonomous vehicle for collision avoidance and stabilisation. Veh. Syst. Dyn. **57**(8), 1163–1187 (2019)
13. www.cranfield.ac.uk, Connected and Autonomous Vehicle Engineering (Automotive) MSc (n.d.), [Online] Available at: https://www.cranfield.ac.uk/courses/taught/connected-and-autonomous-vehicle-engineering-automotive. Accessed 13 July 2020
14. warwick.ac.uk, MSc Smart, Connected and Autonomous Vehicles (n.d.), [Online] Available at: https://warwick.ac.uk/fac/sci/wmg/education/wmgmasters/courses/smart_connected_and_autonomous_vehicles/. Accessed 13 July 2020
15. D. Sperling, E. van der Meer, S. Pike, Vehicle automation: Our best shot at a transportation do-over? in *Three Revolutions*, (Island Press, Washington, DC, 2018), pp. 77–108
16. L. Columbus, Using Patent Analytics To See Why Amazon Bought Zoox. (Forbes, n.d.), [Online] Available at: https://www.forbes.com/sites/louiscolumbus/2020/07/12/using-patent-analytics-to-see-why-amazon-bought-zoox/. Accessed 13 July 2020

17. U.Z.A. Hamid, H. Zamzuri, D.K. Limbu, Internet of vehicle (IoV) applications in expediting the implementation of smart highway of autonomous vehicle: A survey, in *Performability in Internet of Things*, (Springer, Cham, 2019), pp. 137–157
18. C.R. Storck, F. Duarte-Figueiredo, A survey of 5G technology evolution, standards, and infrastructure associated with vehicle-to-everything communications by internet of vehicles. IEEE Access (2020)
19. M. Giordani, M. Polese, M. Mezzavilla, S. Rangan, M. Zorzi, Toward 6g networks: Use cases and technologies. IEEE Commun. Mag. **58**(3), 55–61 (2020)
20. Ericsson.com, A connected car platform for all vehicles (2018), [Online] Available at: https://www.ericsson.com/en/internet-of-things/automotive/connected-vehicle-cloud. Accessed 13 July 2020
21. LuxTurrim5G, Sensible 4 takes all-weather autonomous driving to a new level (n.d.). [Online] Available at: https://www.luxturrim5g.com/new-blog/2020/3/27/sensible-4-takes-all-weather-autonomous-driving-to-a-new-level. Accessed 13 July 2020
22. www.telia.fi, "Look, no hands" – 5G comes in support of self-driving cars | Article | Telia Business (n.d.), [Online] Available at: https://www.telia.fi/business/article/5g-comes-in-support-of-self-driving-cars. Accessed 13 July 2020
23. I. Kotilainen, C. Händel, U.Z.A. Hamid, H. Santamala, A. Schirokoff, M. Autioniemi, et al., Arctic challenge project's final report: Road tranport automation in snowy and icy conditions. *Väyläviraston tutkimuksia* (2019)
24. I. Burch, J. Gilchrist, *Survey of Global Activity to Phase Out Internal Combustion Engine Vehicles* (Center of Climate Protection, Santa Rosa, 2018)
25. I.C. Setiawan, Policy simulation of electricity-based vehicle utilization in Indonesia (electrified vehicle-HEV, PHEV, BEV and FCEV). Automot. Exp. **2**(1), 1–8 (2019)
26. K. Degirmenci, M.H. Breitner, Consumer purchase intentions for electric vehicles: Is green more important than price and range? Transp. Res. D Transp. Environ. **51**, 250–260 (2017)
27. P. Eavis, Rivian Gets $1.3 Billion Investment in Electric Truck Venture. The New York Times (2019). [Online] 23 Dec. Available at: https://www.nytimes.com/2019/12/23/business/rivian-truck-investment.html. Accessed 13 July 2020
28. Beiker, S. and Burgelman, R.A., 2020. The Future of the Automated Mobility Industry: A Strategic Management Perspective.
29. F. Un-Noor, S. Padmanaban, L. Mihet-Popa, M.N. Mollah, E. Hossain, A comprehensive study of key electric vehicle (EV) components, technologies, challenges, impacts, and future direction of development. Energies **10**(8), 1217 (2017)
30. E.C.X. Aw, N.K. Basha, S.I. Ng, M. Sambasivan, To grab or not to grab? The role of trust and perceived value in on-demand ridesharing services. Asia Pac. J. Market. Logist. **31**(5), 1442–1465 (2019)
31. R. Anggriawan, E-hailing transportation and the issue of competition in Indonesia. Indones. Comp. Law Rev. **2**(1) (2020)
32. F. Ciari, H. Becker, How disruptive can shared mobility be? A scenario-based evaluation of shared mobility systems implemented at large scale, in *Disrupting Mobility*, (Springer, Cham, 2017), pp. 51–63
33. G. Laporte, F. Meunier, R.W. Calvo, Shared mobility systems: An updated survey. Ann. Oper. Res. **271**(1), 105–126 (2018)
34. J. Burrieza, New Mobility Options and Urban Mobility Challenges and Opportunities for Transport Planning and Modelling Date -ASAP 24pt, regular D2.1 New Mobility Options and Urban Mobility: Challenges and Opportunities for Transport Planning and Modelling (n.d.). [Online] Available at: https://h2020-momentum.eu/wp-content/uploads/2020/01/MOMENTUM-D2.1-New-Mobility-Options-and-Urban-Mobility.pdf. Accessed 13 July 2020
35. Autonomous Vehicle International, Sensible 4 to test all-weather software in Oslo (2020). [Online] Available at: https://www.autonomousvehicleinternational.com/news/testing/sensible-4-to-test-all-weather-software-in-oslo.html. Accessed 13 July 2020
36. Forum Virium Helsinki, Four consortia enter the prototyping and lab testing phase of the FABULOS pre-commercial procurement for self-driving shuttle services (2019). [Online] Available at: https://forumvirium.fi/en/fabulos-self-driving-bus/

Part II
Connected and Autonomous Vehicle (CAV) as Part of Future Mobility

Cut-ins in Truck Platoons: Modeling Loss of Fuel Savings

Alexey Voronov, Jonas Andersson, and Cristofer Englund

1 Introduction

Platooning is a technology that allows vehicles to drive with short inter-vehicular distance to reduce aerodynamic drag and thus save fuel [1]. As a consequence, when the distance between vehicles decreases, more vehicles can be accommodated on the roads leading to improved road utilization. Platooning is enabled by vehicle automation and vehicle-to-vehicle (V2V) communication. These technologies allow vehicles to collaborate on a short time scale [2], and thus, the control systems of the vehicles will be able to react simultaneously to maintain safety, since they are aware of each other's actions and intentions.

Platooning for fuel savings is previously studied in several projects [3]. Fuel savings for such systems is significant and in some situations reported to be more than 20% [4]. In this paper we investigate the potential loss of fuel savings due to interference from surrounding traffic and in particular cut-ins. The main reason for cut-ins is either vehicles that are joining a highway from an on-ramp or vehicles that suddenly need to exit the highway. Other scenarios include merging due to an upcoming roadwork or courtesy maneuvers from overtaking vehicles to let fast upcoming vehicles from behind to pass. The next section summarizes previous platooning projects and activities.

A. Voronov · J. Andersson
RISE Research Institutes of Sweden, Gothenburg, Sweden
e-mail: jonas.andersson@ri.se

C. Englund (✉)
RISE Research Institutes of Sweden, Gothenburg, Sweden

Center for Applied Intelligent Systems Research (CAISR), Halmstad University, Halmstad, Sweden
e-mail: cristofer.englund@ri.se

U. Z. A. Hamid, F. Al-Turjman (eds.), *Towards Connected and Autonomous Vehicle Highways*, EAI/Springer Innovations in Communication and Computing,
https://doi.org/10.1007/978-3-030-66042-0_2

2 Platooning Overview

Platooning dates back to the 1939 New York World's Fair as General Motors presented their vision of a driverless vehicle that is capable of keeping a safe distance from other vehicles maintained by automatic radio control. Over the years, vehicle automation technology has evolved, with various automated functions enabling both lateral and longitudinal vehicle control systems after the introduction of electronic control units (ECUs) in the 1970s.

The Program for European Traffic of Highest Efficiency and Unprecedented Safety (PROMETHEUS, 1988–1995) was the largest project so far within the field of automated driving. PROMETHEUS had the vision to create intelligent vehicles as a part of an overall intelligent road traffic system. The PROMETHEUS project was structured into several different parts, each dealing with a subtopic, e.g., driving assistance, communications, vehicle control, artificial intelligence, standards, scenarios, and use cases. Industrial partners carried out the research in three of the subtopics, whereas the other four represent basic research areas [5].

The California Partners for Advanced Transportation Technology (PATH) project stretches back until 1986. It is currently divided into six research focus areas: artificial intelligence/deep learning, connected and automated vehicles, connected corridors, human factors research, modal applications, and traffic operations. From a historically perspective, PATH was one of the pioneers in platooning and demonstrated an Automated Highway System (AHS) with automated longitudinal control of a four-car platoon already in 1994. Shortly thereafter, in 1997, at the National Automated Highway System Consortium Demo 1997, a platoon with eight vehicles was demonstrated. The demo proved high technical maturity while driving vehicles in a fully operational platoon with 20 cm RMS positioning error between the vehicles, contributing to a feeling as if the vehicles were connected with a mechanical shaft. In addition, the control system was at the same time able to maintain smooth ride quality for passenger comfort [6].

The COVER [7] project ran between 2006 and 2009 and was a 3-year FP6 project funded by the European Commission. The main focus concerned infrastructure efficiency. The project developed systems that enabled cooperation between different actors in the road network, e.g., vehicles and infrastructure, aiming to achieve intelligent vehicle control, thereby increasing transportation efficiency. One example proposed was an application for advanced cruise assist for highway systems and truck platooning in order to safely and efficiently handle queues and congestion.

The SARTRE project [8], which ran between 2010 and 2012 and was co-funded from the European Commission within FP7, aimed at developing strategies and technologies for allowing vehicle platoons to operate in normal public traffic while significantly reducing the environmental impact and increasing traffic safety and comfort. Within SARTRE, highway platooning with automated vehicles utilizing lateral and longitudinal control was demonstrated in 2012. With the help of V2V communication, local vehicle signals such as speed and sensor data could be shared among the vehicles in the platoon. The control algorithms of the SARTRE platooning system were highly integrated with the vehicles' own control system and can be thought of as a distributed control system. The platooning vehicles were programmed to maintain a safe distance to the preceding vehicle while at the same time following the trajectory of the lead vehicle. The results from SARTRE clearly show the benefit of platooning with measured fuel savings of up to 20% for the platoon members. Savings were even observed for the first vehicle.

The Grand Cooperative Driving Challenge (GCDC) 2011 [9] was organized as a competition where the main goal was to accelerate the development, integration, demonstration, and deployment of cooperative mobility. In GCDC 2011, both highway and urban traffic scenarios were demonstrated. In the urban scenario, traffic coordination in a traffic light controlled intersection was demonstrated where two platoons in the same lane should join. In the highway part, it was demonstrated how traffic shockwaves, which are common on highways, can be attenuated. The prerequisites for being able to participate in the challenge were a number of safety measures that had to be implemented, and the vehicles needed to comply with the local traffic rules. They also had to implement a common communication and interaction protocol to allow safe platooning and intersection control. One of the unique properties of GCDC was that it was a collaboration event to explore automated and cooperative technology, and at the same time, it was a competition with a multi-vendor approach. The participants developed their vehicular control systems around different vehicle makes and types, and both heavy-duty and passenger vehicles took part in the challenge. In GCDC, the vehicles applied automatic longitudinal control, and the driver of the vehicle was responsible for the lateral control. In 2016, the second edition of GCDC was set up [10], this time taking collaboration to a new level by developing a new communication protocol, based on the IEEE 802.11p ITS-G5, which allowed negotiation [11]. GCDC 2016 was demonstrated in two traffic environments, (a) highway platooning and (b) cooperative intersection. In the platooning scenario, the vehicles were driving in two platoons, one in each lane, while approaching a road work where the left lane was closed. By using the interaction protocol [11], the vehicles could automatically open gap for each other and merge. The cooperative intersection scenario demonstrated how an urban intersection could benefit from cooperative vehicles collaborating while approaching an intersection. Besides maintaining traffic safety, the goal of the vehicles was to maintain motion while crossing the intersection and thereby increasing traffic throughput. The interaction protocol [12] allowed the vehicles to achieve the goals, and the scenario was successfully demonstrated during GCDC 2016.

The Japanese project Energy ITS ran between 2008 and 2012 and was financed by the Ministry of Economy, Trade and Industry in Japan. The main goal was to develop energy saving technology in order to reduce CO_2 emissions and other greenhouse gases by means of ITS technology. In the platooning part of the project, there were two tracks, (1) autonomous driving/platooning and (2) evaluation methods to measure the effectiveness of ITS in terms of energy savings. The project demonstrated three automated heavy-duty trucks driving in a platoon at

80 km/h with a gap of 4 m. The result showed that it was possible to reduce energy consumption by up to 15% with the help of the reduction of the aerodynamic drag [13, 14].

The CHAUFFEUR I & II were conducted between 1996 and 1998 and 2000 and 2003, respectively, and funded by the European Commission. CHAUFFEUR I was focused on developing a "tow bar" concept, connecting two consecutive vehicles in a platoon [15] using wireless communication along with an infrared camera system to maintain a safe distance between the trucks. This concept implies that only the first vehicle needs an active driver. CHAUFFEUR II was based on the first project and focused on realization of the platooning maneuvers along with interoperability development of the "tow bar" concept. Generally, the CHAUFFEUR concept can be described as a combination of adaptive cruise control and lane keeping.

In the KONVOI project that ran between 2005 and 2009, the aim was to realize and analyze the use of electronically regulated truck convoys. Besides the technical challenges and demonstrations, one of the goals was to examine what effect convoys have on regular traffic. The evaluation was made in both a driving simulator and in real experimental vehicles. The project developed five vehicles for the experiments with automated longitudinal and lateral control, V2V communication, and a human machine interface (HMI) for interacting with the vehicular system. The vehicles were tested and demonstrated on both test tracks and in real traffic during normal traffic conditions. In addition, they were used by road carrier companies where the functions were tested under realistic traffic and transportation conditions.

The European Truck Platooning Challenge (ETPC) was another event aiming at boosting the introduction of cooperative vehicles. ETPC took place in April 2016 and was initiated by the Dutch government. During the ETPC, six heavy-duty truck platoons were driving from their home base to the port of Rotterdam. Trucks from DAF, Daimler, Iveco, MAN Truck & Bus, Scania, and Volvo Group participated in the challenge and were assessing driver and road user behavior, user acceptance, usability and take up, interaction with other traffic participants and technical assessment, proof of concept, etc.

The Companion project ran between 2013 and 2016 and was co-funded by the European Commission [16]. The main objective of this project was to develop co-operative mobility technologies for supervised vehicle platooning in order to improve fuel efficiency and safety for goods transport. Thus, the focus was on the creation, coordination, and operation of platoons. The project also studied the regulation and standardization of coordinated platooning, together with its acceptance by both end users and the society.

Several commercial companies have tried out platooning in real traffic. For example, Volvo and Scania launched a national multi-brand platooning project in 2017 to test longitudinal control between two different truck brands in real traffic [17]. Together with two Singaporean government agencies, Scania has developed a four-truck platooning system, where only the first truck was driven by a driver. The concept was successfully demonstrated on closed test tracks in 2017 [17].

Recently, truck manufacturer MAN has performed tests with two vehicle platoons on Autobahn between München and Nürnberg which have driven a total of 39,000 km. The first truck is manually driven, whereas the other vehicle is driven automatically, though a driver is in the cabin. The distance between the trucks was 15–20 m, and the plan is to continue making tests with a 10 m distance between the trucks. In the experiments, lower fuel savings than expected was observed. This was mainly due to disturbances such as rain, road works, on-ramps, etc., which forced the drivers to dissolve the platoon [18]. Ford Otosan, a Turkish truck manufacturer co-owned by Ford Motor Company and Koç Holding, has together with AVL made initial tests with platoons in Turkey. The next step is to develop platoons with SAE Level 4 vehicles to realize hub-to-hub autonomous highway transports [19]. Hyundai Motor successfully demonstrated autonomous platooning on a Yeoju Smart Highway in simulated real-world traffic conditions. The trial was made with two trucks driving in 60 km/h and 16.7 m distance between the trucks [20].

Peloton Technology, Inc. is a company focusing solely on platooning technologies for heavy trucks. In their recent FHWA Exploratory Advanced Research project with the title Heavy Truck Cooperative Adaptive Cruise Control: Evaluation, Testing, and Stakeholder Engagement for Near Term Deployment, they have evaluated the commercial feasibility of driver-assisted truck platooning (DATP). DATP can be thought of as a form of cooperative adaptive cruise control (CACC) for heavy trucks, i.e., two-truck platoons. DATP uses V2V communication to achieve low-latency exchange of performance parameters between vehicles to improve freight efficiency, fleet efficiency, safety, and highway mobility as well as reduce emissions [21]. DATP also uses radar to maintain longitudinal distance between trucks, satellite positioning to remain in the lane, and human-machine interfaces to clearly notify the driver if the truck is in leading or following mode. Freight and fleet efficiency is highly dependent on the inter-vehicle gap, and a safe gap setting is dependent on factors such as engine horsepower, estimated mass of the vehicles in the platoon, estimated braking ability, ability to cool the engine with the available air flow, driver acceptance, public acceptance, traffic conditions, and weather conditions [21]. A recent paper from Peloton shows mature platooning technologies are developing, reflecting solutions to challenges they may have come across during their trials [22]. For example, they have control buttons for the driver to dissolve the platoon as well as automatic detection of cut-ins that trigger platoons to dissolve. Other challenges that are brought up are cybersecurity and functional safety, i.e., ISO 26262, which is a challenge as the vehicle may need to react on information obtained over an unreliable wireless communication link.

Platooning has a large potential to improve traffic capacity and efficiency. Despite the field operational tests described above, it has not yet reached the point of realistic deployment. The research and test of the concept of platooning is a very active area. Besides vehicle control, some of the tough requirements that platooning needs include reliable and robust vehicle communications, positioning with high accuracy, as well as the platooning operational logic and coordination [23]. Safety, particularly the coordination of braking maneuvers for trucks with different braking capacity, is a major challenge [24]. There is also a range of human factors related to challenges to avoid interruption of the platoon from the surrounding traffic, e.g., cut-ins [25]. Another challenge is the individual driver's/operator's understanding and interaction with the automation [24].

Among the research findings presented in [26] a platoon coordination methodology for a vehicle to join from the side of the platoon was developed. The coordination of the vehicle's maneuvers is strictly managed by the platoon leader. As the joining vehicle reports to the platoon leader that it is in position to join, the leader commands the other vehicles that are already in the platoon to open up a gap large enough for the merging vehicle to enter. Finally, the joining vehicle waits for a confirmation signal from the platoon leader and distinctly joins the platoon when instructed to do so.

The PATH project proposes a five-layered control architecture. The two highest layers, situated in the infrastructure, monitor and give directives to the different platoons. Control of the maneuvers of the individual vehicles is done by the currently active control law of the vehicle. Consequently, the control law may be altered depending on the position of the vehicle, its current state in the platoon, and its active mode of operation. The system requires timely information exchange between vehicles and infrastructure and is dependent on a well-defined vehicle behavior.

A third method of platooning was demonstrated during the GCDC 2011 where a non-centralized approach to forming the platoons was utilized. With this approach, any vehicle could take on any role in the platoon; thus there was no dedicated vehicle or driver who was acting as platoon leader. This was further elaborated upon to allow negotiation to, e.g., join a platoon or merge between two vehicles in the platoon [11, 27].

Distributed, i.e., non-centralized, platooning is a promising approach when autonomous vehicles become a reality. Thus, the platooning operation can be fulfilled in a completely distributed way through collaboration between the ego vehicle and its neighboring vehicles. Meanwhile, all vehicles within the platoon are aware about each other and have a synchronized platoon status through V2V.

Coordinating vehicles in the platoon maintains freight and fleet efficiency, safety, and highway mobility as well as reduce emissions. With the challenges described above in mind, this paper investigates the potential loss of fuel savings from interference by surrounding traffic and in particular cut-ins.

3 Background on Fuel Consumption

This section summarizes a well-established way to estimate fuel consumption in platooning trucks [28–31].

Fuel consumption for vehicles is proportional to the energy extracted from the fuel:

$$m_{\text{fuel}} = \frac{E_{\text{fuel}}}{c_p^{\text{diesel}}} \tag{1}$$

where E_{fuel} is the total energy stored in the fuel and c_p^{diesel} is the specific energy density of diesel (for diesel-operated vehicles; the same reasoning can be extended to other types of energy storage, e.g., chemical energy in batteries).

To compute the total fuel energy, consideration needs to be taken of how this energy is used. The following are four major ways energy is consumed:

- More than 50% is dissipated by the engine, encoded into the engine efficiency η (for modern diesel engines, engine efficiency is about 35–45%).
- Air drag $F_{\text{air}} = \frac{1}{2}\rho_a C_d A_f v^2$, where ρ_a is the air density, C_d is the drag coefficient, and A_f is the reference area.
- Rolling resistance $F_{\text{roll}} = m\,g\,c_r \cos\alpha$, where m is the vehicle mass, c_r is the rolling resistance coefficient, and α is the slope (gradient) of the road.
- Fuel energy is also used to overcome gravitational force on the uphill $F_g = m\,g \sin\alpha$, resulting in higher potential energy. On the downhill, this potential energy is converted back to kinetic energy (vehicle speed). However, to stay within speed limits, excess kinetic energy is irreversibly converted into waste heat by the friction of disc brakes.

Other energy consumers include gear box friction, air conditioning, and other auxiliary systems.

3.1 Fuel Consumption Reduction in Platoon

Fuel consumption reduction in platoons is mainly related to vehicle aerodynamics and air drag coefficient C_d. A general reference for aerodynamics of vehicles is the book *Aerodynamics of Road Vehicles* published by SAE [32]. The reduction in air drag coefficient is usually estimated either using *computation fluid dynamics* or in a wind tunnel [32–34]. Figure 1 illustrates typical reduction in air drag coefficient depending on position in the platoon and the headway distance. Fuel consumption reduction is sometimes measured directly for the complete set of vehicles, typically on a test track [33, 34]. Field tests on real roads and real traffic are generally challenging for estimating fuel consumption, since the traffic situations vary, as well as pavement properties, weather, wind direction, etc.

Since the literature provides fuel consumption reduction figures for a homogeneous platoon with varying gap, in what follows, we will assume that cutting-in vehicles are similar in shape to the platoon vehicles and that the air resistance increase is only due to the increased gap.

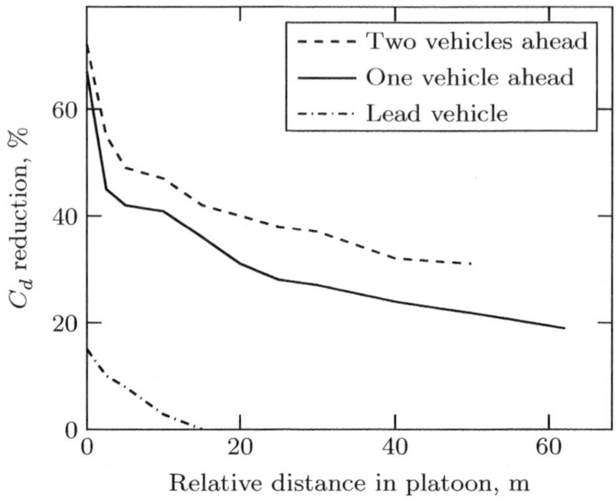

Fig. 1 Reduction or air drag coefficient C_d depending on the position in the platoon and headway distance. (Adapted from [32, 33, 35])

3.1.1 Example

As shown in Fig. 1, at a 10-m inter-vehicular distance, the air drag reduction is about 40%. Assuming that air resistance stands for approximately 1/3 of the fuel consumption for a 40-ton vehicle, and fuel consumption is about 30 L/100 km for an ordinary topographic profile, then air resistance consumes about 10 L/100 km. Driving at a 10 m distance saves 4 L/100 km. This is consistent with 10–15% total fuel savings for a complete vehicle.

3.1.2 Observations on Platoon Fuel-Savings for Vehicles of Different Mass

In the fuel consumption formula, both rolling resistance and gravity/braking depend on vehicle mass, while air resistance is independent of mass. Platooning only reduces air resistance, while rolling resistance remains unaffected. Thus, the relative fuel consumption reduction will be much larger for light vehicles where air resistance plays a larger role. Conversely, for heavy vehicles, where fuel consumption is dominated by the rolling resistance, gravity, and braking, the relative fuel consumption would be lower.

3.1.3 Platooning on a Non-flat Road

On uphill, the first vehicle often naturally slows down due to engine power limitations (normal behavior). The second vehicle in the platoon, however, has to use brakes, not to crash into the first one. Such braking dissipates energy and should not occur. In naturalistic fuel consumption studies, such behavior was linked to severe reduction in fuel savings [36].

Braking can be avoided by increasing the gap when the platoon is approaching uphill, optimizing trade-off between increased air resistance while minimizing the use of friction brakes [37, 38]. Such gap opening poses new questions: "How much does the air resistance increase?" "Does the increased gap invite additional cut-ins?"

4 Cut-ins

At a cut-in, air resistance increases. Even if the vehicle cutting in is a truck with a shape similar to the original preceding vehicle, due to the absence of the V2V platoon communication, the safety distance has to be increased compared to the platoon operation. Moreover, if the shape of the vehicle cutting in is not platooning-optimized, like that of a passenger car, the air resistance is increased even further. Also, the truck behind the cutting-in vehicle will increase its distance to its platooning companion even further.

4.1 Loss of Fuel Savings

Let us say a cut-in removes all fuel savings. We introduce the following formula for estimating the loss of fuel savings:

$$\gamma_{\text{lost}} = \gamma_{\text{platoon}} \cdot \gamma_{\text{air}} \cdot \frac{n_{\text{cut}} \cdot d_{\text{cut}}}{100\,\text{km}} \tag{2}$$

where:

- γ_{lost} is the proportion of fuel lost due to cut-in,
- γ_{platoon} is the air drag reduction in percent, e.g., 40% at a gap of 10 m (see Fig. 1),
- γ_{air} is the proportion of fuel that is used to overcome air resistance, e.g., 1/3 for a 40-ton truck, depending on parameters of vehicle and environment,
- n_{cut} is the number of cut-ins per 100 km,
- $d_{\text{cut}} = v \cdot t_{\text{cut}}$ is the average length of a cut-in,
- v is the platoon speed,
- t_{cut} is the average duration of a cut-in.

The next section describes a detailed estimation of γ_{air} parameter.

4.2 Detailed Parameters Setting Estimation

One way to model fuel consumption is by considering the instantaneous total tank-to-wheel fuel efficiency over the driving mission. Integration can be performed either over time or over distance:

$$E_{\text{fuel}} = \int_{t_1}^{t_2} P_{\text{fuel}}\, dt = \int_{x_1}^{x_2} F_{\text{fuel}}\, dx \tag{3}$$

where $P_{\text{fuel}} = \frac{dE_{\text{fuel}}}{dt}$ is power, $F_{\text{fuel}} = \frac{dE_{\text{fuel}}}{dx}$ is force, t is time, and x is position.

Fuel energy (accounted for engine efficiency η) is converted to the force on the wheels, together with the brakes force, as follows:

$$F_{\text{wheel}} = \eta \cdot F_{\text{fuel}} - F_{\text{brk}}. \tag{4}$$

Force on the wheels is used to overcome resistance as described by the vehicle dynamics, which can be written as follows in the time domain:

$$m\dot{v} = F_{\text{wheel}} - F_{\text{air}} - F_{\text{roll}} - F_g \tag{5}$$

where:

- m is the equivalent mass that includes the actual vehicle mass and terms reflecting inertia of rotational components,
- v is the longitudinal speed,
- F_{wheel} is the total wheel force,
- $F_{\text{air}} = \frac{1}{2}\rho_a C_d A_f v^2$ is the aerodynamic drag,
  - ρ_a is the air density,
  - C_d is the drag coefficient,
  - A_f is the reference area,
- $F_{\text{roll}} = mgc_r \cos\alpha$ is the rolling resistance,
- $F_g = mg \sin\alpha$ is the gravitational force,
- α is the slope (gradient) of the road.

In space domain, the left-hand side of Eq. (5) can be rewritten as follows:

$$m\frac{dv}{dt} = m\frac{dx}{dt}\frac{dv}{dx} = mv\frac{dv}{dx} = \frac{d}{dx}(\frac{mv^2}{2}) = \frac{dE}{dx} \tag{6}$$

where $E = \frac{mv^2}{2}$ is the kinetic energy. The whole vehicle dynamics equation (5) becomes:

$$\frac{dE}{dx} = \eta\, F_{\text{fuel}} - F_{\text{brk}} - b_{\text{air}} \cdot E - F_{\text{roll}} - F_g \tag{7}$$

where $b_{\text{air}} = \frac{\rho_a C_d A_f}{m}$.

Then the fuel consumption can be calculated by combining equations (1), (3), and (7).

4.3 Note on Braking

If a cut-in is sudden, the platooning vehicle might be forced to brake heavily. For such braking, a mechanical friction brake might be used, which irreversibly dissipates energy, increasing fuel consumption even further. This might be avoided through using recuperating brakes on hybrid or fully electric vehicles. If only soft braking is required, it can be achieved instead by engine braking or even coasting, keeping parasitic energy dissipation minimal.

5 Data Collection: Interviews and Workshop with Truck Drivers

To estimate the parameters for the fuel savings calculations, and to make realistic calculations on the effect that cut-ins have on fuel consumption, we interviewed both regular truck drivers and truck drivers with experience from platooning (European Truck Platooning Challenge, ETPC). Eight interviews with truck drivers from Sweden and two from IVECO and DAF were made. In addition, a group of six truck drivers from the Transport Lab at Scania and four truck drivers from Volvo Trucks were interviewed. Moreover, six car drivers were interviewed to obtain the opinion from surrounding traffic. The results are summarized in Table 1, and together with assumptions listed in Table 2, they give rise to estimations in Table 3.

5.1 Input from ETPC

The discussions in the workshops with ETPC drivers identified that the following scenarios are the most affected by cut-ins:

- When driving on the highway approaching an on-ramp where other vehicles will join the highway next to the platoon.

Table 1 Summary on cut-in related information extracted from interviews

Scenario	Swedish roads	European roads
Number of cut-ins n_{cut}	3 per 100 km	10 per 100 km
Duration of cut-in d_{cut}	45 s	90 s
Fraction of cut-ins by trucks	5%	10%

Table 2 Assumptions used in this paper for estimating cut-in losses according to Eq. (2). The values are typical for a modern 40-ton truck

Vehicle mass	40 tons
Total fuel consumption m_{fuel}	30 L per 100 km
Proportion of fuel to overcome air drag	$\gamma_{\text{air}} = 30\%$
Platooning gap	10 m
Air drag reduction	$\gamma_{\text{platoon}} = 40\%$
Fuel savings from platooning	13%
Platoon speed	$v = 90$ km/h
Daily driving distance	500 km
Aerodynamic drag coefficient	$c_d A_f = 5.2$
Air density	$\rho_a = 1.184$ kg/m^3
Rolling resistance coefficient	$c_r = 0.005$

Table 3 Estimated cut-in losses according to Eq. (2) with assumptions in Tables 1 and 2

	Swedish roads	European roads
Percentage of time spent in a cut-in	3%	22%
Estimated cut-in losses γ_{lost}	0.5%	3%

- When driving on the highway approaching an off-ramp where overtaking vehicles suddenly want to exit.
- When driving on a highway (that may have oncoming traffic) and there are vehicles behind that want to overtake. Also, if a slow vehicle overtakes the platoon, a queue will form behind, creating an incentive for the slow vehicle to cut-in.
- When driving on a highway and one lane is closed for, e.g., roadwork, the merging vehicles should be aware of the platooning vehicles and that their inter-vehicular distance is short due to their platooning activity.

The workshops revealed that there are large regional differences in the behavior of surrounding traffic between countries. In Sweden, for example, other traffic seldom interferes with trucks that drive after each other. However, this is not the case in other countries in Europe, and the reasons for this are mainly because of the higher traffic density, there is simply not enough room on the roads and cars, and other trucks may need to cut in between trucks that drive close to each other. The other reason is that on-ramps south of the Nordic countries are shorter, typically 400 m in Sweden compared to 250 m Germany, making it harder to make timely entries on the highway when there is heavy traffic.

ETPC data show that cut-ins occur on highways (overtaking) 47%, on-ramps 44%, off-ramps (exits) 7%, and lane merge 2%. In the ETPC, the number of km between cut-ins was 164.2 km in Sweden, 75.5 km in Denmark, 21.9 km direction south of the Netherlands-Belgium, and 14.5 km direction north Belgium-the Netherlands. The mean time a car is between two trucks is 15 s.

It should be noted that ETPC lasted for only a few days and the trucks that generated this data drove one return trip from Gothenburg to Rotterdam; however, the numbers indicate that the traffic situation is completely different in Sweden compared to the Netherlands and Belgium. Most of the data was collected while driving with ACC with 1 s time headway resulting in a following distance of about 22 m. The drivers did not notice any particular behavior difference among the surrounding traffic when the time headway was changed to 1.3 s. It should also be noted that the data includes several experiments made during the ETPC, where the system was evaluated while escort cars made cut-ins.

5.2 *Input from Truck Drivers*

From the interviews with Swedish truck drivers who use adaptive cruise control with time headway of 2 s (approximately 45 m), we estimate that there are approximately 3 cut-ins/100 km on a regular Swedish highway, with each cut-in lasting for about 45 s. This corresponds to a fuel loss of 0.1 L/100 km. Assuming 500 km of driving per day, the total loss is about 0.75 L per day.

In the central European countries, e.g., Germany and the Netherlands, the number of cut-ins is higher due to the generally higher traffic density on the roads. Interview results from European drivers indicate about 10 cut-ins/100 km. This is in line with naturalistic driving data [39], where 4974 cut-ins were found in 44,000 km of driving data. Given the higher traffic density, we assume it is more difficult to leave the pocket between two trucks entailing that each cut-in lasts for about 90 s, and we obtain 0.6 L/100 km. Assuming 500 km of driving per day, the total loss is about 4.5 L per day.

6 Estimating the Effect of Cut-ins

To put those daily losses into perspective, consider that in 2016, Volvo AB sold 190,424 trucks [40, p. 89]. Assuming 250 working days a year, fuel loss of 0.5 L per vehicle per day gives about 24,000,000 L in lost fuel savings per year for a fleet comparable to the one sold by Volvo AB in a year.

In a study from the US Department of Transportation [39], it is concluded that for headway distances <30 m (<1.5 s), no cut-ins were observed when following another truck, and very few cut-ins were observed when following a light vehicle. In interviews with Swedish truck drivers in the ETPC, they report that when driving with a 22 m gap (1.0 s at 80 km/h), Swedish passenger cars avoid going in between the trucks. However, while driving in the Netherlands and Belgium, cut-ins were occurring despite the short distance. Clearly, there are differences between countries and regions in when, where, and how cut-ins occur. Whether the differences in behavior are due to contextual (traffic density, infrastructure design) or cultural factors remains unclear and is a subject for further research.

7 Discussion

The model to assess fuel savings reductions is based on the physical properties of a platooning scenario. To use the model, several assumptions have to be made. The assumptions made in our example are mainly taken from subjective assessments made by experienced drivers and empirical data from a small number of tests in real traffic. They give an estimate but are not as precise as one could wish for, indicating a need for further studies to collect data on real traffic behavior. Even though the figures in Sect. 5 only give very rough impression of the magnitude of possible future losses of fuel savings, the model can be valuable for use in future studies of cut-in behavior when additional data of high quality has been collected.

With the conservative estimate on the number of cut-ins and the duration of the interference, there is still a potential to increase fuel savings for trucks – if they are allowed to drive in a platoon without being disturbed by surrounding traffic.

8 Conclusion

We presented a method to estimate potential fuel loss due to cut-ins in platoons. The method is based on a model of vehicle fuel consumption in platoons. To estimate the potential fuel savings loss due to cut-ins, we used both interviews with truck drivers with experience from platooning in Europe and naturalistic data from the USA. The results show that platoons spend as much as 20% of their time with cut-ins on typical European roads, reducing fuel savings in platooning from 13% down to 10%. Consequently, by avoiding cut-ins, there is a positive environmental effect worth considering. Future work includes verification of the assumptions made through quantitative data collection. Another consideration for future work is the cultural difference, i.e., the frequency of cut-ins observed in the different parts of Europe, which may exist, and how this can be incorporated in the modelling of loss of fuel savings.

Acknowledgments This project was funded by Strategic Innovation Program Drive Sweden.

References

1. K.Y. Liang, J. Mårtensson, K.H. Johansson, Fuel-saving potentials of platooning evaluated through sparse heavy-duty vehicle position data, in *2014 IEEE Intelligent Vehicles Symposium Proceedings*, June 2014, pp. 1061–1068
2. M. Aramrattana, T. Larsson, J. Jansson, C. Englund, Dimensions of cooperative driving, its and automation, in *2015 IEEE Intelligent Vehicles Symposium (IV)*, June 2015, pp. 144–149
3. C. Bergenhem, S. Shladover, E. Coelingh, C. Englund, S. Tsugawa, Overview of platooning systems, in *Proceedings of the 19th ITS World Congress, Oct 22–26, Vienna* (2012)

4. S. Tsugawa, An overview on an automated truck platoon within the energy ITS project, in *IFAC Proceedings Volumes*, Vol. 46, no. 21, pp. 41–46 (2013). [Online]. Available http://www.sciencedirect.com/science/article/pii/S1474667016383409
5. W.J. Gillan, PROMETHEUS and DRIVE: their implications for traffic managers, in *Vehicle Navigation and Information Systems Conference* (1989), pp. 237–243
6. R. Rajesh, T. Han-Shue, L. Boon Kait, W.-B. Zhang, Demonstration of integrated longitudinal and lateral control for the operation of automated vehicles in platoons. IEEE Trans. Control Syst. Technol. **8**(4), 695–708 (2000)
7. J. Sun, Z.-H. Wu, G. Pan, Context-aware smart car: from model to prototype. J. Zhejiang Univ. Sci. A **10**(7), 1049–1059 (2009)
8. C. Bergenhem, H. Pettersson, E. Coelingh, C. Englund, S. Shladover, S. Tsugawa, M. Adolfsson, Overview of platooning systems, in *Proceedings of the 19th ITS World Congress*, October, Vienna (2012), pp. 1–7. [Online]. Available http://publications.lib.chalmers.se/publication/174621
9. E. van Nunen, M.R.J.A.E. Kwakkernaat, J. Ploeg, B.D. Netten, Cooperative competition for future mobility. IEEE Trans. Intell. Transp. Syst. **13**(3), 1018–1025 (2012). [Online]. Available http://ieeexplore.ieee.org/document/6226874/
10. C. Englund, L. Chen, J. Ploeg, E. Semsar-Kazerooni, A. Voronov, H.H. Bengtsson, J. Didoff, The grand cooperative driving challenge 2016: boosting the introduction of cooperative automated vehicles. IEEE Wirel. Commun. **23**(4), 146–152 (2016). [Online]. Available http://ieeexplore.ieee.org/document/7553038/
11. H. Bengtsson, L. Chen, A. Voronov, C. Englund, Interaction protocol for highway platoon merge, in *Intelligent Transportation Systems (ITSC), 2015 IEEE 18th International Conference on* (2015)
12. E.S. Kazerooni, J. Ploeg, Interaction protocols for cooperative merging and lane reduction scenarios, in *Intelligent Transportation Systems (ITSC), 2015 IEEE 18th International Conference on*, Sept 2015, pp. 1964–1970
13. T. Hanai, Intelligent transport systems. J. Soc. Autom. Eng. Japan **67** (2013). [Online]. Available https://www.jsae.or.jp/en/publications/yearbook_e/2013/docu/26_intelligent_transport_systems.pdf
14. S. Tsugawa, S. Kato, Energy ITS: another application of vehicular communications. IEEE Commun. Mag. **48**, 120–126 (2010)
15. T. Benz, et al., CHAUFFEUR. TR 1009. User safety and operational requirements, DaimlerChrysler and Community Research and Development Information Service (CORDIS), Technical Report (1999)
16. S. Eilers, J. Mårtensson, H. Pettersson, M. Pillado, D. Gallegos, M. Tobar, K.H. Johansson, X. Ma, T. Friedrichs, S.S. Borojeni, et al., Companion–towards co-operative platoon management of heavy-duty vehicles, in *2015 IEEE 18th International Conference on Intelligent Transportation Systems* (IEEE, 2015), pp. 1267–1273
17. J. Hellåker, J. Gunnarson, P. King, Drive Sweden: an update on Swedish automation activities, in *Road Vehicle Automation 5* (Springer, Cham, 2019), pp. 41–49
18. P. Maahn, M. Reichhardt, Platooning: MAN stellt Testergebnisse vor, 2019-11-26
19. C. Tara, Ford Otosan and AVL take next AV step, Autonomous Vehicle International, Nov 2019. [Online]. Available https://www.autonomousvehicleinternational.com/news/concept-vehicles/ford-otosan-and-avl-take-next-av-step.html
20. J. Cha, Hyundai conducted the company's first platooning of trailer trucks, Hyundai News Release, Nov 2019. [Online]. Available https://www.hyundai.news/uk/brand/hyundai-conducted-the-companys-first-platooning-of-trailer-trucks/
21. R. Bishop, D. Bevly, L. Humphreys, S. Boyd, D. Murray, Evaluation and testing of driver-assistive truck platooning: phase 2 final results. Transp. Res. Rec. J. Transp. Res. Board **2615**(1), 11–18 (2017). [Online]. Available https://doi.org/10.3141/2615-02
22. J.P. Switkes, R. McLane, S. Laws, M. Luckevich, Truck platooning: connectivity enabled, grounded in safety, properly tested, in *Automated Vehicles Symposium* (Springer, 2019), pp. 92–99

23. C. Bergenhem, Q. Huang, A. Benmimoun, Challenges of platooning on public motorways, in *Proceedings of the 17th ITS World Congress* (2010), pp. 1–12
24. J. Engström, R. Bishop, S.E. Shladover, M.C. Murphy, L. O'Rourke, T. Voege, B. Denaro, R. Demato, D. Demato, Deployment of automated trucking: challenges and opportunities, in *Road Vehicle Automation 5* (Springer, Cham, 2019), pp. 149–162
25. J. Andersson, C. Englund, A. Voronov, Study of communication needs in interaction between trucks and surrounding traffic in platooning. *Gothenburg* (2017)
26. M. McGurrin, Vehicle information exchange needs for mobility applications: version 2.0, FHWA-JPO-12-021, ITS Joint Program Office, Research and Innovative Technology Administration, U.S. Department of Transportation, Technical Report (2012)
27. C. Englund, L. Chen, J. Ploeg, E. Semsar-Kazerooni, A. Voronov, H.H. Bengtsson, J. Didoff, The grand cooperative driving challenge 2016: boosting the introduction of cooperative automated vehicles. IEEE Wirel. Commun. **23**(4), 146–152 (2016)
28. L. Guzzella, A. Sciarretta, *Vehicle Propulsion Systems* (Springer, Berlin/Heidelberg, 2013). [Online]. Available http://link.springer.com/10.1007/978-3-642-35913-2
29. L. Eriksson, L. Nielsen, *Modeling and Control of Engines and Drivelines* (Wiley, Chichester, 2014)
30. D. Crolla, *Automotive Engineering: Powertrain, Chassis System and Vehicle Body* (Butterworth-Heinemann, Burlington, 2009)
31. D. Crolla, B. Mashadi, *Vehicle Powertrain Systems: Integration and Optimization* (John Wiley & Sons Ltd, The Atrium, Southern Gate, Chichester, West Sussex, United Kingdom 2012)
32. T.C. Schuetz, *Aerodynamics of Road Vehicles*, 5th edn (SAE International, 2015). [Online]. Available http://books.sae.org/r-430/
33. C. Bonnet, H. Fritz, Fuel consumption reduction in a platoon: experimental results with two electronically coupled trucks at close spacing, in *Future Transportation Technology Conference & Exposition*, no. 724, 8 (2000). [Online]. Available http://papers.sae.org/2000-01-3056/
34. R. Bishop, D. Bevly, L. Humphreys, S. Boyd, D. Murray, Evaluation and testing of driver assistive truck platooning: final project results, in *ITS World Congress 2016* (2016)
35. A.A. Alam, A. Gattami, K.H. Johansson, An experimental study on the fuel reduction potential of heavy duty vehicle platooning, in *13th International IEEE Conference on Intelligent Transportation Systems* (IEEE, 2010), pp. 306–311. [Online]. Available http://ieeexplore.ieee.org/document/5625054/
36. A. Alam, Fuel-Efficient Heavy-Duty Vehicle Platooning (Ph.D. Thesis) (KTH, Stockholm, 2014). [Online]. Available http://kth.diva-portal.org/smash/record.jsf?searchId=1&pid=diva2:719084
37. N. Murgovski, B. Egardt, M. Nilsson, Cooperative energy management of automated vehicles. Control Eng. Pract. **57**, 84–98 (2016)
38. W.D. Herrera, N. Lidander, Lateral Control of Heavy Duty Vehicles in Platooning Using Model Predictive Control, Master's Thesis 2016:EX-038, Chalmers University of Technology (2016)
39. E. Nodine, A. Lam, M. Yanagisawa, W. Najm, Naturalistic study of truck following behavior. Transp. Res. Rec. J. Transp. Res. Board **2615**, 35–42 (2017). [Online]. Available http://trrjournalonline.trb.org/doi/10.3141/2615-05
40. Volvo Group, Annual and Sustainability Report 2016, Gothenburg, pp. 208, 2 (2017). [Online]. Available http://www.volvogroup.com/en-en/investors/reports-and-presentations.html

The City Adaptation to the Autonomous Vehicles Implementation: Reimagining the Dubai City of Tomorrow

Dalia Hafiz and Ismail Zohdy

1 Introduction: Architectural Approach to the New Smart City

One of the main emerging transportation technologies of the future is "autonomous vehicles" (or as they are usually called: self-driving, unmanned vehicles). Autonomous vehicle is set to be as life changing as the invention of the motor vehicle itself. Modern-day cities have been shaped largely by the mobility that is achieved through motor vehicles, providing transport services to people and goods alike and supported by major road networks worldwide [1].

Today, our vehicles are parked nearly 95% of the time. There are millions of parking spaces worldwide, which soon will not be needed when the new autonomous and connected vehicles technology takes place; these cars can function throughout the day to transport other passengers, they can travel and park beyond the city perimeters, or they can simply travel at a low speed with low fuel consumption. There will be fewer vehicles on the road. Besides, autonomous vehicles can travel very close to each other at high speeds without threatening humans, which means roads themselves can be narrower. This will free up a huge amount of land that is currently congested with vehicles. All these possibilities and iterations can have considerable implications on how cities and buildings are designed. Researchers and transportation professionals claimed that such technologies are the most profound shift in urban land use any architect, urban planner, transportation, and traffic

D. Hafiz (✉)
Al-Ghurair University, Dubai, UAE
e-mail: d.hafiz@agu.ac.ae

I. Zohdy
Roads and Transport Authority (RTA), Dubai, UAE

U. Z. A. Hamid, F. Al-Turjman (eds.), *Towards Connected and Autonomous Vehicle Highways*, EAI/Springer Innovations in Communication and Computing,
https://doi.org/10.1007/978-3-030-66042-0_3

professionals alive today have seen, arguably since the transition from horses to cars a century ago [2].

"We could take 90% of the cars off the road without any reduction in-transit time, while also reclaiming 20% more space inside cities currently occupied by parking and roads." By *Michael* Schmidt.

When closely observing the effect of AVs on the city, people and places are the two key aspects that architects and planners need to examine in the city.

1.1 People

Pedestrians need to feel safe and comfortable, but they also need to be entertained. This represents a very important role that architects, and urban planners must take. According to Steve Mouzon's theory of "walk appeal," holding that how far we will walk is all about what we experience along the way. Stores and businesses with street-level windows and architectural details were found to be the most encouraging walkability features in cities, while lined parking and vertical building lines discourage it, they increase feelings of discomfort, danger, and doubt [3, 4].

Transportation that prioritizes active transportation in walkable, bikeable communities oriented around mass transit. Several Utopian urban planning projects failed not because of their architecture, but because they forgot the most important aspect of any community: the people.

A very famous example of such failure is the radiant city (Ville Radieux), by the famous architect and urban planner Le Corbusier, where he imagined the city planning with a Cartesian grid separating cars, pedestrians, and commercial spaces. Le Corbusier saw the Radiant City's geometric layout, standardization, and repetitive towers as "the perfect form" of urbanism. His designs were based on his theory "A city made for speed is made for success" [5].

1.2 Spaces

To achieve successful spaces, architects need to make both vehicle- and human-friendly buildings with unique designs, while eliminating the empty spaces between them. Such a process can be done through transitional spaces or passages. Such spaces are essential links to sustainable mobility. They can be represented in tunnels, footbridges, escalators, urban funiculars, atriums, courtyards, or corridors that facilitate the transition between various transportation modes, or between urban ambiances.

These transitional spaces are neither classified as indoors nor outdoors and share properties with both indoor and outdoor space. They can incorporate landmarks and remarkable elements that enhance pedestrians' experience and wayfinding [6].

Consequently, the purpose of this chapter is to apply modern techniques of future cities on a use case (Dubai), taking into consideration the new mobility solutions (autonomous vehicles) while capturing the main aspects of successful cities and elements of livability for citizens.

2 Dubai: The Future City and 2030 Vision

Dubai has long been known for its futuristic innovations, and the city's plans to become a world-leading smart and sustainable city are only expanding. In its 2021 plan, the city aims to become one of the best-connected in terms of telecoms, tourism, utilities, education, real estate, public safety, healthcare, and transportation [7]. Also, Dubai is committed to implementing smart and self-driving transport. The goal is for 25% of all trips in Dubai to be smart and driverless by 2030 [8].

While autonomous vehicles are considered an important step towards the future of transportation, such a technology application may repeat previous urban planning mistakes of prioritizing traffic efficiency over walkability and community vitality.

In research conducted by MIT, the simulations demonstrated cities without traffic lights with cars that are easily traveling through an intersection when compared to traditional streets. However, there is a key component missing in such simulations: humans; the simulations did not show pedestrians, bus riders, or cyclists, while intersections represent the most walkable locations in the world [9].

Recent research on connected and autonomous vehicles (CAVs) mainly focuses on the vehicle's design, mechanism, and manufacturing process. Studies on the impact of CAVs are focused on the highway. Very little research examined the impact on buildings and urban design and how the city and buildings can adapt to accommodate CAVs, while soon, there will be a transition period where CAVs will be on the roads with human-driven cars [10].

3 Reimagining Downtown Dubai

The selected case study is located on Sheikh Zayed Road, the longest road in the Emirates; it stretches from Al-Silah in the Emirate of Abu Dhabi and ends in Ras al-Khaimah emirate, running roughly parallel to UAE's coastline along the Persian Gulf.

The case study is reimagination of a block on Sheikh Zayed road. It focuses on human interaction, space-making, and community in the city of Dubai, which may be more challenging with the hot summer weather of Dubai.

The key elements in the examined block are the new museum (Museum of The Future) which will be opened early 2020, a Metro station, and the existing commercial activities. From the literature review, there are *six* main pillars for the design of future cities to be considered: safety, comfort, interaction, and

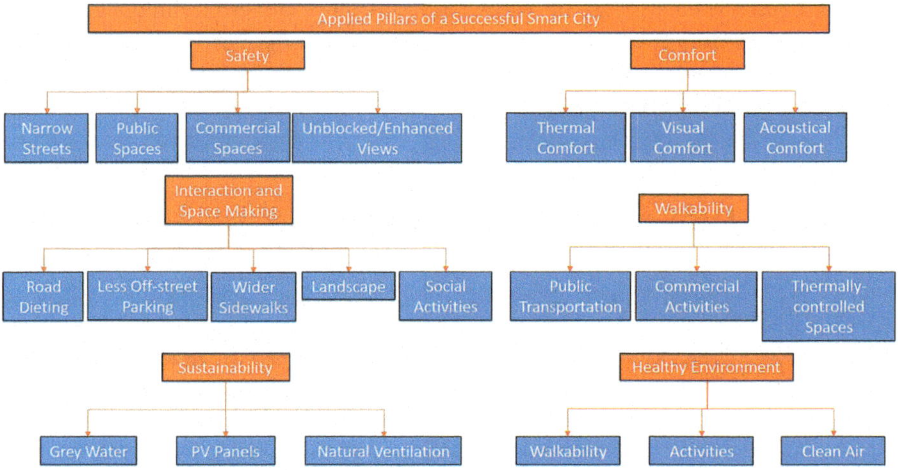

Fig. 1 Pillars of the successful smart city

space-making, walkability, sustainability, and a healthy environment [11, 12], as summarized in Fig. 1.

There are sets of guidelines for each pillar that needs to be maintained when applying AVs technology, as presented in the following sections.

3.1 Pillar 1: Safety

According to the United Nations' International Crime Victim Survey (ICVS), *"urbanization is the strongest predictor for crime and victimization"* [13]. Although several recent studies concluded that the UAE is one of the safest places to live, especially the city of Abu Dhabi and Dubai, it is important to consider the city urban safety with the country's great growth and multicultural approach. Previous research based on real cities showed that environmental and urban design are the best opportunities for safety. The AVs will increase safety with reduced parking spaces. The proposed design focuses on narrower streets (made possible with AVs) surrounded by public spaces, enhanced views to public transportation stops, lighting, and vegetation (Fig. 2).

3.2 Pillar 2: Comfort

Comfort is defined as "the condition of mind which expresses satisfaction with the environment" [14], and "when individuals have the psychological, social and

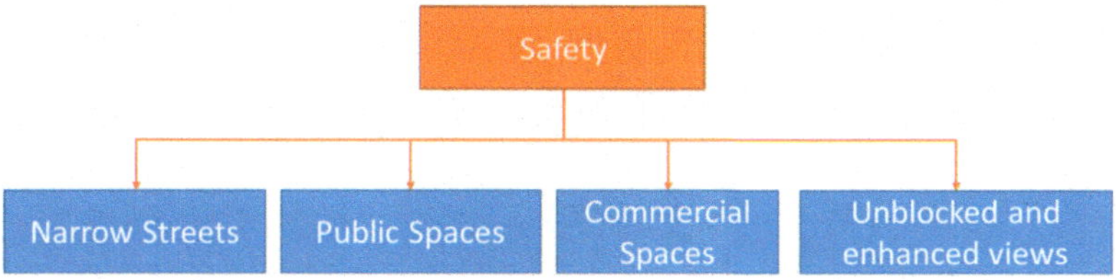

Fig. 2 Safety aspects of smart cities

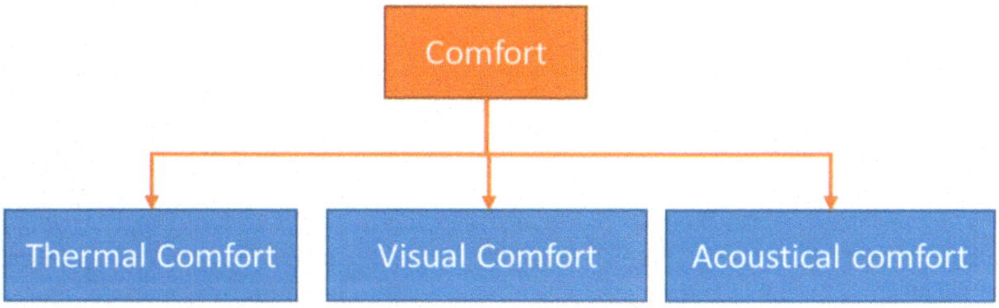

Fig. 3 Comfort criteria for smart cities

physical resources they need to meet a particular psychological, social and/or physical challenge" [15]

Although Comfort is very subjective, it can vary from one person to another based on their comfort level and individual experience. Architects, designers, and urban planners aim at reaching comfort levels to the maximum number of occupants and minimize as much discomfort as possible. Several factors can affect the level of comfort which are mainly related to human senses, levels of activities, and ambience. Generally, the key aspects of comfort can be subdivided into: visual comfort, thermal comfort, and acoustical comfort. The examined case study will focus on achieving such aspects using urban and architecture solutions, as shown in Fig. 3 and as will be discussed in the following subsections:

Thermal Comfort People achieve thermal comfort when they are satisfied with their thermal environment, which affects their ability to function effectively. There are several parameters other than air temperature that can affect thermal comfort (perceived temperature) including direct/diffused sun, wind, and humidity.

In the proposed case study, thermal comfort was enhanced in outdoor spaces using radiant cooling that can be used in the warm season (summer). In addition to thermally controlled shopping and recreational spaces (where energy is achieved from PV panels), it can be used in the hot and humid summer season (Figs. 4 and 5).

Visual Comfort This is to be able to perform the tasks safely and comfortably, in a visually pleasing environment. To achieve visual comfort, there must be an efficient quality and quantity of light [15]. In the selected case study, visual comfort was achieved through the proposal of maintaining natural lighting in the whole urban fabric, based on the theory of the Biophilia hypothesis, which suggests that

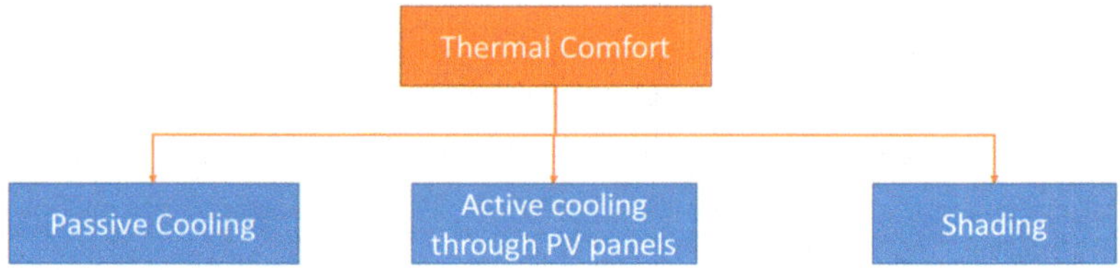

Fig. 4 Thermal comfort aspects of smart cities

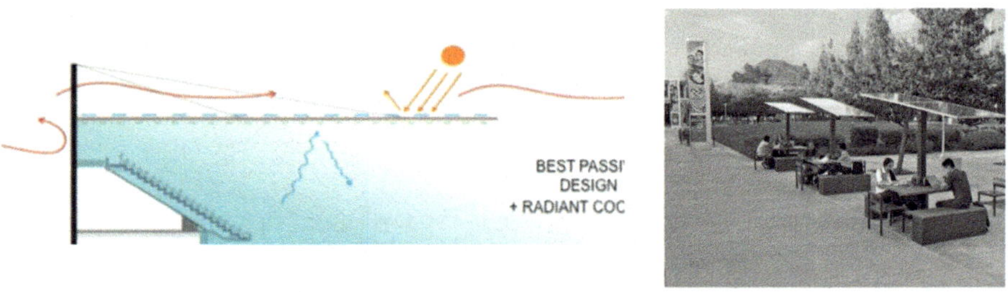

Fig. 5 Passive and active cooling

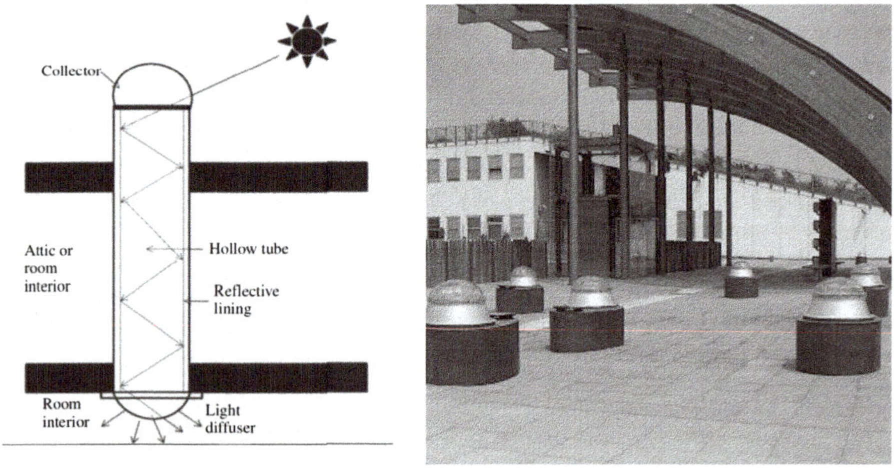

Fig. 6 Daylighting penetration for underground public and commercial spaces

"humans possess an innate tendency to seek connections with nature and other forms of life" [16]. While daylighting is mainly introduced in ground levels. Innovative systems are needed to bring natural daylight in underground spaces using light tubes and fiber optics, which can reach down to 40 meters underground to maintain daylighting to the deepest zones, Fig. 6.

Acoustical Comfort Acoustical comfort aims at reducing noise and disturbance that might affect the city citizens' health and quality of life. The proposed design

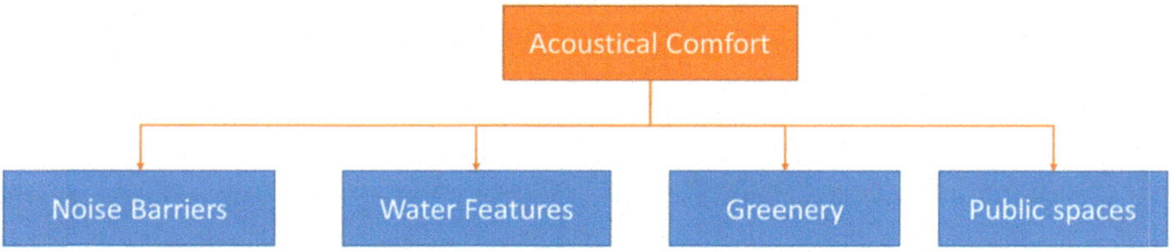

Fig. 7 Acoustical comfort aspects of smart cities

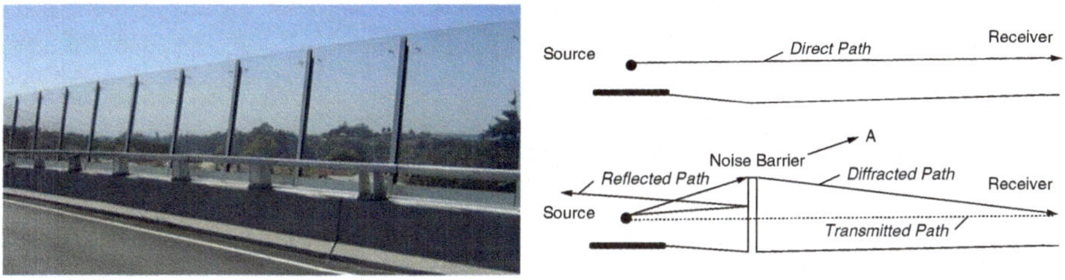

Fig. 8 Smart noise barriers

shifts to cut off traffic noise; it was developed to offer some solutions to minimize noise problems generated by the main-high speed-road. An innovative approach was used by involving a combination of elements including (1) water features, which previously showed high rates of success to minimize street noise, absorb pollution, and enhance the city phenomenon; (2) greenery, which can enhance both air and sound pollution; and (3) smart noise barrier, with advanced transparent materials that work as a sound barrier (e.g., Vinyl, Plexiglass, Polycarbonate) while maintaining the visual connection to the street, as shown in Fig. 8, and (4) local community and businesses, which provide community sounds, which also help in the process by holding public discussions [12]. The design of the continuous walkway has also included more attractive playgrounds, sports areas, and green spaces which have increased the level of lively sounds from human activities, [11] Figs. 7 and 8.

3.3 *Pillar 3: Interaction and Space-Making*

Research studies show that when all vehicles are fully automated, the capacity on freeways will double [17]. Previous research suggests the threshold for designing road diets to taking four-lane streets down to two-lane streets without impacting the capacity. Since the new autonomous vehicles are better able to stay in lanes, regular passenger lanes (none truck or busses) can be reduced to 8 or 9 feet (2.4 or 2.7meters) in width. These narrower lanes are also more pedestrian-friendly, as they are easier to cross. The autonomous vehicles road dieting is also represented

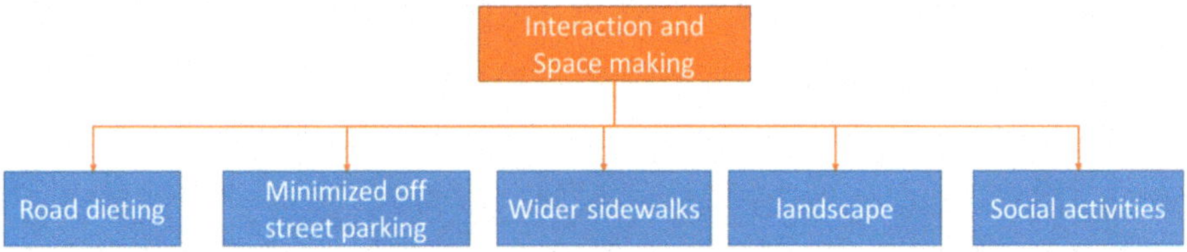

Fig. 9 Interaction and space-making aspects in smart cities

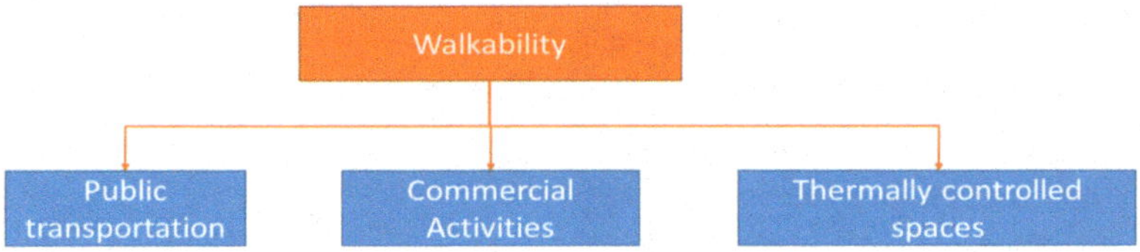

Fig. 10 Walkability aspects in smart cities

in different ways including lanes size and number reduction, bi directions, in addition to the unwanted need for off-street parking. Such a diet gives a vast opportunity in the proposed case study for other urban and social activities, including protected and improved bike lanes, wider sidewalks, street furniture, street vending, public art, fountains, streetscape and landscape, outdoor dining, jogging paths, and playgrounds [6, 12, 11].

To avoid previous urban planner's mistakes where cities were planned based on speed and isolation of activities, continuous visual and physical interaction was maintained between the pedestrians and the streets with AVs and motorcycles. Such interaction aims to minimize the "bubble" effect where activities are differentiated into motors and people, Fig. 9.

3.4 Pillar 4: Walkability

The transportation profession has long called for more walkable communities for many reasons: address traffic congestion, minimize sources of pollution, and enhance the residents' health with the increase in obesity and decreased level of human physical activities among the city residents [3]. Walkability was achieved in the case study through the connection with public transportation (Dubai Metro), cultural experience (The Museum of the Future), commercial experience (stores and restaurants), in addition to the underground thermally controlled, however, daylit spaces (Fig. 10).

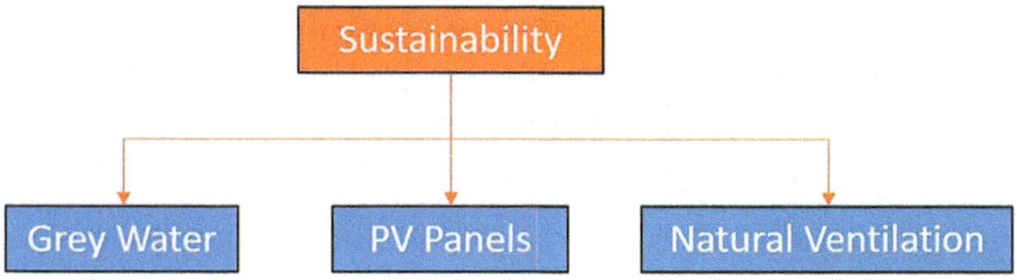

Fig. 11 Sustainability aspects of smart cities

3.5 Pillar 5: Sustainability

A sustainable city approach can be defined as "improving the quality of life in a city, including ecological, cultural, political, institutional, social and economic components without leaving a burden on the future generations" [18]. Recently, several features of sustainability, especially sustainable urban planning where climate change, clean air and water, renewable energy, and land use, are making a large impact on global sustainability, where architecture and transportation are two key ones. In the proposed case study, sustainable urban planning involves several disciplines including (1) recycled irrigation graywater (from nearby buildings), (2) shaded pedestrian paths/walkways covered with photovoltaic (PV) panels which produce enough energy to ventilate the walkways, (3) daylighting in the underground surface using light tubes. Besides, mixed-use developments, walkability, greenways, and open spaces are found to be key features of a healthy city and residents' well-being. Such solutions can help to improve urban city sustainability and create more efficient spaces that contribute more to the Dubai, 2030 sustainability vision, Fig. 11.

3.6 Pillar 6: Healthy Environment

In times of rapid urbanization, the health and well-being of citizens are increasingly recognized as a challenge and a remarkable amount of research on the potential associations between urban areas and health or well-being has been conducted [19].

Higher percentages of green-spaces are generally associated with better health-related aspects and increased quality of life. Fewer vacant land and architectural features that facilitate visual and social contacts are found to be associated with healthier environments. Attractive cities were often characterized by neighborhood safety, aesthetics, walkability, increased services, and access to cultural, shopping, or sport amenities [20].

4 Dubai: Sheikh Zayed Road: Block Reimagination

Dubai, like several other cities, is considered futuristically innovative. The city plans to become one of the world's smart and sustainable cities. One of its key goals is to convert 25% of all trips in the city of Dubai to be smart and driverless by 2030 [21, 8]. The selected case study is located on Sheikh Zayed Road, the longest road in the Emirates, stretching along the Persian Gulf.

The case study is reimagination of a block on Sheikh Zayed road. It focuses on human interaction, space-making, and community in the city of Dubai, which may be more challenging with the hot summer weather of Dubai.

The key elements in the examined block are the Museum of the Future, the Metro station, and the existing commercial activities. From the literature, the six previously examined main pillars for the design of future cities, which are safety, comfort, interaction, and space-making, walkability, and sustainability, and a healthy environment are explored.

While reimagining one arterial UAE road (Sheikh Zayed Road) and applying all the aspects of smart livable cities, especially while implementing AVs is important, the proposed examination aims to achieve the maximum levels of comfort, safety, sustainability, and well-being as shown in Table 1, Figs 12, 13, and 14.

5 Conclusion and Future Research

It is very important to understand how to integrate the AVs technology in the proposed city planning and transportation goal rather than completely embracing AVs or rejecting them entirely. Such a step is a key phase to fully benefit from the opportunities such technology can offer and also avoid fully embracing it without considering its consequences and effect on the urban environment and the life of the city residences while turning it into an automobile-dominated city with AVs.

Further investigations are needed in terms of the structural stability of the planned city, feasibility studies for the additional services provided, the safety of interaction between the pedestrians and the autonomous vehicles, the altered and modified infrastructure needed to support such deviation in the city structure and design.

Table 1 Existing vs imagined features of Sheikh Zayed road with autonomous vehicles

	Existing	Reimagined
1. *Safety*		
Streets		
Lane width (Avg)	3.6	2.7
Lane no (Avg – 2 ways)	12	8
Motorcycle lanes	N/A	1 each way
Bike lanes	N/A	2 each way (isolated from vehicles and motorcycles)
Public/commercial spaces		
	Sidewalks	Wider sidewalks, continuous commercial spaces, underground commercial spaces, seating areas
Views		
Unblocked and enhanced views	Buildings	Greenery, walkways, commercial spaces, transparent smart noise barriers, water surfaces
2. *Comfort*		
Thermal	Shaded/thermally controlled metro stations	Passive cooling, radiant cooling with summer heat, passive cooling (power from PV panels), also PV panels for shading, water features
Visual		
	N/A	Shaded walkways, daylighting underground (light tubes and fiber optics)
Acoustical		
	Solid noise barriers (in residential zones) smart noise barriers for the metro.	Smart noise barriers (streets), water features (fountains), commercial spaces (cafes and shops), greenery (trees), public spaces (walkways, bike lanes, and public seating)
3. *Interaction and space-making*		
	N/A	Commercial and public spaces, water features, AVs drop-off/pick-up zones, greenery
4. *Walkability*		
	Unprotected sidewalks	Shaded and cooled sidewalks, continuous commercial/social and cultural activities, shaded bike lanes
5. *Sustainability*		
	Partial graywater irrigation	PV panels for cooled walkways, daylighting, graywater recycling for irrigation, bike lanes, no off-road parking
6. *Healthy environment*		
	Some commercial activities	Enhanced walkability, fewer vehicles= less emissions, continuous access to activities, thermally controlled outdoor spaces

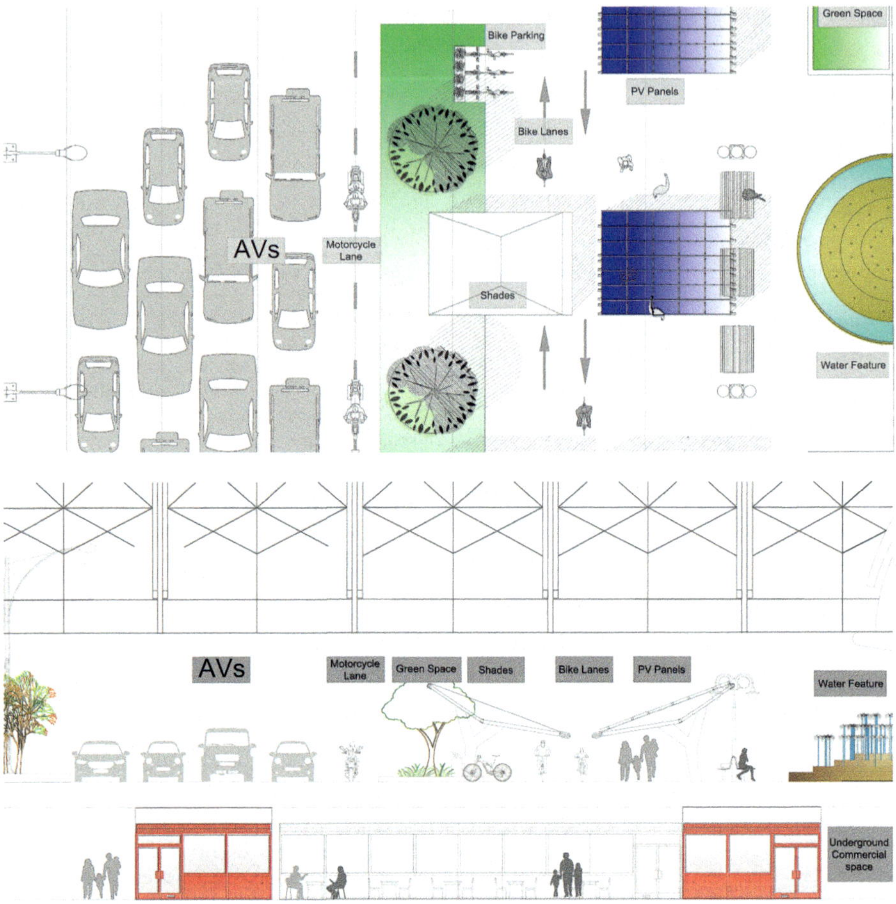

Fig. 12 Reimagined block plan and section of Sheikh Zayed road

AVs technologies should be examined in a livable city where public transit, walkability, public space, and, most importantly, people can thrive.

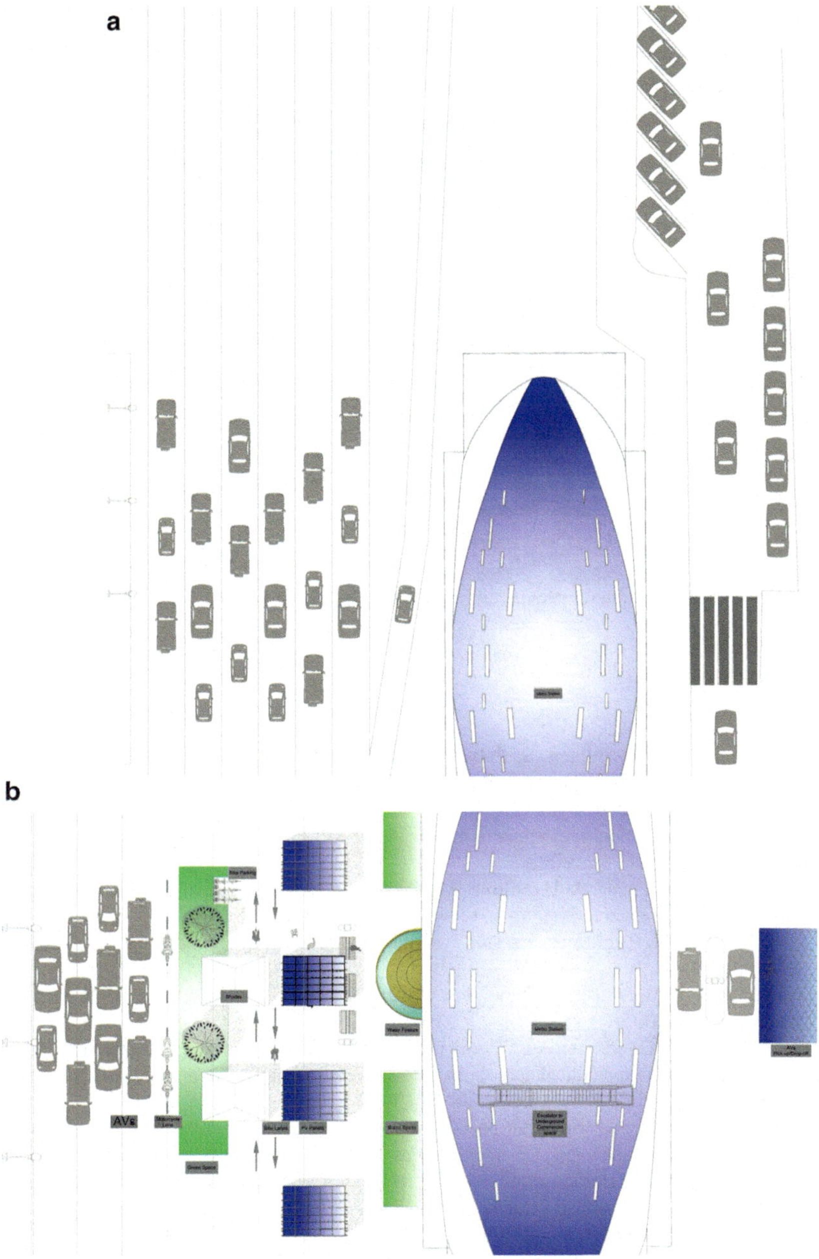

Fig. 13 Existing (Plan-A) versus reimagined (Plan B) overview

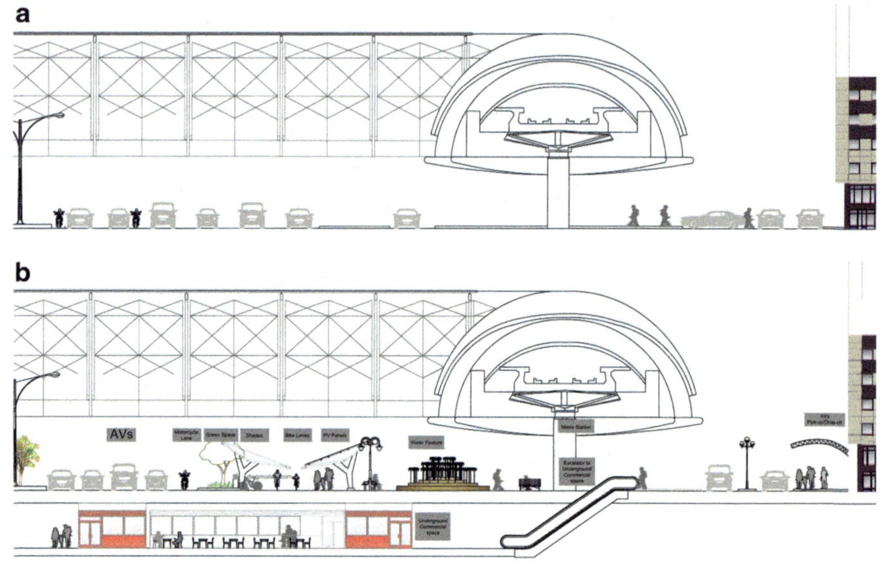

Fig. 14 Existing (Section-A) versus reimagined (Section B) overview

References

1. R. Kulmala, J. Jääskeläinen, S. Pakarinen, The impact of automated transport on the role, operations and costs of road operators and authorities in Finland EU-EIP Activity 4.2 Facilitating automated driving (2019)
2. C. A. V. C. o. Excellence, *Preparing for the Autonomous Vehicles in Canada* (Government of Canada, Toronto, 2015)
3. D. Ogilvie, M. Egan, V. Hamilton, M. Petticrew, Promoting walking and cycling as an alternative to using cars: Systematic review. BMJ **329**(7469), 763 (2004)
4. E. K. S. a. S. S. B. Patel, A review study on the walkable city. Int. Res. J. Eng. Technol. (IRJET) **12**, 06 (2019)
5. A. Tungare, Le Corbusier's Principles of City Planning and Their Application in Virtual Environments (2001)
6. S. Sahlqvist, A. Goodman, A.R. Cooper, D. Ogilvie, Change in active travel and changes in recreational and total physical activity in adults: Longitudinal findings from the iConnect study. Int. J. Behav. Nutr. Phys. Act. **10**(1), 28 (2013)
7. Dubai's Autonomous Transportation Strategy – Dubai Future Foundation, [Online]. Available: https://www.dubaifuture.gov.ae/our-initiatives/dubais-autonomous-transportation-strategy/
8. 2021-2030 – The Official Portal of the UAE Government, [Online]. Available: https://government.ae/en/more/uae-future/2021-2030
9. L. Fridman, Human-Centered Autonomous Vehicle Systems: Principles of Effective Shared Autonomy, 3 10 (2018)
10. Autonomous vehicles: Hype and potential|CNU, [Online]. Available: https://www.cnu.org/publicsquare/2016/09/06/autonomous-vehicles-hype-and-potential
11. E. a. A. K. I. N. O. Kürkçüoğlu, The effects of water elements in urban space perception: A case study in Üsküdar Municipality Square. A|Z ITU J. Fac. Archit. **10**(1), 159–175 (2013)
12. B. a. R. T. Vernon, Place-making through water sensitive urban design. Sustainability **1**(4), 789–814 (2009)

13. E. CPTED, *Secure Cities Through Urban Design and Planning: Standardizing the Process* (European CPTED, 2013)
14. J.A. Jakubiec, *Comfort and Perception in Architecture* (Springer, Berlin, 2016)
15. R. Dodge, A. Daly, J. Huyton, L. Sanders, The Challenge of Defining Wellbeing (2012). internationaljournalofwellbeing.org
16. E.O. Wilson, *Biophilia. 1984* (Harvard University Press, Cambridge, MA, 2003)
17. D. S. a. X.-Y. L. Steven Shladover, Impacts of cooperative adaptive cruise control on freeway traffic flow, in *Transportation Research Record Journal of the Transportation Research Board*, (2012)
18. J.L. Austin, O. Jütersonke, Understanding the Grammar of the City: Urban Safety and Peacebuilding Practice through a Semiotic Lens Paper Series of the Technical Working Group on the Confluence of Urban Safety and Peacebuilding Practice
19. J. Hosking, A. Macmillan, J. Connor, C. Bullen, S. Ameratunga, Organisational travel plans for improving health, in *Cochrane Database of Systematic Reviews*, ed. by A. Macmillan, (Wiley, Chichester, 2010)
20. Smart City Challenge|US Department of Transportation, [Online]. Available: https://www.transportation.gov/smartcity
21. Mohammed bin Rashid approves the Traffic and Transport Plan 2030 (2016)

Combating Road Traffic Congestion with Big Data: A Bibliometric Review and Analysis of Scientific Research

Eustace M. Dogo, Tebogo Makaba, Oluwatobi J. Afolabi, and Augustine C. Ajibo

1 Introduction

Urbanization is a global phenomenon that would have tremendous effects on urban areas now and in the near future; hence, it is rightly gaining lots of attention. Urbanization refers to the movement of people from rural areas to urban areas. Developing regions such as Africa are considered to have the highest urban growth rate in comparison to other regions of the world. For instance, Africa is projected to have an annual average of about 3.9 % with lots of cities having a growth rate of between 5 and 7.8% [104]. About 325 million people are estimated to live in the urban settlement in sub-Saharan Africa, and the UN projects that the number would be three times in the next few decades reaching up to 1 billion by 2050 [41]. It is also projected in Dargay, Gately, and Sommer [31] that the rate of vehicle ownership and use is estimated to rise by 56% in emerging economies and developing countries to over 1 billion units in 2030. This is due to the rise in income levels and the emergence of the middle class [32], but this comes with

E. M. Dogo (✉) · O. J. Afolabi
Department of Electrical & Electronic Engineering Science, University of Johannesburg, Johannesburg, South Africa
e mail: eustaced@uj.ac.za

T. Makaba
Department of Applied Information Systems, University of Johannesburg, Johannesburg, South Africa
e-mail: tmakaba@uj.ac.za

A. C. Ajibo
Department of Systems Innovation, Graduate School of Engineering Science, Osaka University, Osaka, Japan
e-mail: ajibo.augustine@irl.sys.es.osaka-u.ac.jp

U. Z. A. Hamid, F. Al-Turjman (eds.), *Towards Connected and Autonomous Vehicle Highways*, EAI/Springer Innovations in Communication and Computing,
https://doi.org/10.1007/978-3-030-66042-0_4

considerable implication to the current transport infrastructure and services. One of the consequences of urbanization is traffic congestion, especially in emerging economies where there is heavy dependence on road transportation. This has health, economic, and environmental implications, such as physical stress, increased fuel consumption, CO_2 emission, excessive noise, increased road traffic accidents, increased vehicle operating cost, and loss of valuable time and energy. These named consequences have led to the need for efficient mobility and optimization of the flow of traffic in urban cities, as it is vital to the proper functioning of any city due to its effect on the socioeconomic activities and development of any nation [109].

Road traffic congestion is defined in [109] as *"disproportion between the inflow and the outflow of vehicles into and out of a particular space,"* with an average speed flow breakdown occurring between 30 and 40 mph [19]. Road traffic congestion in most big cities is the consequence of inadequate infrastructure, poor traffic management systems, work zone, emergencies, special occasions, etc. This impacts greatly on the daily commuting cost of residents, the quality of life they experience in these cities, their safety, the quality of the air they inhale, and the volume of fuel consumed by automobiles which directly translates to more carbon emission [53]. In many big cities, this has become a herculean challenge for city planning and management. In a bid to ameliorate this challenge and improve the quality of services and transportation systems, most urban cities have adopted advanced traffic control and management systems which leverage on IoT and cloud computing technologies in tandem with data analytics [63].

One of these emerging fields is connected and autonomous vehicles (CAVs), sometimes referred to as driverless vehicles. CAVs are data-driven vehicles that are capable of adaptively sensing their environment and navigating with little or no human intervention [38]. CAV has the potential to revolutionize the global transportation system as well as a solution for combating road traffic congestion with IoT and big data analytics playing a pivotal role [38].

The cause of traffic congestion is broadly categorized into three types [10, 19]:

1. Recurrent congestion (RC) is associated with the regular occurrence of an event usually anticipated by frequent users of the route, such as rush-hour congestion and bad sections of the road such as potholes.
2. Nonrecurrent congestion (NRC) is linked with irregular or unexpected occurrence of events such as road traffic accidents, mechanical faults, bad weather condition, and planned special events, such as political rallies and concerts.
3. Precongestion condition (PCC) is borderline congestion that occurs at the beginning or at the end of congestion occurring. It could be in places with either RC or NRC.

Although the causes of traffic congestion are numerous, they differ from one city to another. Nonetheless, the major root causes of road traffic congestion are summarized in Table 1.

Among several concepts that have been adopted to combat traffic congestion is using big data analytics to predict traffic congestion, provide alternative routes, and monitor environmental conditions [102], in addition to providing several other

Table 1 Major causes of road congestion in most cities [109]

#	Causes	Description
1	Vehicular density	Vehicular density exceeding designed road capacity
2	Urbanization	Urbanization leading to increased motorization on existing road capacity
3	Over-reliance in road mode of transportation and heavy vehicles	Most developing regions rely on road transportation, due to the nonexistence of alternative forms of public transport systems
4	Improper land use and unplanned cities	Lack of proper land-use planning and poor traffic impact analysis before locating facilities in cities
5	Frequency of vehicle breakdown and traffic accidents	Used vehicles dominate in most African cities coupled with unskilled mechanics
6	Poor road infrastructure and conditions	Low capacity roads, failed and collapsed roads
7	Security checks	The stationing of security checkpoints due to security challenges
8	VIP movements and street parties/events	Closure of major roads due to passage of political and senior government officials, and planned social events.
9	Construction activities	Work zone, road repairs activities, and traffic diversion
10	Low-speed limit and speed bumps	Low speed around public places such as markets, schools, and residential and commercial areas
11	Traffic law violation, indiscipline by road users, and road rage	Violation of traffic laws and regulations by motorists and impatient behavior and bad driving habits of drivers
12	Inadequate parking space	Lack of or improper planning of parking space
13	Seasonal festivities	The mass movement of people and motorization during religious festivals and public holidays
14	Poor traffic management system	Manual traffic control by traffic wardens/traffic lights breakdown

real-time traffic situational information [125]. This concept is generally studied under the term intelligence transportation systems (ITS, which encompasses using information and communication technology alongside other enabling technologies in managing efficiently the entire road transportation ecosystem).

The International Data Corporation (IDC) defines big data as *"a new generation of technologies and architectures that have been designed to economically extract value from the huge volume of a wide variety of data by enabling high-velocity capture, discovery and/or analysis"* ([22], p. 6). The influx and huge growth of information which are mainly unstructured (i.e., they come in different formats) in all fields of life have created new and interesting ways of solving problems. This information comes in high volumes, different varieties, and is being generated at high velocity from everything that surrounds us [17, 26, 39, 77, 92, 148]. Information comes from things like wristwatches, smart phones, computer networks, vehicular networks, social networks, web, cellular infrastructures, wireless sensor networks, building automation sources, and so much more, and it has been estimated that the number of devices connecting everything around us will be about 50–100 billion by 2020 [29, 43]. The reality we live in today is that everybody and everything is networked.

Leveraging on big data has created new ways to solve traffic problems in large cities around the world by making both qualitative and quantitative traffic information available to commuters. In the past two decades, most countries of the world have undergone three digital revolutions (DR) that are key drivers to the adoption of smart city initiatives, namely:

- DR 1.0 – Mobile communication
- DR 2.0 – Cloud computing
- DR 3.0 – Artificial intelligence and big data analytics

Developed and developing countries have different approaches to combating traffic congestion. This section discusses the various methods that have been in use before the use of big data to combat congestion in developed and developing (also applicable to underdeveloped) countries.

1.1 Developed Countries

- The use of traffic controls, especially traffic lights, is probably the most adopted means of combating traffic in developed countries. However, these controls are being optimized to allow free flow with minimum delay. In addition, some lanes are actively prioritized at intersections allowing for nonstop and free movement of vehicles on reaching such intersections (used mostly for bus and emergency vehicle lanes).
- Providing commuters with some traffic information. However, this traffic information is collected through some inefficient means, such as the use of clipboards and clickers; traffic enumerators stand by the roads with clipboards and take note

of traffic behavior. The information obtained through this means is then analyzed and made available to the commuters.
- The use of sensors to collect data has also been in vogue. Inductive loops, piezo-electric strips, pneumatic tubes, cameras, infrared sensors, passive acoustic, microwave, and RFID are the sensors that are being used to record traffic activities. However, they have the disadvantages of wear and tear, the constant need for maintenance, being costly, and not effective for predictive analysis.
- Other methods adopted include encouraging people to live at locations nearest to their place of work, encouraging workers to work from home, encouraging the use of bicycles and walking, providing other modes of transport like trains, and encouraging citizens to use large-scale transportation systems like Bus Rapid Transit (BRT) and Light Rail Transit (LRT).

1.2 Developing Countries

- The most intuitive and most common method adopted by developing countries in combating traffic is increasing the capacity of the road by widening the road, constructing new roads, or repairing a previously dilapidated road. This is a very cost-intensive approach and would not fully solve the congestion problem in the future.
- The use of traffic wardens to control and direct traffic flow is also a major method to combat traffic.
- Words of mouth: commuters pass on to other commuters their previous unpleasant traffic experience, thereby warning other commuters to avoid the heavy traffic or to travel via another route.

Following the introductory section, the rest of the chapter is structured as follows: Section 2 provides a brief overview of background concepts that are complementary to big data analytics research. Section 3 describes the methodology for search and data extraction from the examined repositories. The results obtained from the bibliometric analysis are presented in Sect. 4. Section 5 gives a discussion, analysis, and insights from the results obtained. Finally, Sect. 6 concludes the chapter and provides perspective on future works in this research domain.

2 Fundamental Background Concepts

This section provides a summary of fundamental background concepts which are interlinked and key to the research in combating road traffic congestion (RTC) using data analytics.

2.1 Internet of Things

The internet has tremendously evolved in the last decade, creating new technologies for connectivity such as the Internet of Things (IoT) [28]. This technology has transformed internet communications such as machine-to-machine (M2M) and people-to-machine (P2M). According to the European Commission, the *"Future Internet"* is envisioned to be characterized with *"Things having identities and virtual personalities operating in smart spaces using intelligent interfaces in connecting and communicating within social, environmental, and user contexts"* [73]. In other words, the future internet is a network of networks in which an enormous number of objects, devices, or sensors are connected through the information communication technology (ICT) infrastructure to provide value-added services [73, 129]. In this regard, the IoTs is currently a key enabler, as it promises innovative business models and novel user experiences through strong connectivity and effective use of the next-generation networks (4G, 5G, and future 6G) with embedded devices. Currently, IoT models find application in virtually all areas of human activities, including the economy, transportation, energy, water services, and environment and health sectors [93, 114, 129]. Readers may refer to Minerva, Biru, and Rotondi [99] for a detailed perspective on IoT.

2.2 Cloud Computing

The last decade has been marked by the tremendous growth in data processing and storage technologies attributable to the technological achievements of the Internet [69]. The advent of the internet has made computing resources more ubiquitously available, cheaper, and powerful [52]. This technological trend is popularly referred to as cloud computing. Cloud computing *"is a model for allowing ubiquitous, convenient, and on-demand network access to several configured computing resources that can be rapidly provisioned and released with minimal management effort or service provider interaction."* It currently provides a robust architecture for performing large-scale and complex computing as well as providing a flexible online setting with the capability of handling large volumes in minimum execution time [52]. It also avails for virtualized resource sharing, parallel processing, security, and data service integration with scalable data storage. Overall, it minimizes the cost to individuals and enterprises in terms of automation and computerization, as well as provides reduced infrastructure maintenance cost, efficient management, and user access. Given the numerous potentials inherent with this technology, several applications that leverage various cloud platforms have been developed and deployed in various sectors and domains, such as in interrelated smart city concepts like smart health system, intelligent transportation system, smart financial and economic systems, and many others. This has remarkably increased the amount of data generated (big data) and utilized by these applications and services [6,

34, 35, 52]. The fast growth of data is, however, limiting the ability of cloud computing in terms of latency and good network connectivity. Cloud computing is, therefore, evolving into innovative and complementary concepts such as edge and fog computing, and recently, mist and dew computing, to address these bottlenecks. These new and complementary concepts extend cloud computing closer to the sensing devices, which are incorporated with a level of computing intelligence [35].

2.3 Big Data

The expeditious surge in the adaptation of IoT models and future internet technologies for sustainability and resilience of most urban cities has overtime resulted in the generation of an enormous volume of data referred to as "Big data" [73]. These data, which are managed and analyzed using structured and integrated ICT techniques, in turn, find various usages as ICT tools for smart cities [129]. Thus, big data entitles discovering actionable information and insights in structured, semi-structured, or unstructured large raw dataset from heterogeneous sources emanating at high speed and frequency, and leveraging cloud computing technology for storage and processing capabilities using advanced computational techniques such as artificial intelligence and machine learning [6, 8]. Of recent, big data is used in addressing real-life challenges confronting urban cities through real-time, cross-thematic data collection, processing, and integration and sharing over inter-operable services deployed in a cloud environment. The United Nations has championed big data as a catalyst that will foster international development. In the transportation sector, it has been utilized in optimizing routes and schedules and lately, in CAV domain. Researchers and experts have characterized big data by volume, velocity, variety of data types generated, as well as value and veracity, which explains the importance of true value and reliability of data. This is commonly referred to as the big data 5Vs [6, 52, 124], which forms the bedrock in big data definition by researchers and experts. Due to the rapidly evolving nature of big data, other big data "Vs" is continually evolving to better understand and define the big data phenomena, such as complexity, immutability, verification, and validation. It is envisaged that big data and data mining, when harmonized, could also help in rebranding the transportation system, especially in developing countries. They could also be pivotal tools in addressing some of the major challenges affecting interrelated smart city domains such as healthcare, power grid, telecommunication, and water and financial services.

2.4 Smart Cities

A smart city is composed of numerous private and public infrastructures such as in electricity network, gas distribution system, intelligent transportation systems,

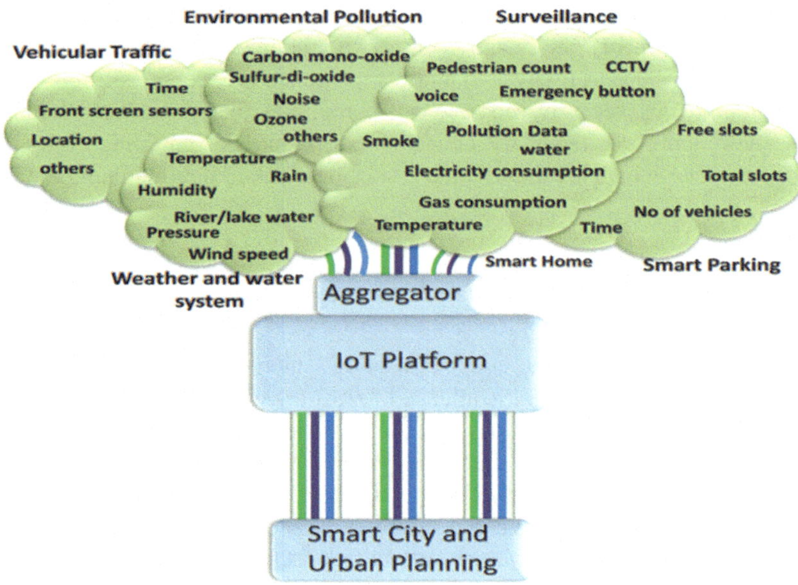

Fig. 1 IoT as the main source of data in smart cities [117]

telecommunication networks, and water management system. The core aim of smart city concept is to improve the quality of life of residents in such city through improved services and offering better planning of a city in the future [69, 159]. IoT technologies and the ever-increasing amount of big data generated from various sources are key enablers for the gradual evolving smart city initiatives [6]. Some key smart city application areas include smart energy and utilities, smart transportation, smart healthcare, smart education, and smart retail and business [6, 129]. Fig. 1 depicts the IoT deployment scenario where sensors generate real-time data for immediate decision support.

Addressing issues of mobility, congestion, low fuel consumption, and emission is the key concern in smart city strategies and targets toward reducing traffic congestion across the globe by employing sensors and data analytics in what is now termed intelligent transportation system (ITS).

2.5 Data Generation Sources

Before the advent of innovations in technology, traditional methods of data collection were employed such as official statistics and survey. Digital innovations have brought about the massive amount of data generated from a plethora of sources that will complement the traditional methods that are often not reliable, especially in most developing countries. These data can potentially provide useful and new insights using data mining techniques such as machine learning in the context of combating road traffic congestion in smart cities and communities [117]. Some of the data generation sources, which are broadly clustered under IoT technology, are briefly discussed in the subsequent subsections.

- Mobile phone devices: Over the years, numerous mobile phone applications have been developed for traffic management and control. Some of these Apps are briefly outlined in Table 2.

- Social media: The use of social media platforms such as Facebook and Twitter is growing rapidly globally, especially in Africa, with an estimated 170 million Facebook user-base accessing the platform majorly through mobile devices, with a corresponding 557 million mobile users [123]. This is evident in high internet traffic, with the bulk coming from mobile devices [34]. Africa is the second-highest population in the world of about 1.25 billion next to Asia but has about 9.6% Facebook subscribers (internetworldstats, 2020).
- Physical sensors, drones, and satellite images: Images from satellites are increasingly becoming an alternative source of huge data collection [16], especially in developing countries for policy-decision support because of challenges associated with reliability of data, lack of open data sources, and limitation of internet infrastructure. Improved technology has made cheaper orbiting of satellites possible, and these satellites could take pictures around the earth for up to 60 times a day. Data from the satellites are analyzed with machine learning techniques to give useful information. Data from these satellite images could also be exploited for road traffic congestion management. The use of CCTV and video cameras has also been a source of data. CCTV and video cameras are widely used for conducting basic traffic flow analysis to enhance safety on roads. This research area is broadly captured in the computer vision domain applied to road traffic management [128].
- Crowdsourced data: Crowdsourced road traffic data are becoming a more viable candidate over the manual and traditional means of traffic data collection [106]. However, its accuracy and reliability have been a subject of debate [60, 83]. Nevertheless, with regard to traffic congestion, digital data from social media messaging platforms, Google, and smart navigation Apps could be sourced from citizens and utilized to analyze traffic flow, user location, and travel speed and patterns, in tandem with other data sources.

2.6 Security and Privacy Issues

Data generation, collection, and processing ascribable to success in technologies such as IoT, big data analytics, and cloud computing are currently overwhelming

Table 2 Example of some smart mobile application platforms [125]

#	Application domain	Description	Examples
1	Route planning	Applications use maps from Google, GPS manufacturers and Transport agencies to assist drivers and commuters to establish the best route from origin to destination	Driving Route Finder, BestRoute
2	Ridesharing/carpooling	Applications match drivers of vehicles with passengers in real time both having common destination and origin	Uber and Taxify
3	Traffic safety	Applications save important data about the vehicle in real time. Data such as the speed of vehicle, location, date, time, and vehicle incidents such as deceleration and acceleration are saved to improve the vehicle's safety	ZoomSafer and SafeCar
4	Parking information	Applications provide real-time information about parking space availability and also include an option that allows payment by phone	Find my Car, central parking, parkmobile
5	Transportation data collection	Applications use GPS and/or mobile cameras to analyze location, speed of vehicles, and traffic counts of both pedestrians and vehicles	Tumcount, R&Scounts, speedlimit
6	Vehicle fuel consumption and emission	Applications use the mobile device's accelerometer to quantify the amount of fuel consumed, cost of fuel consumed, and the distance covered by the vehicle	Gas manager, Fuel economy calculator, Carbon emission calculator
7	Travel information	Applications display real-time information about the traffic. They offer information such as accident alerts, updates on road construction, and travel time between origin and destination	Beat the traffic plus, UK bus checker, CDOT mobile
8	Speed and hazard warning	Applications alert drivers when they exceed the prescribed speed limit for a particular road Application alerts by giving a beep sound	CamSam, Radardroid Lite, TomTom Speed Cameras & Speed Camera Radar Pro
9	Drivers' smart advisory systems	Applications alert drivers when approaching an un-signalized intersection, a stop sign or yield sign by sending a warning message much earlier before reaching the intersection	DSAS (deployed in Houston, Texas)

in today's digital world, and there exists the possibility of unauthorized access to vital information. The overall objective of information security consists of measures put in place to protect confidentiality, integrity, and availability of data from unauthorized access. Hence, big data is currently a great concern to individuals, researchers, governments, and private enterprises not only because of its complex nature and the computational demands but also because the conventional security and privacy systems and tools are no longer adequate in dealing with security and privacy issues in the current big data era. Traditional techniques for data security and privacy such as encryption schemes, access permissions, firewalls, transport layer security could easily be disrupted, thus rendering the source of data to be unidentified. Similarly, data whose source are originally unspecified can be made known. Given these challenges, advanced techniques and technologies are constantly being developed to protect, monitor, and audit big data processes as they pertain to infrastructure, application, and data. The security and privacy issues associated with big data can be categorized into five titles, namely Hadoop security, cloud security, monitoring and auditing, key management, and anonymization. Similarly, according to Cloud Security Alliance, the security and privacy challenges associated with big data can be categorized into four, namely infrastructure security, data privacy, data management, and integrity and reactive security. Infrastructure security entails secure distributed programming and security practices in nonrelational data stores. Data privacy refers to privacy-preserving analytics, encrypted data center, and granular access control. Data management involves secure data storage and transaction logs, auditing, and data provenance. In addition, integrity and reactive security encompass validation, filtering and real-time monitoring. Based on these proposed issues, authorization and authentication techniques are composed of both users and applications, and encryption and data masking are deployed for both data at rest and in motion or streaming. Big data is faced with privacy challenges owing to the global use of numerous cloud computing services and applications. Additionally, exchanging datasets or information between organizations across different geographical jurisdiction with conflicting laws also abounds with big data. Hence, data security can be compromised in the absence of adequate security measures. In an effort to remedy these challenges, researchers have suggested the enforcement of laws and regulations with obvious frontiers as regards unlawful access, data sharing and abuse, and replication of personal information [95, 130, 133, 134]. Since IoT is the key technology through which data are generated, implementing intelligent security approaches supported by IoT technology is crucial to addressing smart cities' security concerns [7, 9]. A promising solution to mitigate security and privacy issues is incorporating blockchain technology into IoT for intelligent transportation systems, as has been proposed for other smart city domain concepts such as in intelligent water management system [36].

3 Research Methodology

3.1 Database Search Criteria and Extraction

The research methodology relates to search criteria and bibliometric scientific data extraction similar to previous studies [85, 105, 110, 119], and involves the following:

- The search criteria in this research were conducted between 2011 and March 2020, using "Big data" AND "Traffic congestion" OR "Traffic flow" as the search index-words query.
- The search was performed on March 24, 2020, using Scopus and Web of Science (WoS), these being the two largest academic and scientific databases [101]. The choice of Scopus and WoS is aligned with the objective of this research study, which is to analyze and review wide interconnected research patterns and gain an observed perspective on research utilizing big data for combating road traffic congestion within the past ten years in published scientific articles.
- Our search criteria only considered journal articles published in English language for both Scopus and WoS databases because a high percentage of articles published in these two databases are in English language.

Even though this chapter has provided cogent reasons for using Scopus and WoS indexed-articles for the analysis, this has prompted the exclusion of articles not indexed in Scopus and WoS. Therefore, this research work does not claim to study and analyze the entire works of scholarly articles in the field of big data for combating road traffic congestion. However, the findings of this research work provide an adequate representation of studies in this domain globally.

This research leverages on *"Preferred Reporting Items for Systematic Reviews and Meta-Analyses (PRIMA)"* reporting approach studied in Moher, Liberati, Tetzlaff, and Altman [100], for performing systematic scholarly research reviews. For both Scopus and WoS database search engines, the following criteria were used:

- Investigated period: 2011–March 2020.
- The subject areas considered: Engineering, Energy, Computer Science, Environmental Science, and Decision Sciences. This information is in Table 4.
- The document types: Conference and journal articles, conference review papers, books, chapters in books, reviews papers, editorials, letters, and short surveys as listed in Table 5.
- The Source document types: Proceedings in conference, journal articles, books and book chapters, and other specialized publications.
- The languages excluded: Chinese, French, Russia, and Portuguese.

The publications extracted from the subject areas and document types were further scrutinized to establish their relevance to this study. After scrutinizing, the filtering data from the Scopus and WoS repositories were transferred into Microsoft Excel.

Consequently, the Scopus search yielded 653 publications comprising 643 published articles and 10 articles still in press, while the WoS search returned a total of 1799 publications. The data obtained are composed of article information such as the name of authors, the affiliation of authors, titles of articles, abstracts, indexed words/keywords, and other key citation data. The tags retrieved based on the Scopus and WoS search criteria are outlined in Table 5 (subject area) and Table 6 (document types). The document types shown in Table 6 are linked with the total number of publications (TNP) and the total number of citations (TNC), adopting the same abbreviations used in [119].

4 Bibliometric Results Analysis

This section presents useful insights based on results extracted on the research knowledge distribution in big data for combating road traffic congestion in Scopus and WoS database repositories.

4.1 Research Trends

Figures 2 and 3 show the research trends in terms of the TNP and TNC in Scopus and WoS, respectively.

For Scopus, as depicted in Fig. 2, there is a slow but steady growth of research in big data for combating road traffic congestion in terms of TNP from 2011, peaking in 2019 with 175 publications. A sharp decline is observed from 2019 until March 2020, which is attributed to a considerable drop in publication activities

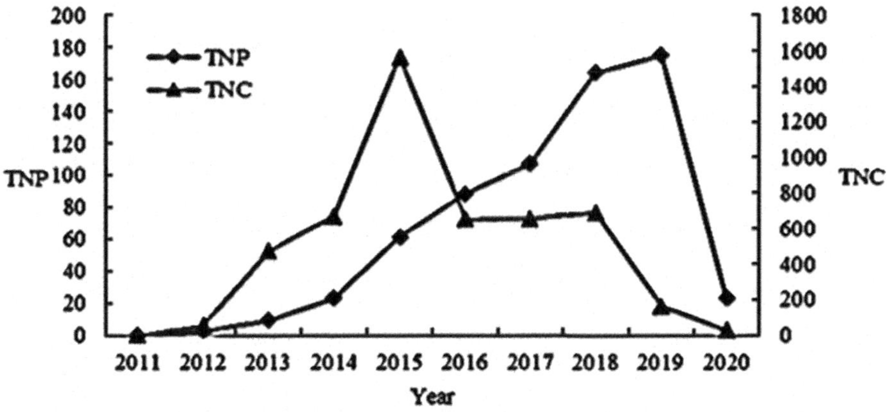

Fig. 2 Trend in the total number of publications and citations for Scopus

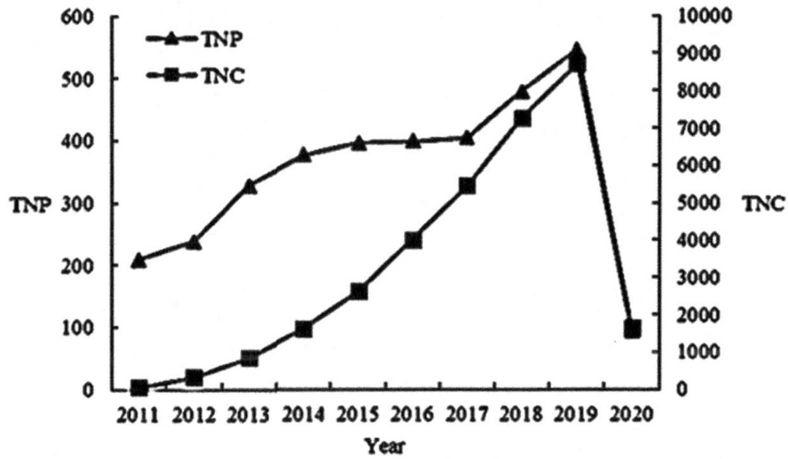

Fig. 3 Trend in the total number of publications and citations for WoS

within that period. On the other hand, a steady rise in TNCs is observed from 2011 peaking in 2015 with cumulative citation of 1562. There is a sharp decline observed between 2015 and 2016, which is attributed to a considerable drop in publication activities within that period. There is, however, a marginal rise from 649 to 690 TNC between 2016 and 2018. From 2018, TNC steadily declined until March 2020. This is attributed to newer publications that have not yet been cited. Based on the analysis on the TNP and TNC, a cumulative 653 publications with a TNC count of 4933 were published between 2011 and March 24, 2020. Even though there is an observed steady decline in TNP and TNC for Scopus as shown in Fig. 2, nonetheless, we anticipate that TNP and TNC counts are most likely to rise between April and December 2020, as newer papers are published.

For WoS, as shown in Fig. 3, we observe a gradual but steady growth of research in big data for combating road traffic congestion in terms of TNP from 2011, but with a slight decline between 2014 and 2017, and thereafter peaking in 2019 with about 300 publications. A sharp decline is observed from 2019 until March 2020, which is also attributed to a considerable drop in publication activities within that period. However, TNC has a steady rise from 2011, peaking in 2019 with TNC at over 4500 citations. Afterwards, TNC declines steadily after 2019 until March 2020. This is attributed to newer publications that have not yet been cited. From analysis on the TNP and TNC, it is observed that 1799 publications with TNC count of 32550 were published between 2011 and March 24, 2020. We also anticipate TNP and TNC to most likely rise between March and December of 2020, as newer papers are published.

In summary, we observed, based on the results obtained from our query and search criteria, that there are more research activities in WoS in comparison to Scopus.

4.2 The Topmost 20 Influential and Productive Authors

The list of the most influential and productive authors is presented in Table 3. The authors' ranking is based on the TNP.

For Scopus, Costa R., Figueiras P., and Guerreiro G. are the top three most productive authors with TNP = 5 and a corresponding TNC counts of 18 each. Interesting to note that author Wang, F.-Y., ranked 10th with four publications, is overall first in terms of TNC counts of 963. The entire top 20 most productive authors have at least three publications with varying TNC.

For WoS, Goatin, P. is the topmost productive author with 29 publications and TNC = 259. Hoogendoorn S.P. is the second with TNP = 22 and TNC = 383. Papageorgiou M. is the third most productive author with TNP = 19 with TNC = 420, as the most-cited author among the top 20 authors. The outstanding third to twentieth authors have at least 11 TNP with corresponding variations in TNC.

Table 3 Top 20 productive authors in terms of publications in Scopus and WoS

#	Scopus				WoS			
	Author	TNP	TNC	TNC/TNP	Author	TNP	TNC	TNC/TNP
1.	Costa, R.	5	18	3.6	Goatin, P.	29	259	8.9
2.	Figueiras, P.	5	18	3.6	Hoogendoorn, S.P.	22	383	17.4
3	Guerreiro, G.	5	18	3.6	Papageorgiou, M.	19	420	22.1
4	Mahgoub, I.	5	52	10.4	Herty, M.	17	164	9.6
5	Dubey, A.	4	31	7.8	Piccoli, B.	17	342	20.1
6	Fan, X.	4	9	2.3	Van Arem, B.	17	346	20.4
7	Hong, B.	4	31	7.8	Wang, H.	17	118	6.9
8	Paik, I.	4	8	2.0	Gottlich, S.	16	65	4.1
9	Wang, F.-Y.	4	963	240.8	Wang, W.	16	94	5.9
10	Bhuiyan, M.Z.A.	3	25	8.3	Ran, B.	14	105	7.5
11	Chen, Y.	3	7	2.3	Bressan, A.	13	136	10.5
12	Chen, Y.	3	4	1.3	Knoop, V.L.	13	68	5.2
13	He, Z.	3	53	17.7	Ngoduy, D.	13	338	26.0
14	Imawan, A.	3	16	5.3	Lil, L.	12	369	30.8
15	Jardim-Goncalves, R.	3	17	5.7	Papamichail, I.	12	244	20.3
16	Kanasugi, H.	3	2	0.7	Wang, Y.	12	134	11.2
17	Kwon, J.	3	16	5.3	Bhuyan, P.K.	11	34	3.1
18	Liu, K.	3	19	6.3	Gaeavello, M.	11	146	13.3
19	Liu, L.	3	7	2.3	Tosin, A.	11	174	15.8
20	Liu, S.	3	25	8.3	Wang, M.	11	226	20.5
Total		73	1339	345.4	Total	303	4165	279.6

4.3 Analysis of the Subject Area

The dominant subject areas in both Scopus and WoS repositories are enumerated in Table 4. For Scopus, Computer Science leads in the first position with TNP = 532, followed by Engineering in the second place having TNP = 275. Decision Science occupies the third place with TNP = 136. For WoS, Transportation Science and Technology is occupying the leading position with TNP = 1411, followed by Transportation having TNP = 584 in the second position, while Computer Science Information Systems is the third with TNP = 442. The vast amount of work in other disciplines and subject areas supports the interdisciplinary approach of research in the domain of combating road traffic congestion with big data as evident in Table 4.

4.4 Analysis of Top Document Types and Sources

Tables 5 lists the document types and sources distribution in both Scopus and WoS repositories. For Scopus, the majority of published papers are in conference proceedings (408) and journal papers (181). Books and book chapters and special issues are the other document types and sources adding to the whole TNP counts.

Whereas in WoS, overwhelming numbers of documents are published in journal articles (1760), which is in contrast to Scopus. Conference proceedings are in a distant second with 60 documents, with the rest of documents being review, book chapters, etc. (Table 6)

Table 4 Subject areas in Engineering and Physical Science disciplines in Scopus and WoS

Scopus			WoS		
Subject area	TNP	%	Subject area	TNP	%
Computer Science	532	46.8	Transportation Science Technology	1411	32.1
Engineering	275	24.2	Engineering Electrical Electronic	288	6.6
Decision Sciences	136	12.0	Transportation	584	13.3
Mathematics	136	12.0	Computer Science Information Systems	442	10.1
Environmental Science	27	2.4	Computer Science Interdisciplinary Applications	152	3.5
Energy	26	2.3	Engineering Multidisciplinary	325	7.4
Multidisciplinary	5	0.4	Mathematics Applied	305	6.9
–	–	–	Computer Science Software Engineering	215	4.9
–	–	–	Mathematics Interdisciplinary Applications	311	7.1
–	–	–	Computer Science Artificial Intelligence	252	5.7
–	–	–	Mathematics	109	2.5

Table 5 Document types and sources in Scopus and WoS

Scopus			WoS		
Document type	TNP	%	Document type	TNP	%
Conference papers	408	54.2	Article	1760	93.1
Articles	181	24.0	Proceedings paper	60	3.2
Conference review	134	17.8	Review	29	1.5
Book chapter	14	1.9	Early access	27	1.4
Review	7	0.9	Editorial material	9	0.5
Book	2	0.3	Book chapter	5	0.3
Editorial	2	0.3	Correction	1	0.1
Short survey	1	0.1	–	–	–
Others	4	0.5	–	–	–
Total	753	100	Total	1891	100

Table 6 Criteria for selecting countries thresholds in VoSViewer

#	Criteria	Scopus	WoS
1	Minimum number of documents of a country	5	5
2	Minimum number of citations of a country	20	0
3	Countries thresholds meeting criteria	20	20
4	Number of countries retrieved	20	20

4.5 *Topmost Countries and Regions*

Table 7 lists the topmost 20 influential and productive countries publishing scholarly works in both Scopus and WoS in this field of study. To represent graphically the results on VoSViewer software, filtering criteria are set as listed in Table 5.

In both repositories, China and the United States are the leading competing countries, with China occupying the first position. The two countries have a combined TNP of 390 in Scopus. India is ranked distant third with TNP = 48. Noticeably, each of the topmost 20 countries has at least 6 articles to their credit. Using VOSviewer software, the top 20 countries publishing in this domain are also shown in Fig. 4 for Scopus. In Fig. 4, the inter-network research strengths between the publishing countries are clearly shown.

For WoS, USA is the topmost country publishing in this domain with TNP = 158. This time, China is occupying the second position with TNP = 144. England occupies the third position with TNP of 40 slightly above Italy with 39. It is worthy to note that each of the top 20 countries has a minimum of 7 published articles to their credit. Similarly, using VOSviewer software, the topmost 20 publishing countries are visually represented in Fig. 5 displaying inter-networking research strength among the publishing countries.

In Fig. 5, we observe a strong inter-networking research connection between China and the USA (the leading publishing countries) in both the Scopus and the

Table 7 Topmost20 productive countries in Scopus and WoS

#	Scopus			WoS		
	Country	TNP	%	Country	TNP	%
1	China	259	38	USA	158	25
2	USA	131	19	China	144	23
3	India	48	7	England	40	6
4	South Korea	37	5	Australia	29	5
5	Japan	29	4	Germany	22	3
6	United Kingdom	23	3	Italy	39	6
7	Australia	20	3	Netherlands	32	5
8	Spain	18	3	France	29	5
9	Italy	16	2	Greece	23	4
10	Canada	15	2	Singapore	12	2
11	Taiwan	14	2	Canada	13	2
12	Germany	12	2	Taiwan	19	3
13	Hong Kong	11	2	Switzerland	7	1
14	Portugal	10	1	Iran	11	2
15	Malaysia	9	1	Spain	17	3
16	France	6	1	Brazil	6	1
17	Indonesia	6	1	South Korea	7	1
18	Israel	6	1	Sweden	7	1
19	Netherlands	6	1	India	13	2
20	Saudi Arabia	6	1	Japan	9	1
Total		682	100	Total	637	100

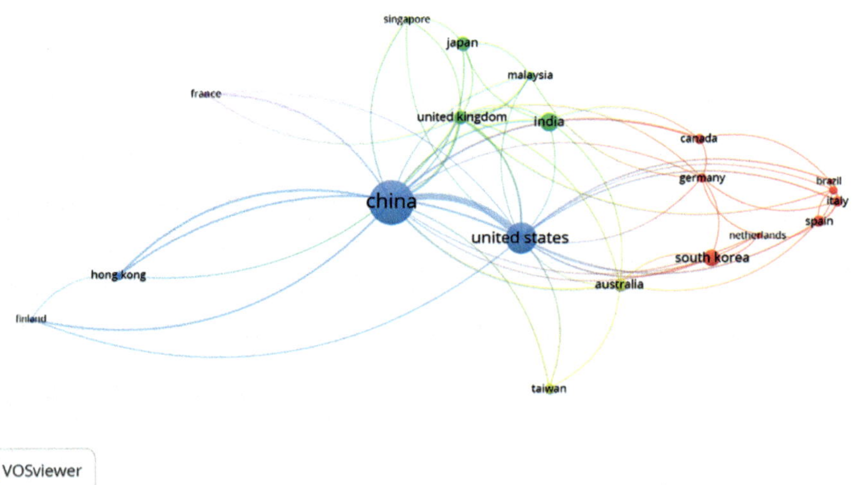

Fig. 4 Graphical representation of the top 20 countries in terms of publications in Scopus

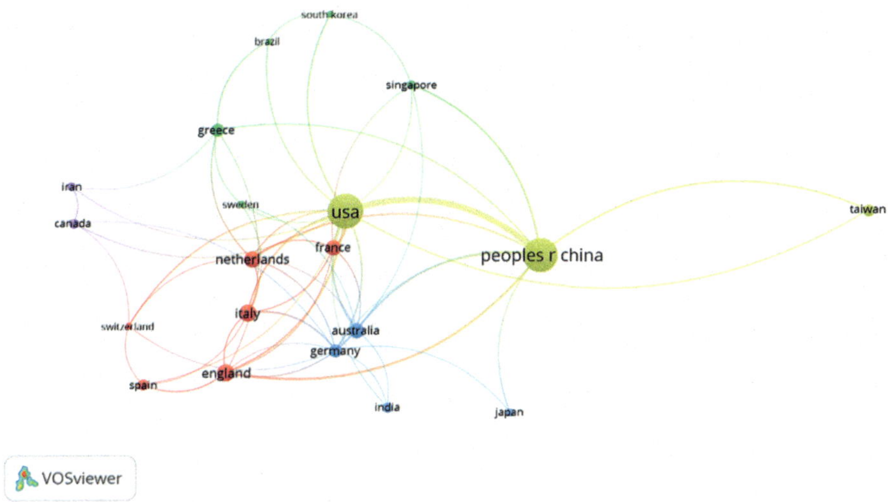

Fig. 5 Graphical representation of the top 20 countries in terms of publications in WoS

WoS repositories. This is an indication of solid research collaborations and in terms of co-authorship between these two countries.

4.6 *Topmost 20 Research Institutions/Affiliations in Scopus and WoS*

The topmost 20 research establishments for Scopus and WoS listed in Table 8.

For Scopus, Chinese Academy of Sciences is ranked number one, with a TNP = 18 and TNC = 1090, making this institution the most influential and productive in combating road traffic congestion with big data research. Beijing Jiaotong University (TNP = 18 and TNC = 130) and Tsinghua University (TNP = 14 and TNC = 97) occupy the second and third positions, respectively. Interestingly, Shanghai Jiao Tong University, ranked 17th with TNP = 6, has TNC = 231. In contrast, Instituto de Desenvolvimento de Novas Tecnologias with TNP = 5 has not yet recorded any citation. Each of the top 20 institutions/affiliations has a minimum of 5 published articles to their credit.

For WoS, Delft University of Technology is ranked number one, with TNP = 65 published articles and a corresponding TNC = 1063. The Delft University of Technology is thus the most influential and productive institution in combating road traffic congestion with big data research area. While Beijing Jiaotong University (TNP = 60 and TNC = 632) and Southeast University (TNP = 49 and TNC = 432) are in second and third positions, respectively. Each of the top 20 institutions/affiliations has a minimum of 17 published articles.

Table 8 Topmost 20 productive institutions and affiliations in Scopus and WoS

#	Scopus			WoS		
	Institution/affiliation	TNP	TNC	Institution/affiliation	TNP	TNC
1	Chinese Academy of Sciences	18	1090	Delft University of Technology	65	1063
2	Beijing Jiaotong University	14	130	Beijing Jiaotong University	60	632
3	Tsinghua University	14	97	Southeast University	49	432
4	Wuhan University of Technology	10	18	Tongji University	40	269
5	Beihang University	9	86	Tsinghua University	37	564
6	Tongji University	9	4	Zhejiang University	35	505
7	Beijing University of Posts and Telecommunications	8	4	University Wisconsin	24	247
8	University of Tokyo	8	8	Dalian University Technology	23	187
9	University of Chinese Academy of Sciences	8	60	Hong Kong Polytech University	22	296
10	Wuhan University	8	69	University California Berkeley	22	392
11	Ministry of Education China	7	11	Changan University	21	85
12	Southeast University, Nanjing	7	78	Tech University Crete	21	421
13	Pusan National University	7	47	Penn State University	20	299
14	Sun Yat-Sen University	7	16	Indian Institute of Technology	19	194
15	Dalian University of Technology	6	24	Rutgers State University	19	315
16	Beijing University of Technology	6	4	University Mannheim	19	77
17	Shanghai Jiao Tong University	6	231	Beihang University	18	150
18	Zhejiang University	5	41	Jiangsu Prov Collaborat Innovat Ctr Modern Urban	18	202
19	Instituto de Desenvolvimento de Novas Tecnologias	5	0	University of Maryland	18	379
20	Hong Kong University of Science and Technology	5	93	Consiglio Nazionale delle Ricerche, Italy	17	207
Total		368	2111	Total	567	6916

4.7 Top Funding Institutions and Affiliations

Table 9 shows the top 20 funding institutions for Scopus and WoS. They are ranked in terms of their TNPs.

For Scopus, the National Natural Science Foundation in China is ranked in the first position with TNP = 94 and TNC = 1323. The National Science Foundation is in the second position with TNP = 17 and TNC = 160, while Fundamental Research Funds for the Central Universities is occupying the third position having TNP = 16 and TNC = 7. Interestingly, there are 5 institutions in total, namely National Research Foundation of Korea (TNP = 9), Institute for Information and Communications Technology Promotion (TNP = 5), Japan Society for the Promotion of Science (TNP = 4), Beijing Municipal Science and Commission (TNP = 3), and Ministerio de Cienciay TecnologÃa (TNP = 3) with no citations.

In WoS, once again, the National Natural Science Foundation in China is occupying the first position, this time with TNP = 375 and TNC = 2945. The National Science Foundation, NSF, with TNP = 78 and TNC = 970 and the Fundamental Research Funds for the Central Universities with TNP = 71 and TNC = 601 are second and third, respectively. Noticeably, all the top 20 institutions have at least 11 publications to their credit.

4.8 Top 20 Indexed Terms and Keywords

The topmost 20 popularly indexed-terms or keywords used by majority authors in publications are enumerated in Table 10. Besides, the visual representation using VOSviewer software is depicted in Figs. 6 and 7 for Scopus and WoS, respectively. Big data (TNP = 524), traffic congestion (TNP = 344), and forecasting (TNP = 117) are the leading top three keywords in Scopus. Whereas traffic flow (TNP = 116), model (TNP = 82), and waves (TNP = 48) are the top three commonly used keywords by all authors in WoS.

4.9 The Topmost 50 Significant Authors and Articles

In Tables 11 and 12 are enumerated the topmost 50 cited papers, their authors, publication years, the title of the articles, and their publishing sources titles.

For Scopus, the review paper by Lv et al. 2015 is the most cited with TNC = 923. [166] with TNC = 414, which is a review paper that studies concepts, methodology, and applications of urban computing, is ranked second. Noticeably, all the 50 papers have citation counts of at least 18. Majority of the papers with high citation counts are typically concentrated on the overview of key concepts, research challenges,

Table 9 The topmost 20 funding establishments in Scopus and WoS based on TNP

#	Scopus			WoS		
	Name of the funding body	TNP	TNC	Name of the funding body	TNP	TNC
1	National Natural Science Foundation of China	94	1323	National Natural Science Foundation of China	375	2945
2	National Science Foundation	17	160	National Science Foundation NSF	78	970
3	Fundamental Research Funds for the Central Universities	16	7	Fundamental Research Funds for the Central Universities	71	601
4	National Research Foundation of Korea	9	0	National Basic Research Program of China	46	940
5	National Basic Research Program of China (973 Program)	8	60	European Union EU	35	561
6	European Commission	6	17	China Postdoctoral Science Foundation	25	241
7	Ministry of Science and Technology, Taiwan	6	6	Engineering Physical Sciences Research Council EPSRC	22	434
8	Ministry of Science, ICT and Future Planning	6	41	German Research Foundation DFG	21	79
9	China Postdoctoral Science Foundation			National High Technology Research And Development Program of China	21	264
10	Institute for Information and Communications Technology Promotion	5	3	National Science Council of Taiwan	21	708
11	China Scholarship Council	5	0	Hong Kong Research Grants Council	17	191
12	European Regional Development Fund	4	11	Natural Science Foundation of Jiangsu Province	17	11
13	Japan Society for the Promotion of Science	4	3	Natural Science Foundation Of Zhejiang Province	17	189
14	Ministry of Education, China	4	0	French National Research Agency ANR		124
15	Natural Science Foundation of Liaoning Province	4	1	Beijing Natural Science Foundation	16	126
16	Seventh Framework Programme	4	7	China Scholarship Council	15	119
17	Beijing Municipal Science and Commission	4	22	European Research Council ERC	14	302
18	Ministerio de Ciencia y TecnologÃa	3	0	National Key Research and Development Program of China	13	28
19	Ministry of Science and ICT, South Korea	3	0	Deutscher Akademischer Austausch Dienst Daad	12	138
20	Natural Science Foundation of Shandong Province	3	10	Ministry Of Education Culture Sports Science And Technology Japan MEXT	12	82
Total		205	1671	Total	848	9053

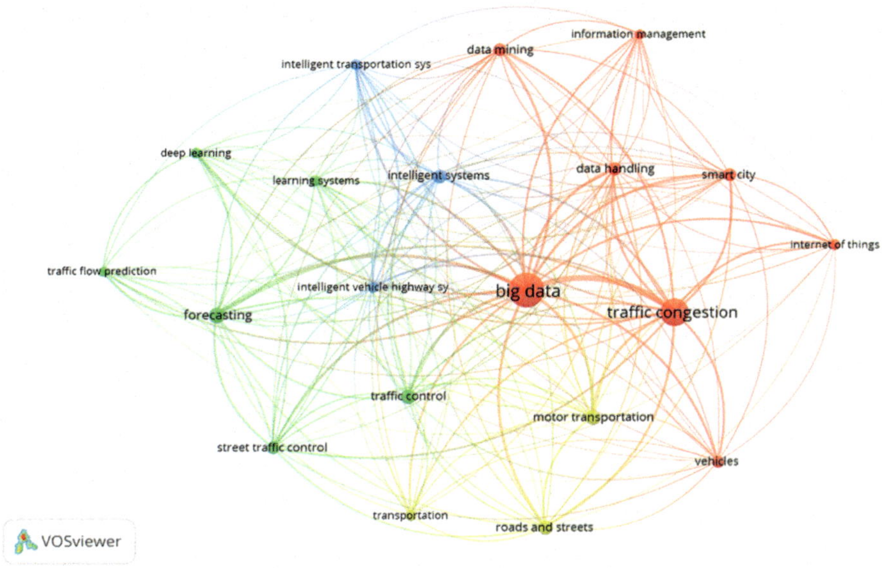

Fig. 6 Graphical visuals of the topmost 20 keywords/indexed-terms used by authors in Scopus

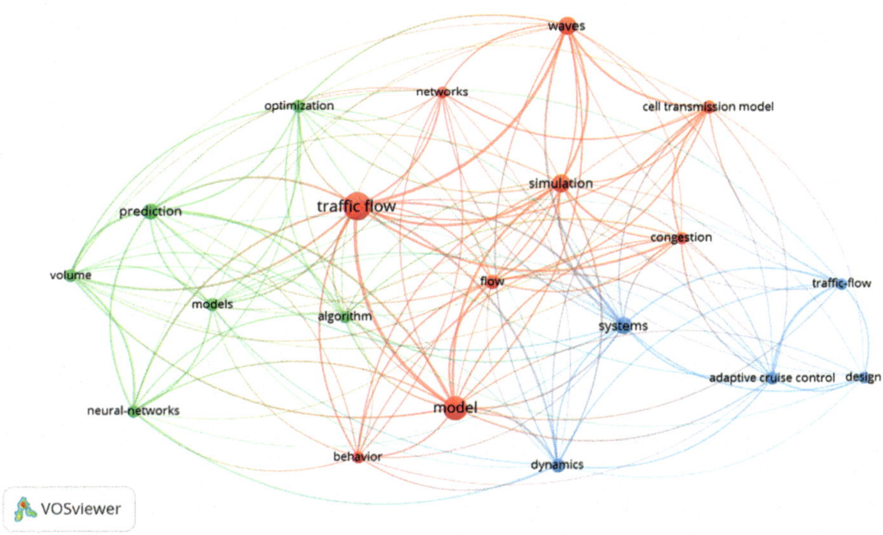

Fig. 7 Graphical visuals of the topmost 20 keywords/indexed-term used by authors in WoS

Table 10 Topmost 20 used keywords/indexed-terms by authors in Scopus and WoS

#	Scopus		WoS	
	Keywords/indexed-terms	TNP	Keywords/indexed-terms	TNP
1	Big Data	524	Traffic flow	116
2	Traffic Congestion	344	Model	82
3	Forecasting	117	Systems	43
4	Traffic Control	96	Simulation	46
5	Intelligent Systems	94	Waves	48
6	Motor Transportation	88	Prediction	36
7	Roads and Streets	84	Dynamics	31
8	Street Traffic Control	82	Volume	26
9	Data Mining	75	Adaptive cruise control	26
10	Data Handling	74	Neural networks	23
11	Smart City	73	Flow	28
12	Vehicles	70	Cell transmission model	24
13	Intelligent Transport Systems	59	Optimization	27
14	Intelligent Vehicle Highway Systems	57	Congestion	23
15	Internet of Things	55	Traffic-flow	20
16	Deep Learning	54	Algorithm	21
17	Learning Systems	54	Behavior	21
18	Transportation	53	Models	25
19	Information Management	47	Networks	20
20	Traffic Flow	45	Design	23
Total		2145	Total	709

practical applications, as well as in applying deep learning algorithms in road traffic flow prediction and forecasting.

For WoS, a review paper by [166] is the most cited with 407 counts, overlapping with Scopus. A paper by [68], that studied the difference between statistical and neural network methods in transportation research is ranked second with TNP = 300. All the top 50 authors have at least 47 publications to their credit.

4.10 Topmost Used Dataset in RTC Research

This section presents the commonly used dataset in combating road traffic congestion using big data; they are presented in Table 13. We observed that data generated from GPS mobile sensors and social media tweets are the most used dataset in smart transportation system because of the advantage in terms of better coverage of the cities' road traffic network and ease of accessing the publicly available dataset.

Table 11 The topmost 50 cited authors and articles in Scopus

#	Authors	Title	Source title	TNC	%
1	Lv et al. [91]	Traffic flow prediction with big data: A deep learning approach	IEEE Transactions on Intelligent Transportation Systems	923	29.4
2	Zheng et al. [166]	Urban computing: Concepts, methodologies, and applications	ACM Transactions on Intelligent Systems and Technology	414	13.2
3	Shi and Abdel-Aty [124]	Big Data applications in real-time traffic operation and safety monitoring and improvement on urban expressways	Transportation Research Part C: Emerging Technologies	123	3.9
4	Yongxue Tian and Li Pan [135]	Predicting short-term traffic flow by long short-term memory recurrent neural network	Proceedings – 2015 IEEE International Conference on Smart City, SmartCity 2015, Held Jointly with 8th IEEE International Conference on Social Computing and Networking, SocialCom 2015, 5th IEEE International Conference on Sustainable Computing and Communications, SustainCom 2015, 2015 International Conference on Big Data Intelligence and Computing, DataCom 2015, 5th International Symposium on Cloud and Service Computing, SC2 2015	119	3.8
5	Mehmood et al. [96]	Internet-of-Things-based smart cities: Recent advances and challenges	IEEE Communications Magazine	118	3.8
6	Cai et al. [20]	Siting public electric vehicle charging stations in Beijing using big-data informed travel patterns of the taxi fleet	Transportation Research Part D: Transport and Environment	109	3.5
7	Wu et al. [149]	A hybrid deep learning based traffic flow prediction method and its understanding	Transportation Research Part C: Emerging Technologies	94	3.0
8	Dawei Chen [24]	Research on traffic flow prediction in the big data environment based on the improved RBF neural network	IEEE Transactions on Industrial Informatics	73	2.3

9	Hao-Fan Yang et al. [156]	Optimized structure of the traffic flow forecasting model with a deep learning approach	IEEE Transactions on Neural Networks and Learning Systems	70	2.2
10	Yao et al. [157]	Deep multi-view spatial-temporal network for taxi demand prediction	32nd AAAI Conference on Artificial Intelligence, AAAI 2018	68	2.2
11	Li et al. [82]	Robust causal dependence mining in big data network and its application to traffic flow predictions	Transportation Research Part C: Emerging Technologies	51	1.6
12	He et al. [54]	Mapping to cells: A simple method to extract traffic dynamics from probe vehicle data	Computer-Aided Civil and Infrastructure Engineering	49	1.6
13	Wang et al. [141]	Joint virtual machine assignment and traffic engineering for green data center networks	Performance Evaluation Review	45	1.4
14	Jalali et al. [62]	Smart city architecture for community-level services through the internet of things	2015 18th International Conference on Intelligence in Next Generation Networks, ICIN 2015	43	1.4
15	Fusco et al. [42]	Short-term speed predictions exploiting big data on large urban road networks	Transportation Research Part C: Emerging Technologies	41	1.3
16	Aujla et al. [12]	Optimal decision making for big data processing at edge-cloud environment: An SDN perspective	IEEE Transactions on Industrial Informatics	39	1.2
17	Wibisono et al. [147]	Traffic big data prediction and visualization using Fast Incremental Model Trees-Drift Detection (FIMT-DD)	Knowledge-Based Systems	38	1.2
18	Yi et al. [158]	Deep neural networks for traffic flow prediction	2017 IEEE International Conference on Big Data and Smart Computing, BigComp 2017	34	1.1
19	Zhu et al. [168]	Big data analytics in intelligent transportation systems: A survey	IEEE Transactions on Intelligent Transportation Systems	32	1.0
20	Hou and Xingyi Li [57]	Repeatability and similarity of freeway traffic flow and long-term prediction under big data	IEEE Transactions on Intelligent Transportation Systems	32	1.0

(continued)

Table 11 (continued)

#	Authors	Title	Source title	TNC	%
21	Chen et al. [27]	Cyber-physical system enabled nearby traffic flow modelling for autonomous vehicles	2017 IEEE 36th International Performance Computing and Communications Conference, IPCCC 2017	30	1.0
22	Kim et al. (2016)	A deep learning approach to flight delay prediction	AIAA/IEEE Digital Avionics Systems Conference – Proceedings	30	1.0
23	Ota et al. [111]	A scalable approach for data-driven taxi ride-sharing simulation	Proceedings – 2015 IEEE International Conference on Big Data, IEEE Big Data 2015	29	0.9
24	St-Aubin et al. [128]	Large-scale automated proactive road safety analysis using video data	Transportation Research Part C: Emerging Technologies	29	0.9
25	Wu et al. [150]	Smart city with Chinese characteristics against the background of big data: Idea, action and risk	Journal of Cleaner Production	28	0.9
26	Ehmke et al. [37]	Data-driven approaches for emissions-minimized paths in urban areas	Computers and Operations Research	27	0.9
27	Jianjun Yu et al. [67]	RTIC-C: A big data system for massive traffic information mining	Proceedings – 2013 International Conference on Cloud Computing and Big Data, CLOUDCOM-ASIA 2013	26	0.8
28	Lu et al. [89]	Big data-driven based real-time traffic flow state identification and prediction	Discrete Dynamics in Nature and Society	24	0.8
29	Liu et al. [88]	Short-term traffic flow prediction with Conv-LSTM	2017 9th International Conference on Wireless Communications and Signal Processing, WCSP 2017 – Proceedings	23	0.7
30	Glatz et al. [46]	Visualizing big network traffic data using frequent pattern mining and hypergraphs	Computing	23	0.7

31	Soua et al. [127]	Big-data-generated traffic flow prediction using deep learning and Dempster-Shafer theory	Proceedings of the International Joint Conference on Neural Networks	22	0.7
32	Xia et al. [151]	A map reduce-based nearest neighbor approach for big data-driven traffic flow prediction	IEEE Access	22	0.7
33	Wu et al. [149]	A novel fuzzy deep-learning approach to traffic flow prediction with uncertain spatial-temporal data features	Future Generation Computer Systems	21	0.7
34	Cárdenas-Benítez et al. [21]	Traffic congestion detection system through connected vehicles and big data	Sensors (Switzerland)	21	0.7
35	Kim et al. [71]	LADS: optimizing data transfers using layout-aware data scheduling	Proceedings of the 13th USENIX Conference on File and Storage Technologies, FAST 2015	20	0.6
36	Al Najada and Mahgoub [4]	Anticipation and alert system of congestion and accidents in VANET using big data analysis for Intelligent Transportation Systems	2016 IEEE Symposium Series on Computational Intelligence, SSCI 2016	19	0.6
37	Al Najada and Mahgoub [5]	Big vehicular traffic data mining: Toward accident and congestion prevention	2016 International Wireless Communications and Mobile Computing Conference, IWCMC 2016	19	0.6
38	Ahn et al. [2]	Highway traffic flow prediction using support vector regression and Bayesian classifier	2016 International Conference on Big Data and Smart Computing, BigComp 2016	19	0.6
39	Ahmed and Hyoungshick Kim [1]	DDoS attack mitigation in internet of things using software defined networking	Proceedings – 3rd IEEE International Conference on Big Data Computing Service and Applications, BigDataService 2017	18	0.6
40	Yang et al. [154]	Spatiotemporal context awareness for urban traffic modeling and prediction: Sparse representation based variable selection	PLoS ONE	18	0.6

(continued)

Table 11 (continued)

#	Authors	Title	Source title	TNC	%
41	Bellemans et al. [13]	An agent-based model to evaluate carpooling at large manufacturing plants	Procedia Computer Science	18	0.6
42	Garg et al. [44]	Edge computing-based security framework for big data analytics in VANETs	IEEE Network	17	0.5
43	Xie et al. [153]	Analysis of traffic crashes involving pedestrians using big data: Investigation of contributing factors and identification of hotspots	Risk Analysis	17	0.5
44	Wang et al. [142]	Estimating online vacancies in real-time road traffic monitoring with traffic sensor data stream	Ad Hoc Networks	16	0.5
45	Jeon and Hong [64]	Monte Carlo simulation-based traffic speed forecasting using historical big data	Future Generation Computer Systems	15	0.5
46	Li Ye et al. [84]	Spatial-temporal analysis on spring festival travel rush in China based on multisource big data	Sustainability (Switzerland)	15	0.5
47	Tang et al. [132]	Travel time estimation at intersections based on low-frequency spatial-temporal GPS trajectory big data	Cartography and Geographic Information Science	15	0.5
48	Nguyen et al. [108]	Deep learning methods in transportation domain: A review	IET Intelligent Transport Systems	14	0.4
49	Deri et al. [32]	Big data computation of taxi movement in New York City	Proceedings – 2016 IEEE International Conference on Big Data, Big Data 2016	14	0.4
50	Ozbayoglu et al. [112]	A real-time autonomous highway accident detection model based on big data processing and computational intelligence	Proceedings – 2016 IEEE International Conference on Big Data, Big Data 2016	14	0.4
Total				3138	100

Table 12 Top 50 most-cited authors and articles in WoS

#	Authors	Title	Source title	TNC	%
1	Zheng et al. [166]	Urban computing: Concepts, methodologies, and applications	ACM Transactions On Intelligent Systems And Technology	407	9
2	Karlaftis and Vlahogianni [68]	Statistical methods versus neural networks in transportation research: Differences, similarities and some insights	Transportation Research Part C-Emerging Technologies	300	7
3	Talebpour and Mahmassani [131]	Influence of connected and autonomous vehicles on traffic flow stability and throughput	Transportation Research Part C-Emerging Technologies	179	4
4	Wei and Chen [145]	Forecasting the short-term metro passenger flow with empirical mode decomposition and neural networks	Transportation Research Part C-Emerging Technologies	177	4
5	Polson and Sokolov [116]	Deep learning for short-term traffic flow prediction	Transportation Research Part C-Emerging Technologies	170	4
6	Guo et al. [49]	Adaptive Kalman filter approach for stochastic short-term traffic flow rate prediction and uncertainty quantification	Transportation Research Part C-Emerging Technologies	134	3
7	Wang et al. [139]	Traffic-constrained multi-objective planning of electric-vehicle charging stations	IEEE Transactions on Power Delivery	133	3
8	Guler et al. [48]	Using connected vehicle technology to improve the efficiency of intersections	Transportation Research Part C-Emerging Technologies	132	3
9	Milanés and Shladover [98]	Modeling cooperative and autonomous adaptive cruise control dynamic responses using experimental data	Transportation Research Part C-Emerging Technologies	128	3
10	Fei et al. [40]	A Bayesian dynamic linear model approach for real-time short-term freeway travel time prediction	Transportation Research Part C-Emerging Technologies	123	3

(continued)

Table 12 (continued)

#	Authors	Title	Source title	TNC	%
11	Luettel et al. [90]	Autonomous ground vehicles-concepts and a path to the future	Proceedings of the IEEE	116	3
12	Li et al. [79]	Efficient missing data imputing for traffic flow by considering temporal and spatial dependence	Transportation Research Part C-Emerging Technologies	101	2
13	Zheng et al. [164]	Mining travel patterns from geotagged photos	ACM Transactions On Intelligent Systems And Technology	100	2
14	Lin et al. [86]	Efficient network-wide model-based predictive control for urban traffic networks	Transportation Research Part C-Emerging Technologies	98	2
15	Mahmassani et al. [94]	Urban network gridlock: Theory, characteristics, and dynamics	Transportation Research Part C-Emerging Technologies	94	2
16	Zhang et al. [163]	A hybrid short-term traffic flow forecasting method based on spectral analysis and statistical volatility model	Transportation Research Part C-Emerging Technologies	91	2
17	Zhang and Haghani [162]	A gradient boosting method to improve travel time prediction	Transportation Research Part C-Emerging Technologies	90	2
18	Hong [56]	Traffic flow forecasting by seasonal SVR with chaotic simulated annealing algorithm	Neurocomputing	87	2
19	Chen et al. [25]	The retrieval of intra-day trend and its influence on traffic prediction	Transportation Research Part C-Emerging Technologies	80	2
20	Moretti et al. [103]	Urban traffic flow forecasting through statistical and neural network bagging ensemble hybrid modeling	Neurocomputing	79	2

21	Habtemichael and Cetin [51]	Short-term traffic flow rate forecasting based on identifying similar traffic patterns	Transportation Research Part C-Emerging Technologies	76	2
22	Kumar and Vanajakshi [74]	Short-term traffic flow prediction using seasonal ARIMA model with limited input data	European Transport Research Review	73	2
23	Jia and Ngoduy [65]	Platoon based cooperative driving model with consideration of realistic inter-vehicle communication	Transportation Research Part C-Emerging Technologies	72	2
24	Khondaker and Kattan [70]	Variable speed limit: A microscopic analysis in a connected vehicle environment	Transportation Research Part C-Emerging Technologies	71	2
25	Pascale et al. [113]	Wireless sensor networks for traffic management and road safety	IET Intelligent Transport Systems	69	2
26	Wu et al. [149]	A hybrid deep learning based traffic flow prediction method and its understanding	Transportation Research Part C-Emerging Technologies	67	1
27	Levin and Boyles [78]	A multiclass cell transmission model for shared human and autonomous vehicle roads	Transportation Research Part C-Emerging Technologies	67	1
28	Chang et al. [23]	Dynamic near-term traffic flow prediction: system-oriented approach based on past experiences	IET Intelligent Transport Systems	65	1
29	Yu et al. [160]	A bi-level programming for bus lane network design	Transportation Research Part C-Emerging Technologies	64	1

(continued)

Table 12 (continued)

#	Authors	Title	Source title	TNC	%
30	Liu and Chang [87]	An arterial signal optimization model for intersections experiencing queue spillback and lane blockage	Transportation Research Part C-Emerging Technologies	62	1
31	Bressan et al. [18]	Flows on networks: recent results and perspectives	Ems Surveys in Mathematical Sciences	61	1
32	Yang et al. [155]	Isolated intersection control for various levels of vehicle technology: Conventional, connected, and automated vehicles	Transportation Research Part C-Emerging Technologies	60	1
33	Zegeye et al. [161]	Integrated macroscopic traffic flow, emission, and fuel consumption model for control purposes	Transportation Research Part C-Emerging Technologies	59	1
34	Di Francesco et al. [33]	On the Hughes' model for pedestrian flow: The one-dimensional case	Journal of Differential Equations	59	1
35	Ngoduy [107]	Multiclass first-order traffic model using stochastic fundamental diagrams	Transportmetrica	58	1
36	Antoniou et al. [11]	Dynamic data-driven local traffic state estimation and prediction	Transportation Research Part C-Emerging Technologies	57	1
37	Zheng et al. [165]	The effects of lane-changing on the immediate follower: Anticipation, relaxation, and change in driver characteristics	Transportation Research Part C-Emerging Technologies	57	1
38	Roncoli et al. [118]	Traffic flow optimisation in presence of vehicle automation and communication systems - Part II: Optimal control for multi-lane motorways	Transportation Research Part C-Emerging Technologies	55	1
39	Blandin et al. [15]	A general phase transition model for vehicular traffic	Siam Journal On Applied Mathematics	54	1
40	Zhu et al. [167]	Traffic volume forecasting based on radial basis function neural network with the consideration of traffic flows at the adjacent intersections	Transportation Research Part C-Emerging Technologies	53	1

41	Xia et al. [152]	A distributed spatial-temporal weighted model on MapReduce for short-term traffic flow forecasting	Neurocomputing	52	1
42	Wang et al. [140]	New Bayesian combination method for short-term traffic flow forecasting	Transportation Research Part C-Emerging Technologies	52	1
43	Hu et al. [58]	A Short-term traffic flow forecasting method based on the hybrid PSO-SVR	Neural Processing Letters	51	1
44	Płaczek [115]	Selective data collection in vehicular networks for traffic control applications	Transportation Research Part C-Emerging Technologies	51	1
45	Li et al. [80]	Urban traffic flow forecasting using Gauss-SVR with cat mapping, cloud model and PSO hybrid algorithm	Neurocomputing	50	1
46	Sewall et al. [121]	Interactive hybrid simulation of large-scale traffic	ACM Transactions on Graphics	49	1
47	Heydecker and Addison [55]	Analysis and modelling of traffic flow under variable speed limits	Transportation Research Part C-Emerging Technologies	49	1
48	Jiang et al. [66]	Eco approaching at an isolated signalized intersection under partially connected and automated vehicles environment	Transportation Research Part C-Emerging Technologies	48	1
49	Bellouquid et al. [14]	Toward the modeling of vehicular traffic as a complex system: A kinetic theory approach	Mathematical Models & Methods in Applied Sciences	48	1
50	Li et al. [81]	Missing traffic data: comparison of imputation methods	IET Intelligent Transport Systems	47	1
Total				4575	100

Table 13 The most popular dataset used in RTC research

#	Study	Datasets
1	Wei and Hong-ying [146]	Image (video, CCTV camera)
2	Gu et al. [47]	Social media (tweet)
3	An et al. [10], D'Andrea and Marcelloni [30], Gidófalvi and Yang [45], Jabbarpour et al. [61], Kong et al. [72], Kwoczek et al. [75], Shan and Zhu [122] and Wang et al. [144]	GPS Mobile sensors
4	Aissaoui et al. [3] and Wang et al. [143]	Vehicular Network and V2X
5	Lana et al. [76] and Melnikov et al. [97]	Fixed traffic sensors data
6	Huang and Xiao [59], Toole et al. [136], Tosi and Marzorati [137], Wan et al. [138] and Guo et al. [50]	Mobile cellular network and crowdsourced
7	Al Najada and Mahgoub [5]	Recorded traffic accident

4.11 The Most Popularly Used Algorithms and Simulation Tools in RTC Research

This section provides the most used models, algorithms, and research simulation tools. The search criteria used are based on articles with at least 20 citations given in Tables 11 and 12. Generally, the research is conducted in the following broad areas: traffic congestion, traffic, security, and privacy mitigation using big data techniques. Specifically, machine learning, deep learning, computational intelligence, and variants of their hybrid approaches. Several big data mining simulation researchers that are broadly categorized as either open-source, for example, are: Anaconda Python and R distribution, Kaggle, Weka, etc., or commercial, for example, Matlab, Tableau, Oracle, and IBM SPSS modeler [126].

5 Analysis and Discussions

This section offers analysis and implications of the main findings and limitations of this current study.

This chapter employed bibliometric and science mapping as techniques for documenting and examining the research knowledge amassed in big data for combating road traffic congestion between 2011 and March 24, 2020. This study analyzed 653 documents published in Scopus and 1799 in WoS repositories. Conference proceedings and journal articles are the prominent document sources and types targeted by research scholars for publication in this field. The papers by Costa et al. 2012 is the most influential in Scopus, while Goatin et al. is the most influential in WoS. The main subject areas for research in Scopus are in Computer Science and Engineering. On the other hand, Transportation, Science and Technology are the most targeted subject areas in WoS. China and the USA are the leading countries researching this domain. Chinese Academy of Sciences leads other Institutions with the most TNPs and TNCs in Scopus. National Natural Science Foundation in China is the foremost funding institution using the TNPs as criteria in both Scopus and WoS. Big data and traffic congestion are the most common keywords in Scopus, whereas traffic flow and model are the most common keywords in WoS. Most significant author/article is Lv et al. 2015 in Scopus, while Zheng et al. in WoS. Most commonly used dataset in both Scopus and WoS repositories includes GPS mobile sensor data, CCTV and image, social media tweets, fixed sensors, etc. Machine learning, deep learning, nature-inspired algorithms, computational intelligence, and their hybrids are the most commonly used models/algorithms in both Scopus and WoS. Research in this domain is clustered around the following themes: Traffic flow prediction, traffic congestion detection, traffic speed/flow forecasting alert system, optimization of vehicular routing, and security and privacy mitigation. Other research theme areas include analysis of traffic crashes with pedestrians, analysis of aging bridges, vehicle emission profiles, travel time estimation, journey planning and congestion prediction, travel and parking guidance, traffic volume count using RFID technology. It is worth noting also that the publications and research projects on big data in combating traffic congestion reported in this current research do not necessarily result in impactful research and development outputs, and a detailed analysis is still required in this regard.

6 Conclusion and Future Research Direction

Cloud computing and IoT and the concept of smart city are technological advancements that are transforming practical design of big data analytical models in support of combating road traffic congestion in the overall concept of smart cities phenomena. Combating road traffic congestion is an essential part of the smart cities concept because it has socioeconomic implications to the well-being of a city. This chapter conducted a systematic study that highlights the accumulated key research works in combating road traffic congestion using big data. Specifically, this chapter employed bibliometric analysis and VOSviewer science mapping software tool to analyze scholarly publications indexed in two large scientific databases (SCOPUS and WoS) from 2011 to March 24, 2020. This research systematically examined and interpreted the resulting findings of the analysis based on the intellectual structure of research in this domain, to carve out useful and interpretable outcomes. Additionally, this chapter identifies and highlights needed research directions and propositions that could be of applicability in facilitating future research in this domain, and by extension, in the CAV field.

The following are limitations of this study that will be of interest to the authors of future work:

- Possible inclusion of non-English publications to capture regions that are publishing impactful and groundbreaking research in other languages.
- Inclusion of and considering impactful non-Scopus and non-WoS publications.
- However, the findings in this research indicate that the National Natural Science Foundation in China is the foremost funding institution, using TNP and TNC as criteria. This may not entirely be a true representation regarding funding capabilities at the disposal of this funding body. Hence, further investigation is required to give credence to this accession.

In conclusion, the authors hope that this study will assist new researchers and application engineers by providing a quick overview of research activities in this domain, as well as in assisting them to develop new and future research directions in emerging and transformative technologies, such as connected autonomous vehicle domain. Additionally, the collaborative information generated from the semi-autonomous vehicles (Advanced Driver Assistance Systems), future fully autonomous vehicles, people, and infrastructure will provide vital insight to policymakers in fashioning out strategies for intelligent transportation systems and smart city initiatives toward finding a lasting solution to combating road traffic congestion in cities around the world.

Acknowledgments The authors would like to thank the University of Johannesburg, South Africa for affording the resources to complete this research work.

References

1. M.E. Ahmed, H. Kim, DDoS attack mitigation in Internet of Things using software defined networking, in *2017 IEEE Third International Conference on Big Data Computing Service and Applications (BigDataService)*, (IEEE, 2017), pp. 271–276
2. J. Ahn, E. Ko, E.Y. Kim, Highway traffic flow prediction using support vector regression and Bayesian classifier, in *2016 International Conference on Big Data and Smart Computing (BigComp)*, (IEEE, 2016), pp. 239–244
3. R. Aissaoui, H. Menouar, A. Dhraief, F. Filali, A. Belghith, A. Abu-Dayya, Advanced real-time traffic monitoring system based on V2X communications, in *2014 IEEE International Conference on Communications (ICC)*, (IEEE, 2014), pp. 2713–2718
4. H. Al Najada, I. Mahgoub, Anticipation and alert system of congestion and accidents in VANET using Big Data analysis for Intelligent Transportation Systems, in *2016 IEEE Symposium Series on Computational Intelligence (SSCI)*, (IEEE, 2016a), pp. 1–8
5. H. Al Najada, I. Mahgoub, Big vehicular traffic data mining: Towards accident and congestion prevention, in *2016 International Wireless Communications and Mobile Computing Conference (IWCMC)*, (IEEE, 2016b), pp. 256–261
6. E. Al Nuaimi, H. Al Neyadi, N. Mohamed, J. Al-Jaroodi, Applications of big data to smart cities. J. Internet Serv. Appl. **6**(1), 1–15 (2015)
7. F. Al-Turjman, *Intelligence in IoT-Enabled Smart Cities* (CRC Press, Boca Raton, 2019a)
8. F. Al-Turjman, *Artificial Intelligence in IoT* (Springer, Cham, 2019b)
9. F. Al-Turjman, *Smart Cities Performability, Cognition, & Security* (Springer, Cham, 2020)
10. S. An, H. Yang, J. Wang, N. Cui, J. Cui, Mining urban recurrent congestion evolution patterns from GPS-equipped vehicle mobility data. Inf. Sci. **373**, 515–526 (2016)

11. C. Antoniou, H.N. Koutsopoulos, G. Yannis, Dynamic data-driven local traffic state estimation and prediction. Transp. Res. Part C **34**, 89–107 (2013)
12. G.S. Aujla, N. Kumar, A.Y. Zomaya, R. Ranjan, Optimal decision making for big data processing at edge-cloud environment: An SDN perspective. IEEE Trans. Ind. Informatics **14**(2), 778–789 (2018)
13. T. Bellemans, S. Bothe, S. Cho, F. Giannotti, D. Janssens, L. Knapen, et al., An agent-based model to evaluate carpooling at large manufacturing plants. Proc. Comput. Sci. **10**, 1221–1227 (2012)
14. A. Bellouquid, E. De Angelis, L. Fermo, Towards the modeling of vehicular traffic as a complex system: A kinetic theory approach. Math. Models Method Appl. Sci. **22**(supp01), 1140003 (2012)
15. S. Blandin, D. Work, P. Goatin, B. Piccoli, A. Bayen, A general phase transition model for vehicular traffic. SIAM J. Appl. Math. **71**(1), 107–127 (2011)
16. J.E. Blumenstock, Fighting poverty with data. Science **353**(6301), 753–754 (2016)
17. D. Boyd, K. Crawford, Critical questions for big data. Inf. Commun. Soc. **15**(5), 662–679 (2012)
18. A. Bressan, S. Čanić, M. Garavello, M. Herty, B. Piccoli, Flows on networks: Recent results and perspectives. EMS Surv. Math. Sci. **1**(1), 47–111 (2014)
19. J. Brownfield, A. Graham, H. Eveleigh, F. Maunsell, H. Ward, S. Robertson, R. Allsop, Congestion and accident risk. Department for Transport Road Safety Research Report **44** (2003)
20. H. Cai, X. Jia, A.S.F. Chiu, X. Hu, M. Xu, Siting public electric vehicle charging stations in Beijing using big-data informed travel patterns of the taxi fleet. Transp. Res. D **33**(C), 39–46 (2014)
21. N. Cárdenas-Benítez, R. Aquino-Santos, P. Magaña-Espinoza, J. Aguilar-Velazco, A. Edwards-Block, A. Medina Cass, Traffic congestion detection system through connected vehicles and big data. Sensors (Basel, Switzerland) **16**(5), 599 (2016)
22. P. Carter, Big data analytics: Future architectures, skills and roadmaps for the CIO, in *White Paper, IDC Sponsored by SAS*, (2011), pp. 1–16
23. H. Chang, Y. Lee, B. Yoon, S. Baek, Dynamic near-term traffic flow prediction: System-oriented approach based on past experiences. IET Intell. Transp. Syst. **6**(3), 292 (2012)
24. D. Chen, Research on traffic flow prediction in the big data environment based on the improved RBF neural network. IEEE Trans. Ind. Informatics **13**(4), 2000–2008 (2017)
25. C. Chen, Y. Wang, L. Li, J. Hu, Z. Zhang, The retrieval of intra-day trend and its influence on traffic prediction. Transp. Res. Part C **22**, 103–118 (2012a)
26. H. Chen, R.H.L. Chiang, V.C. Storey, Business intelligence and analytics: From big data to big impact. MIS Q. **36**(4), 1165–1188 (2012b)
27. B. Chen, Z. Yang, S. Huang, X. Du, Z. Cui, J. Bhimani, et al., Cyber-physical system enabled nearby traffic flow modelling for autonomous vehicles, in *2017 IEEE 36th International Performance Computing and Communications Conference (IPCCC)*, (IEEE, 2017), pp. 1–6
28. L. Coetzee, J. Eksteen, The Internet of Things-promise for the future? An introduction, in *2011 IST-Africa Conference Proceedings*, (IEEE, 2011), pp. 1–9
29. Commonwealth of Kentucky, The Kentucky transportation cabinet (2014), Retrieved November 28, 2017, from http://transportation.ky.gov/sites/GoKY/Pages/home.aspx
30. E. D'Andrea, F. Marcelloni, Detection of traffic congestion and incidents from GPS trace analysis. Expert Syst. Appl. **73**, 43–56 (2017)
31. J. Dargay, D. Gately, M. Sommer, Vehicle ownership and income growth, worldwide: 1960–2030. Energy J. **28**(4), 143–170 (2007)
32. M. Davies, T. Schiller, Deloitte Africa automotive insights - Navigating the African automotive sector: Ethiopia, Kenya and Nigeria (2018), Retrieved February 2, 2020, from https://www2.deloitte.com/content/dam/Deloitte/za/Documents/manufacturing/za_Africa-Auto-2016-Report-28-May-2018.pdf

32. J.A. Deri, F. Franchetti, J.M. Moura, Big data computation of taxi movement in New York City, in *2016 IEEE International Conference on Big Data (Big Data)*, (IEEE, 2016), pp. 2616–2625
33. M. Di Francesco, P.A. Markowich, J. Pietschmann, M. Wolfram, On the Hughes' model for pedestrian flow: The one-dimensional case. J. Differ. Equ. **250**(3), 1334–1362 (2011)
34. E.M. Dogo, A. Salami, S. Salman, Feasibility analysis of critical factors affecting cloud computing in Nigeria. Int. J. Cloud Comput. Serv. Sci. **2**(4), 276 (2013). https://doi.org/10.11591/closer.v2i4.4162
35. E.M. Dogo, A.F. Salami, C.O. Aigbavboa, T. Nkonyana, Taking cloud computing to the extreme edge: A review of mist computing for smart cities and industry 4.0 in Africa, in *Edge Computing*, ed. by F. Al-Turjman, (Springer, Cham, 2019a)
36. E.M. Dogo, A.F. Salami, N.I. Nwulu, C.O. Aigbavboa, Blockchain and internet of things-based technologies for intelligent water management system, in *Artificial Intelligence in IoT*, ed. by F. Al-Turjman, (Springer, Cham, 2019b)
37. J.F. Ehmke, A.M. Campbell, B.W. Thomas, Data-driven approaches for emissions-minimized paths in urban areas. Comput. Oper. Res. **67**, 34–47 (2016)
38. D. Elliott, W. Keen, L. Miao, Recent advances in connected and automated vehicles. J. Traffic Transp. Eng. **6**(2), 109–131 (2019) (English Edition)
39. B. Emmerson, M2M: the Internet of 50 billion devices. WinWin Magazine **1**, 19–22 (2010)
40. X. Fei, C. Lu, K. Liu, A Bayesian dynamic linear model approach for real-time short-term freeway travel time prediction. Transp. Res. Part C **19**(6), 1306–1318 (2011)
41. S. Fox, Africa's urban transition: Challenges, misconceptions and opportunities. *Africa at LSE* (2012)
42. G. Fusco, C. Colombaroni, N. Isaenko, Short-term speed predictions exploiting big data on large urban road networks. Transp. Res. Part C **73**, 183–201 (2016)
43. J. Gantz, D. Reinsel, Extracting value from chaos. IDC iView **1142**(2011), 1–12 (2011)
44. S. Garg, A. Singh, K. Kaur, G.S. Aujla, S. Batra, N. Kumar, et al., Edge computing-based security framework for big data analytics in VANETs. IEEE Netw. **33**(2), 72–81 (2019)
45. G. Gidófalvi, C. Yang, Scalable detection of traffic congestion from massive floating car data streams, in *Proceedings of the 1st International ACM SIGSPATIAL Workshop on Smart Cities and Urban Analytics*, (2015), pp. 114–121
46. E. Glatz, E. Glatz, S. Mavromatidis, S. Mavromatidis, B. Ager, B. Ager, et al., Visualizing big network traffic data using frequent pattern mining and hypergraphs. Computing **96**(1), 27–38 (2014)
47. Y. Gu, Z. Qian, F. Chen, From twitter to detector: Real-time traffic incident detection using social media data. Transp. Res. Part C **67**, 321–342 (2016)
48. S.I. Guler, M. Menendez, L. Meier, Using connected vehicle technology to improve the efficiency of intersections. Transp. Res. Part C **46**, 121–131 (2014)
49. J. Guo, W. Huang, B.M. Williams, Adaptive Kalman filter approach for stochastic short-term traffic flow rate prediction and uncertainty quantification. Transp. Res. Part C **43**, 50–64 (2014)
50. Y. Guo, J. Zhang, Y. Zhang, A method of traffic congestion state detection based on mobile big data, in *2017 IEEE 2nd International Conference on Big Data Analysis (ICBDA)*, (IEEE, 2017), pp. 489–493
51. F.G. Habtemichael, M. Cetin, Short-term traffic flow rate forecasting based on identifying similar traffic patterns. Transp. Res. Part C **66**, 61–78 (2016)
52. I.A.T. Hashem, I. Yaqoob, N.B. Anuar, S. Mokhtar, A. Gani, S. Ullah Khan, The rise of "big data" on cloud computing: Review and open research issues. Inf. Syst. **47**, 98–115 (2015)
53. F. He, X. Yan, Y. Liu, L. Ma, A traffic congestion assessment method for urban road networks based on speed performance index. Proc Eng **137**, 425–433 (2016)
54. Z. He, L. Zheng, P. Chen, W. Guan, Mapping to cells: A simple method to extract traffic dynamics from probe vehicle data. Comput. Aided Civil Infrastruct. Eng. **32**(3), 252–267 (2017)
55. B.G. Heydecker, J.D. Addison, Analysis and modelling of traffic flow under variable speed limits. Transp. Res. Part C **19**(2), 206–217 (2011)

56. W. Hong, Traffic flow forecasting by seasonal SVR with chaotic simulated annealing algorithm. Neurocomputing **74**(12), 2096–2107 (2011)
57. Z. Hou, X. Li, Repeatability and similarity of freeway traffic flow and long-term prediction under big data. IEEE Trans. Intell. Transp. Syst. **17**(6), 1786–1796 (2016)
58. W. Hu, W. Hu, L. Yan, L. Yan, K. Liu, K. Liu, et al., A short-term traffic flow forecasting method based on the hybrid PSO-SVR. Neural. Process. Lett. **43**(1), 155–172 (2016)
59. J. Huang, M. Xiao, State of the art on road traffic sensing and learning based on mobile user network log data. Neurocomputing, 1–9 (2017)
60. Informatica, Crowdsourced data: Can you trust it? (2017), Accessed 1 December 2017, from https://www.informatica.com/potential-at-work/crowdsourced-data-can-you-trust-it.html#fbid=Nt3BdYMjtEJ
61. M.R. Jabbarpour, A. Jalooli, E. Shaghaghi, R.M. Noor, L. Rothkrantz, R.H. Khokhar, et al., Ant-based vehicle congestion avoidance system using vehicular networks. Eng. Appl. Artif. Intell. **36**, 303–319 (2014)
62. R. Jalali, K. El-khatib, C. McGregor, *Smart City Architecture for Community-Level Services Through the Internet of Things* (2015), pp. 108–113
63. M. Janssen, S. Chattopadhyay, Z. Rehena, A reference architecture for context-aware intelligent traffic management platforms. Int. J. Electron. Gov. Res. (IJEGR) **14**(4), 65–79 (2018)
64. S. Jeon, B. Hong, Monte Carlo simulation-based traffic speed forecasting using historical big data. Futur. Gener. Comput. Syst. **65**, 182–195 (2016)
65. D. Jia, D. Ngoduy, Platoon based cooperative driving model with consideration of realistic inter-vehicle communication. Transp. Res. Part C **68**, 245–264 (2016)
66. H. Jiang, J. Hu, S. An, M. Wang, B.B. Park, Eco approaching at an isolated signalized intersection under partially connected and automated vehicles environment. Transp. Res. Part C **79**, 290–307 (2017)
67. Y. Jianjun, F. Jiang, T. Zhu, RTIC-C: A big data system for massive traffic information mining, in *2013 International Conference on Cloud Computing and Big Data*, (IEEE, 2013), pp. 395–402
68. M.G. Karlaftis, E.I. Vlahogianni, Statistical methods versus neural networks in transportation research: Differences, similarities and some insights. Transp. Res. Part C **19**(3), 387–399 (2011)
69. Z. Khan, A. Anjum, K. Soomro, M. Tahir, Towards cloud-based big data analytics for smart future cities. J. Cloud Comput. **4**(1), 1–11 (2015)
70. B. Khondaker, L. Kattan, Variable speed limit: A microscopic analysis in a connected vehicle environment. Transp. Res. Part C **58**, 146–159 (2015)
71. Y. Kim, S. Atchley, G.R. Vallée, G.M. Shipman, {LADS}: Optimizing data transfers using layout-aware data scheduling, in *13th {USENIX} Conference on File and Storage Technologies ({FAST} 15)*, (2015), pp. 67–80
72. X. Kong, Z. Xu, G. Shen, J. Wang, Q. Yang, B. Zhang, Urban traffic congestion estimation and prediction based on floating car trajectory data. Futur. Gener. Comput. Syst. **61**, 97–107 (2016). https://doi.org/10.1016/j.future.2015.11.013
73. S. Kraijak, P. Tuwanut, A survey on internet of things architecture, protocols, possible applications, security, privacy, real-world implementation and future trends, in *2015 IEEE 16th International Conference on Communication Technology (ICCT)*, (IEEE, 2015), pp. 26–31
74. S. Kumar, L. Vanajakshi, Short-term traffic flow prediction using seasonal ARIMA model with limited input data. Eur. Transp. Res. Rev. **7**(3), 1–9 (2015)
75. S. Kwoczek, S. Di Martino, W. Nejdl, Predicting and visualizing traffic congestion in the presence of planned special events. J. Vis. Lang. Comput. **25**(6), 973–980 (2014)
76. I. Lana, J. Del Ser, I.I. Olabarrieta, Understanding daily mobility patterns in urban road networks using traffic flow analytics, in *NOMS 2016–2016 IEEE/IFIP Network Operations and Management Symposium*, (IEEE, 2016), pp. 1157–1162

77. D. Lazer, R. Kennedy, G. King, A. Vespignani, Big data. The parable of Google Flu: Traps in big data analysis. Science (New York, N.Y.) (6176), 343, 1203–1205 (2014)
78. M.W. Levin, S.D. Boyles, A multiclass cell transmission model for shared human and autonomous vehicle roads. Transp. Res. Part C **62**, 103–116 (2016)
79. L. Li, Y. Li, Z. Li, Efficient missing data imputing for traffic flow by considering temporal and spatial dependence. Transp. Res. Part C **34**, 108–120 (2013a)
80. M. Li, W. Hong, H. Kang, Urban traffic flow forecasting using Gauss–SVR with cat mapping, cloud model and PSO hybrid algorithm. Neurocomputing **99**, 230–240 (2013b)
81. Y. Li, Z. Li, L. Li, Missing traffic data: Comparison of imputation methods. IET Intell. Transp. Syst. **8**(1), 51–57 (2014)
82. L. Li, X. Su, Y. Wang, Y. Lin, Z. Li, Y. Li, Robust causal dependence mining in big data network and its application to traffic flow predictions. Transp. Res. Part C **58**, 292–307 (2015)
83. G. Li, J. Wang, Y. Zheng, M.J. Franklin, Crowdsourced data management: A survey. IEEE Trans. Knowl. Data Eng. **28**(9), 2296–2319 (2016a)
84. J. Li, Q. Ye, X. Deng, Y. Liu, Y. Liu, Spatial-temporal analysis on spring festival travel rush in china based on multisource big data. Sustainability **8**(11), 1184 (2016b)
85. H. Liao, M. Tang, L. Luo, C. Li, F. Chiclana, X. Zeng, A bibliometric analysis and visualization of medical big data research. Sustainability **10**(2), 166 (2018)
86. S. Lin, B. De Schutter, Y. Xi, H. Hellendoorn, Efficient network-wide model-based predictive control for urban traffic networks. Transp. Res. Part C **24**, 122–140 (2012)
87. Y. Liu, G. Chang, An arterial signal optimization model for intersections experiencing queue spillback and lane blockage. Transp. Res. Part C **19**(1), 130–144 (2011)
88. Y. Liu, H. Zheng, X. Feng, Z. Chen, Short-term traffic flow prediction with Conv-LSTM, in *2017 9th International Conference on Wireless Communications and Signal Processing (WCSP)*, (IEEE, 2017), pp. 1–6
89. H.P. Lu, Z.Y. Sun, W.C. Qu, Big data-driven based real-time traffic flow state identification and prediction. Discret. Dyn. Nat. Soc. **2015** (2015)
90. T. Luettel, M. Himmelsbach, H. Wuensche, Autonomous ground vehicles-concepts and a path to the future, in *Proceedings of the IEEE, 100 (Special Centennial Issue)*, (2012), pp. 1831–1839
91. Y. Lv, Y. Duan, W. Kang, Z. Li, F.Y. Wang, Traffic flow prediction with big data: A deep learning approach. IEEE Trans. Intell. Transp. Syst. **16**(2), 865–873 (2014)
92. C. Lynch, Big data how do your data grow? Nature **455**(7209), 28–29 (2008)
93. S. Madakam, R. Ramaswamy, S. Tripathi, Internet of things (IoT): A literature review. J. Comput. Commun. **3**(5), 164–173 (2015)
94. H.S. Mahmassani, M. Saberi, A. Zockaie, Urban network gridlock: Theory, characteristics, and dynamics. Transp. Res. Part C **36**, 480–497 (2013)
95. B. Matturdi, X. Zhou, S. Li, F. Lin, Big data security and privacy: A review. China Commun. **11**(14), 135–145 (2014)
96. Y. Mehmood, F. Ahmad, I. Yaqoob, A. Adnane, M. Imran, S. Guizani, Internet-of-things-based smart cities: Recent advances and challenges. IEEE Commun. Mag. **55**(9), 16–24 (2017)
97. V.R. Melnikov, V.V. Krzhizhanovskaya, A.V. Boukhanovsky, P.M.A. Sloot, Data-driven modeling of transportation systems and traffic data analysis during a major power outage in the Netherlands. Proc. Comput. Sci. **66**, 336–345 (2015)
98. V. Milanés, S.E. Shladover, Modeling cooperative and autonomous adaptive cruise control dynamic responses using experimental data. Transp. Res. Part C **48**, 285–300 (2014)
99. R. Minerva, A. Biru, D. Rotondi, Towards a definition of the Internet of Things (IoT). IEEE Internet Initiat. **1**(1), 1–86 (2015)
100. D. Moher, A. Liberati, J. Tetzlaff, D.G. Altman, Preferred reporting items for systematic reviews and meta-analyses: The PRISMA statement. PLoS Med. **6**(7), e1000097 (2009)
101. P. Mongeon, A. Paul-Hus, The journal coverage of web of science and Scopus: A comparative analysis. Scientometrics **106**(1), 213–228 (2016). https://doi.org/10.1007/s11192-015-1765-5

102. M.V. Moreno, A.F. Skarmeta, A.J. Jara, How to intelligently make sense of real data of smart cities, in *2015 International Conference on Recent Advances in Internet of Things (RIoT)*, (IEEE, 2015), pp. 1–6
103. F. Moretti, S. Pizzuti, S. Panzieri, M. Annunziato, Urban traffic flow forecasting through statistical and neural network bagging ensemble hybrid modeling. Neurocomputing **167**, 3–7 (2015)
104. R. Muggah, D. Kilcullen, These are Africa's fastest-growing cities – And they'll make or break the continent (2016), Retrieved December 1, 2017, from https://www.weforum.org/agenda/2016/05/africa-biggest-cities-fragility/
105. P.K. Muhuri, A.K. Shukla, A. Abraham, Industry 4.0: A bibliometric analysis and detailed overview. Eng. Appl. Artif. Intell. **78**, 218–235 (2019)
106. D.J. Nair, F. Gilles, S. Chand, N. Saxena, V. Dixit, Correction: Characterizing multicity urban traffic conditions using crowdsourced data. PLoS One **14**(4), e0215728 (2019)
107. D. Ngoduy, Multiclass first-order traffic model using stochastic fundamental diagrams. Transportmetrica **7**(2), 111–125 (2011)
108. H. Nguyen, L. Kieu, T. Wen, C. Cai, Deep learning methods in transportation domain: A review. IET Intell. Transp. Syst. **12**(9), 998–1004 (2018)
109. K. Olagunju, Evaluating traffic congestion in developing countries – A case study of Nigeria (2015)
110. H. Ospina-Mateus, L. Quintana Jiménez, F. Lopez-Valdes, K. Salas-Navarro, Bibliometric analysis in motorcycle accident research: A global overview. Scientometrics **121**(2), 793–815 (2019)
111. M. Ota, H. Vo, C. Silva, J. Freire, A scalable approach for data-driven taxi ride-sharing simulation, in *2015 IEEE International Conference on Big Data (Big Data)*, (IEEE, 2015), pp. 888–897
112. M. Ozbayoglu, G. Kucukayan, E. Dogdu, A real-time autonomous highway accident detection model based on big data processing and computational intelligence, in *2016 IEEE International Conference on Big Data (Big Data)*, (IEEE, 2016), pp. 1807–1813
113. A. Pascale, M. Nicoli, F. Deflorio, B. Dalla Chiara, U. Spagnolini, Wireless sensor networks for traffic management and road safety. IET Intell. Transp. Syst. **6**(1), 67 (2012)
114. C. Perera, R. Ranjan, L. Wang, S.U. Khan, A.Y. Zomaya, Big data privacy in the internet of things era. IT Prof. **17**(3), 32–39 (2015)
115. B. Płaczek, Selective data collection in vehicular networks for traffic control applications. Transp. Res. Part C **23**, 14–28 (2012)
116. N.G. Polson, V.O. Sokolov, Deep learning for short-term traffic flow prediction. Transp. Res. Part C **79**, 1–17 (2017)
117. M.M. Rathore, A. Ahmad, A. Paul, S. Rho, Urban planning and building smart cities based on the Internet of Things using big data analytics. Comput. Netw. **101**, 63–80 (2016)
118. C. Roncoli, M. Papageorgiou, I. Papamichail, Traffic flow optimisation in presence of vehicle automation and communication systems – Part II: Optimal control for multi-lane motorways. Transp. Res. Part C **57**, 260–275 (2015)
119. A.F. Salami, E.M. Dogo, T. Makaba, E.A. Adedokun, M.B. Muazu, B.O. Sadiq, A.T. Salawudeen, A decade bibliometric analysis of underwater sensor network research on the Internet of Underwater Things: An African perspective, in *Trends in Cloud-Based IoT*, (Springer, Cham, 2020), pp. 147–182
121. J. Sewall, D. Wilkie, M.C. Lin, Interactive hybrid simulation of large-scale traffic. ACM Trans. Graph. **30**(6), 1–12 (2011)
122. Z. Shan, Q. Zhu, Camera location for real-time traffic state estimation in urban road network using big GPS data. Neurocomputing **169**, 134–143 (2015)
123. T. Shapshak, Facebook has 170 MILLION AFRICAN USERS, mostly on mobile (2017), Retrieved November 8, 2017, from https://www.forbes.com/sites/tobyshapshak/2017/04/05/facebook-has-170m-african-users-mostly-on-mobile/#271fb1f853dc
124. Q. Shi, M. Abdel-Aty, Big data applications in real-time traffic operation and safety monitoring and improvement on urban expressways. Transp. Res. Part C **58**, 380–394 (2015)

125. S. Siuhi, J. Mwakalonge, Opportunities and challenges of smart mobile applications in transportation. J. Traffic Transp. Eng. (English Edition) **3**(6), 582–592 (2016)
126. Software Testing Help, Top 15 big data tools (big data analytics tools) in 2020 (2020), Retrieved April 2, 2020, from https://www.softwaretestinghelp.com/big-data-tools/
127. R. Soua, A. Koesdwiady, F. Karray, Big-data-generated traffic flow prediction using deep learning and Dempster-Shafer theory, in *2016 International Joint Conference on Neural Networks (IJCNN)*, (IEEE, 2016), pp. 3195–3202
128. P. St-Aubin, N. Saunier, L. Miranda-Moreno, Large-scale automated proactive road safety analysis using video data. Transp. Res. Part C **58**, 363–379 (2015)
129. M. Strohbach, H. Ziekow, V. Gazis, N. Akiva, Towards a big data analytics framework for IoT and smart city applications, in *Modeling and Processing for Next-Generation Big-Data Technologies*, (Springer, Cham, 2015), pp. 257–282
130. C. Su, Big data security and privacy protection, in *2019 International Conference on Virtual Reality and Intelligent Systems (ICVRIS)*, (IEEE, 2019), pp. 87–89
131. A. Talebpour, H.S. Mahmassani, Influence of connected and autonomous vehicles on traffic flow stability and throughput. Transp. Res. Part C **71**, 143–163 (2016)
132. L. Tang, Z. Kan, X. Zhang, X. Yang, F. Huang, Q. Li, Travel time estimation at intersections based on low-frequency spatial-temporal GPS trajectory big data. Cartogr. Geogr. Inf. Sci. **43**(5), 417–426 (2016)
133. D.S. Terzi, R. Terzi, S. Sagiroglu, A survey on security and privacy issues in big data, in *2015 10th International Conference for Internet Technology and Secured Transactions (ICITST)*, (IEEE, 2015), pp. 202–207
134. B. Thuraisingham, Big data security and privacy, in *Proceedings of the 5th ACM Conference on Data and Application Security and Privacy*, (2015), pp. 279–280
135. Y. Tian, L. Pan, Predicting short-term traffic flow by long short-term memory recurrent neural network, in *2015 IEEE International Conference on Smart City/SocialCom/SustainCom (SmartCity)*, (IEEE, 2015), pp. 153–158
136. J.L. Toole, S. Colak, B. Sturt, L.P. Alexander, A. Evsukoff, M.C. González, The path most travelled: Travel demand estimation using big data resources. Transp. Res. Part C **58**, 162–177 (2015)
137. D. Tosi, S. Marzorati, Big data from cellular networks: Real mobility scenarios for future smart cities, (2016). Paper presented at the pp. 131–141
138. J. Wan, J. Liu, Z. Shao, A.V. Vasilakos, M. Imran, K. Zhou, Mobile crowdsensing for traffic prediction in internet of vehicles. Sensors (Basel, Switzerland) **16**(1), 88 (2016)
139. G. Wang, Z. Xu, F. Wen, K.P. Wong, Traffic-constrained multi-objective planning of electric-vehicle charging stations. IEEE Trans. Power Deliv. **28**(4), 2363–2372 (2013)
140. J. Wang, W. Deng, Y. Guo, New Bayesian combination method for short-term traffic flow forecasting. Transp. Res. Part C **43**, 79–94 (2014a)
141. L. Wang, F. Zhang, A. Vasilakos, C. Hou, Z. Liu, Joint virtual machine assignment and traffic engineering for green data center networks. ACM SIGMETRICS Perform. Eval. Rev. **41**(3), 107–112 (2014b)
142. F. Wang, L. Hu, D. Zhou, R. Sun, J. Hu, K. Zhao, Estimating online vacancies in real-time road traffic monitoring with traffic sensor data stream. Ad Hoc Netw. **35**, 3–13 (2015)
143. P.-W. Wang, H.-b. Yu, X. Lin, L. Wang, Online traffic condition evaluation method for connected vehicles based on multisource data fusion. J. Sens. **2017** (2017a)
144. Y. Wang, J. Cao, W. Li, T. Gu, W. Shi, Exploring traffic congestion correlation from multiple data sources. Pervasive Mobile Comput. **41**, 470–483 (2017b)
145. Y. Wei, M. Chen, Forecasting the short-term metro passenger flow with empirical mode decomposition and neural networks. Transp. Res. Part C **21**(1), 148–162 (2012)
146. L. Wei, D. Hong-ying, Real-time road congestion detection based on image texture analysis. Proc Eng **137**, 196–201 (2016)
147. A. Wibisono, W. Jatmiko, H.A. Wisesa, B. Hardjono, P. Mursanto, Traffic big data prediction and visualization using fast incremental model trees-drift detection (FIMT-DD). Knowl.-Based Syst. **93**, 33–46 (2016)

148. X. Wu, X. Zhu, G.-Q. Wu, W. Ding, Data mining with big data. IEEE Trans. Knowl. Data Eng. **26**(1), 97–107 (2014)
149. Y. Wu, H. Tan, L. Qin, B. Ran, Z. Jiang, A hybrid deep learning based traffic flow prediction method and its understanding. Transp. Res. Part C **90**, 166–180 (2018a)
150. Y. Wu, W. Zhang, J. Shen, Z. Mo, Y. Peng, Smart city with Chinese characteristics against the background of big data: Idea, action and risk. J. Clean. Prod. **173**, 60–66 (2018b)
151. D. Xia, H. Li, B. Wang, Y. Li, Z. Zhang, A map reduce-based nearest neighbor approach for big-data-driven traffic flow prediction. IEEE Access **4**, 2920–2934 (2016a)
152. D. Xia, B. Wang, H. Li, Y. Li, Z. Zhang, A distributed spatial-temporal weighted model on MapReduce for short-term traffic flow forecasting. Neurocomputing **179**, 246–263 (2016b)
153. K. Xie, K. Ozbay, A. Kurkcu, H. Yang, Analysis of traffic crashes involving pedestrians using big data: Investigation of contributing factors and identification of hotspots. Risk Anal. **37**(8), 1459–1476 (2017)
154. S. Yang, S. Shi, X. Hu, M. Wang, Spatiotemporal context awareness for urban traffic modeling and prediction: Sparse representation based variable selection. PLoS One **10**(10) (2015)
155. K. Yang, S.I. Guler, M. Menendez, Isolated intersection control for various levels of vehicle technology: Conventional, connected, and automated vehicles. Transp. Res. Part C **72**, 109–129 (2016)
156. H.-F. Yang, T.S. Dillon, Y.P. Chen, Optimized structure of the traffic flow forecasting model with a deep learning approach. IEEE Trans. Neural Netw. Learn. Syst. **28**(10), 2371–2381 (2017)
157. H. Yao, F. Wu, J. Ke, X. Tang, Y. Jia, S. Lu, et al., Deep multi-view spatial-temporal network for taxi demand prediction, in *Thirty-Second AAAI Conference on Artificial Intelligence*, (2018)
158. H. Yi, H. Jung, S. Bae, Deep neural networks for traffic flow prediction, in *2017 IEEE International Conference on Big Data and Smart Computing (BigComp)*, (IEEE, 2017), pp. 328–331
159. C. Yin, Z. Xiong, H. Chen, J. Wang, D. Cooper, B. David, A literature survey on smart cities. SCIENCE CHINA Inf. Sci. **58**(10), 1–18 (2015)
160. B. Yu, L. Kong, Y. Sun, B. Yao, Z. Gao, A bi-level programming for bus lane network design. Transp. Res. Part C **55**, 310–327 (2015)
161. S.K. Zegeye, B. De Schutter, J. Hellendoorn, E.A. Breunesse, A. Hegyi, Integrated macroscopic traffic flow, emission, and fuel consumption model for control purposes. Transp. Res. Part C **31**, 158–171 (2013)
162. Y. Zhang, A. Haghani, A gradient boosting method to improve travel time prediction. Transp. Res. Part C **58**, 308–324 (2015)
163. Y. Zhang, Y. Zhang, A. Haghani, A hybrid short-term traffic flow forecasting method based on spectral analysis and statistical volatility model. Transp. Res. Part C **43**, 65–78 (2014)
164. Y. Zheng, Z. Zha, T. Chua, Mining travel patterns from geotagged photos. ACM Trans. Intell. Syst. Technol. (TIST) **3**(3), 1–18 (2012)
165. Z. Zheng, S. Ahn, D. Chen, J. Laval, The effects of lane-changing on the immediate follower: Anticipation, relaxation, and change in driver characteristics. Transp. Res. Part C **26**, 367–379 (2013)
166. Y. Zheng, L. Capra, O. Wolfson, H. Yang, Urban computing: Concepts, methodologies, and applications. ACM Trans. Intell. Syst. Technol. (TIST) **5**(3), 1–55 (2014)
167. J.Z. Zhu, J.X. Cao, Y. Zhu, Traffic volume forecasting based on radial basis function neural network with the consideration of traffic flows at the adjacent intersections. Transp. Res. Part C **47**, 139–154 (2014)
168. L. Zhu, F.R. Yu, Y. Wang, B. Ning, T. Tang, Big data analytics in intelligent transportation systems: A survey. IEEE Trans. Intell. Transp. Syst. **20**(1), 383–398 (2019)

Smart City and Smart Transportation: Intelligent IoT-Based Transportation Objects "Me-Online Mobile Application: A Mutual Practice of Internet of Mobile Things"

Haifaa Omar Elayyan

1 Smart Technologies for Smart Cities

1.1 Introduction

The uprising revolution of creating and manipulating developed urban areas that assures a high quality of daily life activities based on Smart technology is called Smart cities. The basic, direct, and simplest we found to define a Smart city, according to Techopedia dictionary [1], is a designation with predetermined specifications that usually include terms and conditions regarding technologies being illustrated and deployed. The technology comes with real potentials of infrastructure abilities to incorporate information and communication technologies to enhance the quality and performance of urban services. Services can be energy, transportation, and/or any innovations that help to reduce resource consumption and the unjustified wastage we witness these days. As a result, it is expected to manage overall costs and increase productivity by installing Smart technologies [1].

This flexible definition of Smart cities is limited to the technology these cities had installed so far. In practice, we often hear about cities labeling themselves as "Smart," limiting the term to the efforts of deploying and utilizing some types of Smart solutions based on information communication technology (ICT) [2]. The transformation from the old version to the Smart version usually requires the authority to redesign their infrastructures to cope with more digitalization, urbanization, and globalization features [2]. Unfortunately, studying and observing most of current "Smart cities," we find that Smart governments do assign central grants for this huge transformation and spend billions on relatively conventional

H. O. Elayyan (✉)
Middle East University, Amman, Jordan

U. Z. A. Hamid, F. Al-Turjman (eds.), *Towards Connected and Autonomous Vehicle Highways*, EAI/Springer Innovations in Communication and Computing,
https://doi.org/10.1007/978-3-030-66042-0_5

developments and infrastructure projects that are not predominantly "Smart" enough to activate Smart cities and meet the expectations of Smart residents.

Most of these plans and projects are driven by the obsession for installing Smart technologies more than being steered by real needs and expectations of citizens. In other words, citizen experience is overlooked or neglected, which is the opposite of the main purpose of creating Smart cities. The purpose should empower Smart citizens to design and shape their future if the Smart city theme is properly applied.

These scenarios are not dominant as we can find other examples of Smart cities being aware of such crucial factors. For example, Toronto, Canada, is a city with a strong ambition to act like a Smart city [3]. The local authority recognizes the power of citizen participation in producing innovative solutions that can transform the way society functions. This Smart city cares about the citizen's requirements and has been leveraging its "creative class" of financiers, healthcare researchers, artists, corporate strategies, social work pioneers, and lawyers to incorporate and interact together shaping the future of the city in the direction the citizens want and desire. The uncontrolled race, in most scenarios, creates more challenges as the main idea of Smart cities lies behind improving citizens' quality of life, not only exploiting technologies for urban growth that might come with a shortage of energy resources, inadequate and deteriorating energy infrastructure. Most of these Smart city projects were ended without meeting the requirements of Smart citizens.

Then, we simply say that the modest definition of a "Smart city" is subjected to a continuous assessing process to cope with the rapid escalation of Smart technologies being employed for urban services. Excelling multiple crucial keys such as economy and mobility is framed with strong human capital, social capitals, ICT infrastructures, and supportive governance indeed. If we archeologically read history, we find that developing any city is an expected process and spearheaded in contemporary times publicized in different history chapters [4]. This continued to specify and shape five types of sociopolitical actors to develop that city:

1. Agenda-setters, who are usually the governments
2. Experts, who are the planners for the new urban area
3. Sponsors, who invest in this development process
4. Developers, who mature and enhance the development process
5. The citizens, the real influencers, who can be academia leaders, residents, and/or business initiatives

The process of such developing will end to plan and act smarter to increase productivity and reduce costs, which comprehensively explains why Smart cities should be modeled on platforms of Smart technologies. Next, we thoughtfully explain why Smart technologies play the role of pillars in the Smart city concept and what we do expect to develop when we emerge Smart technologies into the different components of a Smart city. This at first can be shown in Fig. 1 [5].

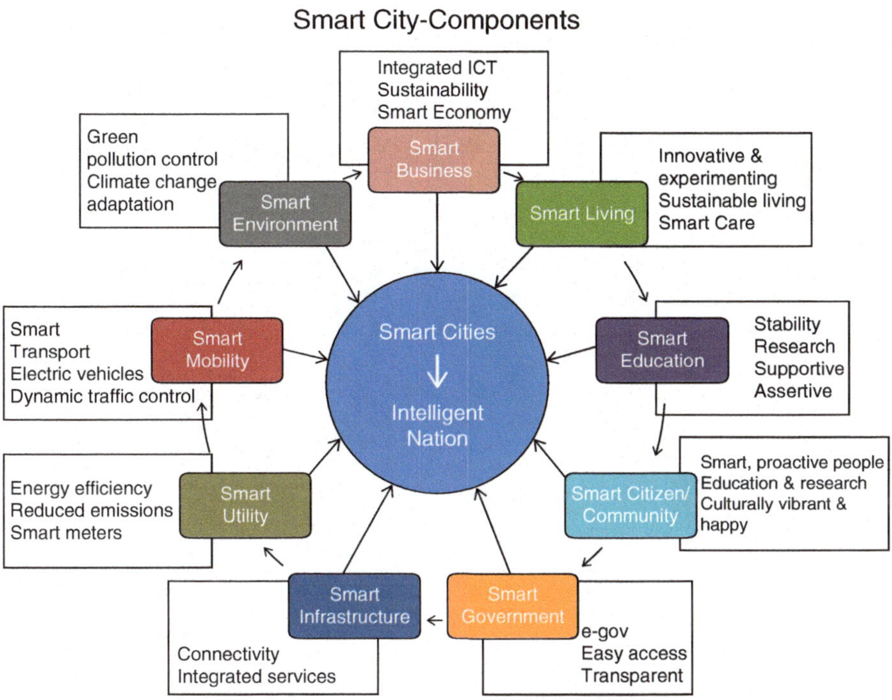

Fig. 1 Smart city components. [URL: https://slideplayer.com/slide/13426791/80/images/15/Smart+City-Components.jpg]

1.2 Smart Technology: Why Is It Needed?

Smart cities around the world are changing in terms of their requirements and components. The Smart city standards are established by organizations such as the International Organization for Standardization (ISO) and forwarded globally with specifications to drive expansion and growth while maintaining and ensuring quality, efficiency, and community safety as well. Agreed standards for all developing and transforming procedures to create Smart cities can play important roles in the enhancing cycles and the construction of these Smart cities, adding more infrastructure layers. Standards can help tackle climate change, address security and transportations problems while ensuring the quality of life and positive impacts on the surrounding environments, and taking into consideration some other important factors such as business practices and resource management platforms. The IEEE has been developing standards for Smart cities for different components of Smart cities as Smart grids, Internet of Things (IoT), e-health, and Intelligent Transportation Systems (ITS) [6].

The sustainability of Smart cities is related to the city infrastructure and governance as they ensure the quality of other interacted components. The quality

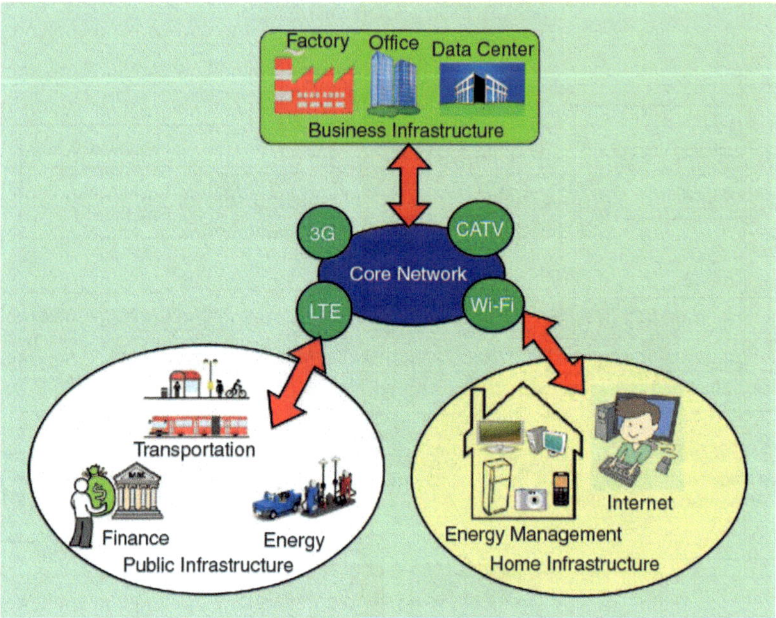

Fig. 2 Smart infrastructure depictions [6]. [URL: https://ai2-s2-public.s3.amazonaws.com/figures/2017-08-08/8fb828364718dd9eff4711dd7f18e342a3366bc9/4-Figure3-1.png]

of life of Smart citizens is not limited to the financial well-being but also exceeds to include multiple emotional aspects and indicators such as technology being deployed, smartness of conceptualizing requirements, and needs of those Smart citizens. The general model of a Smart city is expected to be robust to make decisions based on learning and training procedures using the information and deploying advanced physical ICT infrastructures. In turn, Smart technologies are the backend of Smart infrastructure which makes the physical infrastructure "Smart" enough to handle the information [6]. The Smart infrastructure examples can be sensors, software, and middleware interacting as Smart components as shown in Fig. 2 above.

Smart technologies are promoted to be the right key to designing, implementing, and operating Smart cities. Technologies such as communication infrastructure, Global ICT infrastructures, social networks, cyber-physical systems, and sustainable transportations systems are examples of Smart technologies that help to construct Smart cities and bring them into real-life functioning. It is important to mention that implanting Smart technologies is not that costly if it is compared to the benefits of deploying them [7]. The use of Smart technologies in various forms for different daily activities undoubtedly leads to an increase in the effectiveness of Smart city operations, flexibility, and affordability at one time.

This chapter continues presenting Smart technologies and suggests how Smart transportation should be modeled. The chapter also discusses challenges of digital transportation systems based on the Internet of Things concept (IoT) and presents successful cases when the Smart citizens' experiences are efficiently utilized, such as the case of "Uber" detailing how positively it impacted the Smart environment and refreshed small businesses.

The chapter later proposes a mutual concept of an intelligent "transportation – learning object" designed and scored on the customization of learning – objects and the intelligent tutoring system (ITS) concepts taking into consideration that these concepts have been being employed in virtual learning environments to attain the ultimate quality of the educational systems [8]. The proposal brings to the reader's consideration that Smart technology at first is built on utilizing, analyzing, and customizing digital information into components and frames that act individually or in combination with other components to increase the efficiency of any activity or operation.

The proposed intelligent transportation learning object is reciprocally presented as a collection of semantic analysis and content items of regular users with daily activities and practices in the digital transportation systems. The suggested design reflects the practice items and assessment items that are combined based on a single learning object framed with user's preferences. We mainly suggest customizing the learning object to play an innovation role of positioning numerous other terms, including content objects, chunks, educational objects, information objects, intelligent objects, knowledge bits, knowledge objects, learning components into an IoT platform designed for transporting/relocation purposes.

The core idea of forming these intelligent transportation learning objects is to tame the new applications of IoT and the Internet of Mobile things (IoM) toward having more flexibility, re-usability, efficiency, and sustainability when it comes to the use of information and telecommunication technologies. Besides those mentioned, transportation objects will mature Smart citizens' experience with Smart transportation systems emerged with a collective of many other public services and IoM-based small businesses in order to overcome obstacles in terms of adding more physical infrastructures and translate digital technologies into better public services for inhabitants and better use of resources while reducing the environmental impacts.

The proposed system is mutually designed to enable the user to arrange for a trip digitally from location A to location B. The system should be supported with all Internet of Mobile things technology (IoM) and Geographical Information System (GIS) to characterize media objects, reusable information objects, and reusable learning objects as the system is also built on artificial intelligence units to learn, train, test, and make decisions.

2 Internet of Things (IoT): IoT-Based Smart Transportation

2.1 Introduction

Internet of Things (IoT) stands for embedded systems networking physical objects via devices that can collect and transmit information about a real-time activity, such as sensors or actuators. The deployment of Internet of Things (IoT) technologies in Smart cities has been undeniable, but also the evolution of IoT technologies promotes the idea of Smart cities in general and the transportation section in specific. Emerging IoT technologies with smart transportation systems comes with high expectorations to reconstruct the whole system improving passengers experience safely and efficiently. The Smart technology has the ultimate potentiality to positively impact and transform the transport industry into a more manageable platform by overpoweringly altering how this transportation system gathers and deploys data and information using the major technical and business trends of mobility, automation, and big data analytics. IoT technologies are at the core of processes that have forces to reshape transportation systems into Smart ones functioning at the highest performance can ever be encountered affording more efficacy, improved vehicle, and aircraft maintenance. Moreover, it will be permitted for more well-formed strategic plans to enhance traffic management and quality control actions [9].

More examples of reimbursements of embedding IoT technologies into the transportation system can be detailed as follows:

1. Promoting autonomous vehicles that are able to predict behavior, communicate with other Smart vehicles for decision-making purposes by sensing the surrounding environment, and react promptly to real-life risk scenarios.
2. Promoting frequent monitoring and direct supervising with video surveillance solutions supported by high-resolution CCTV cameras. Streamed data and information are analyzed and automated for early detections and accidental avoidance.
3. Dynamic guidance notifying the system as an example of illustrating a dynamic roadside sign displaying messages to warn drivers of what they should expect while driving. Intelligent transportation systems can display real-time road statuses such as lane changing, lane closures, or even weather broadcasting and next anticipating events. Streamed data are displayed on these dynamic signs and automatically fed from sensors and cameras.
4. Increases system capacity and enhances passenger safety while lowering costs and risks. This means that IoT can enhance transportation performance and efficiency by transmitting big data set to be analyzed and treated for quality control purposes and making the right verdicts at the right time.

The data gathered from these interrelated devices are analyzed by the management system for many purposes that we can mention some according to Alcatel

Lucent online release about solutions brief – IoT in transportation. The online article is listed in the references for further updates and reading [10, 11]:

1. Data are collected through IoT to mainly upgrade the operational performance by proactively monitoring life-threatening infrastructure deficiencies and generating more efficient methods to incorporate system capacity with real-life situations and requirements.
2. A probationary to reduce congestion and energy use by analyzing real-time data and anticipate fast traffic changing patterns.
3. Increases safety by a better understanding of transit architecture and system which in turn improves the travel experience affording a more reliable transportation system.

Many studies and research proposals have presented theoretical and practical IoT-based provisions and applications for Smart and intelligent transportation systems. Sherly and Somasundareswari [12] published an article entitled "Internet of things based smart transportation systems," which described the improvisations that have been done so far to the prevailing transportation systems and presented according to analysis and studies of traditional parking system without deploying the IoT embedded bodies, and explained challenges and difficulties the users have to suffer. The publication later proposed the IoT-based Smart parking assistance system which makes sure of wireless sensors to obtain real-time traffic information, such as traffic conditions on each road, number of vehicles, and average speed. Wireless sensors are appropriate due to their low power consumption and low cost of illustrating. To achieve large-scale network layout, they used the wireless cluster sensor network which has a set of wireless sensors that is represented by the head node. Transmitting data at these head nodes directly to the back-end system through mobile agents takes into consideration that most of the vehicles have been equipped with the Global Position System (GPS) capable of receiving and sending driving information. The paperback in 2015 presented a real-time traffic monitoring system to solve the problem of finding parking lots via the communication between the installed GPS – inside the car – and the satellites. That was a good trial of deploying the concept of IoT systems into the transportation sector [12].

With the revolution of smartphones spreading, Saarika et al. [13] proposed the Smart parking system composed of intelligent sensors along with intelligent signboards. The proposed system is composed of a number of intelligent sensors positioned on-site to monitor and notify about the availability of parking places. A mobile application is launched and installed on the user's smartphone to check cautiously looking for the free spot to park. The signboard with embedded RD module networking with sensors functioning with solar energy will enlighten the desired place, detailing the distance to that spot and give a list of different routes to that spot. Such a proposed system manages parking needs more effectively. If we widen the result of employing such a system, we will shortly be cautious that a developed idea of that proposed system can also manage the traffic in that area as well by getting real-time visibility of surrounding traffic conditions. The IoT e-solution for traffic and parking management comes with a package of Smart sensors to collect real data that can be analyzed and redirected to other drivers and local authorities in case of suspicious feedback. The IoT e-solution increases safety by providing city authorities with important information about the flow of people and resources. It can also broadcast alerts in emergency cases and offer alternative routes to reduce congestion at the end. The intelligent solutions can be embedded in more distributed devices such as traffic lights, highway entrance, and exits to control and monitor traffic usage inside and outside cities.

Another exciting IoT-based application is the lighting application to make city streets safer while reducing power and maintenance costs with connected street lighting. Monitoring and controlling traffic while classifying streets into slots showing which streets are crowded and need lighting and other streets that are with free lanes at specific times that do not need lighting. Such a Smart solution is to reduce the energy being used in these places and consequently reduce costs.

The illustrated intelligent systems into the Smart transportation industry come to cope with the need for intelligent transportation systems being able to optimize movements of users, people, and products improving economies, public safety, and the polishing environment to get expanded over the concept of globalization. The scheming of these intelligent systems can be in many forms, such as being end-to-end intelligent transportation solutions or in the form of developing individual components of Smart transportation systems. Regardless of the form, these solutions should eliminate cost concerns, complexity, and should have built-in compatibility units to get integrated directly into the market for real-life functioning without the anxieties of adopting scalability. Intelligent transportation system openings proliferate across a wide range of industries and market segments, such as the following [14]:

1. Transport logistic solutions that help cargo operators to operate, ship, and deliver products regardless of distance or geographical circumstances as the system is controlled, managed through embedded intelligent units to gather and analyze data from onboard sensors to track containers and packages and to monitor environmental conditions, with the high level of quality management and control. The figure shows an approximate illustration of such a service where a small application can be used for transporting and delivering logistics. As you can imagine, the track condition and location of shipping are automated and in real time. We can conclude that such an example of an intelligent IoT-based Smart transportation system optimizes the transport and delivery of inventory as well as reducing product spoilage and damage, as there will be always a risk management plans. Such Smart technology improves customer satisfaction and profitability and eliminates useless human interventions.

   In addition, such a solution with pre-scheduling, real-time control, and supervision for sure will reduce fuel costs and vehicle maintenance expenses. Figure 3 shows an example of real-time control.
2. Smart train control systems which are an intelligent solution to control train routes and build a reliable collision avoidance system will be designed to operate and cooperate with the rail operators to ensure safety by predicting

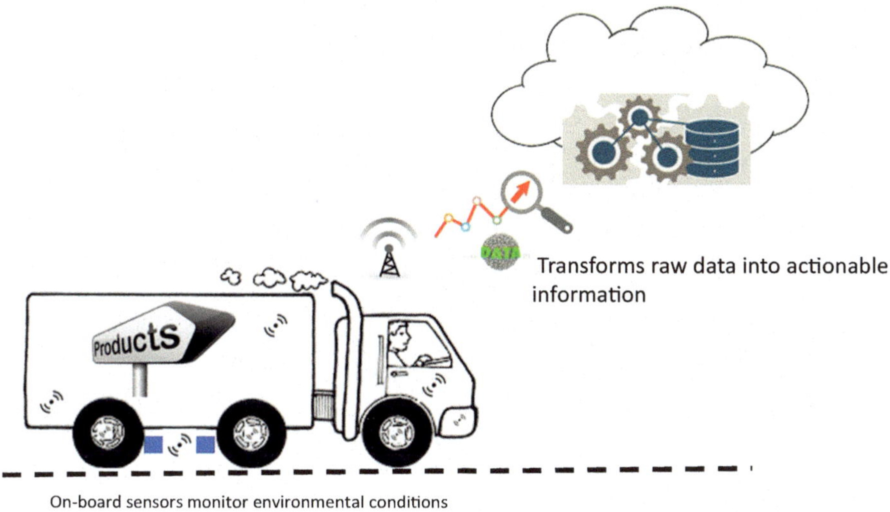

Fig. 3 Transportation logistics example

failures in advance and control the speed as in the case of auto-pilot Smart systems. The intelligent solution will track and guide trains in real time with fully automated control of braking actions and managing routing and relevant costs. An intelligent train system can operate dynamically to avoid any collisions and construction zone incursions. An intelligent train control system improves safety and guarantees on-time arrivals, departures, and reduces fuel consumption. The intelligent system is designed to ensure compliance with government and industry regulations to improve system productivity and mitigate risks. Figure 4 shows an approximate model of a train control system based on intelligent IoT applications and technologies.

3. Figure 5 shows a use case of an IoT-based intelligent solutions that are called fleet telematics. Fleet telematics Smart solutions help businesses and transportation carriers and governments improve the economy, safety, and compliance by intelligently monitoring their vehicles. The Smart solution is extended to monitor the driver's behavior and performance and improve customer satisfaction. The solution is designed to optimize the use of the fleet, dynamically route trams based on real-time conditions as well as ensure that trams are available when needed and not sent out unnecessarily. Example of fleet telematics in the market includes an Italian tram system operator.

In the preceding examples, we have shown and highlighted the importance of collecting big data, analyzing, and making smarter decisions for Smart cities. The Internet of Things (IoT) involves connecting physical objects as we have presented the concept so far to the Internet to build Smart systems and modernize further e-solutions to advance technologies based on universal mobile accessibility

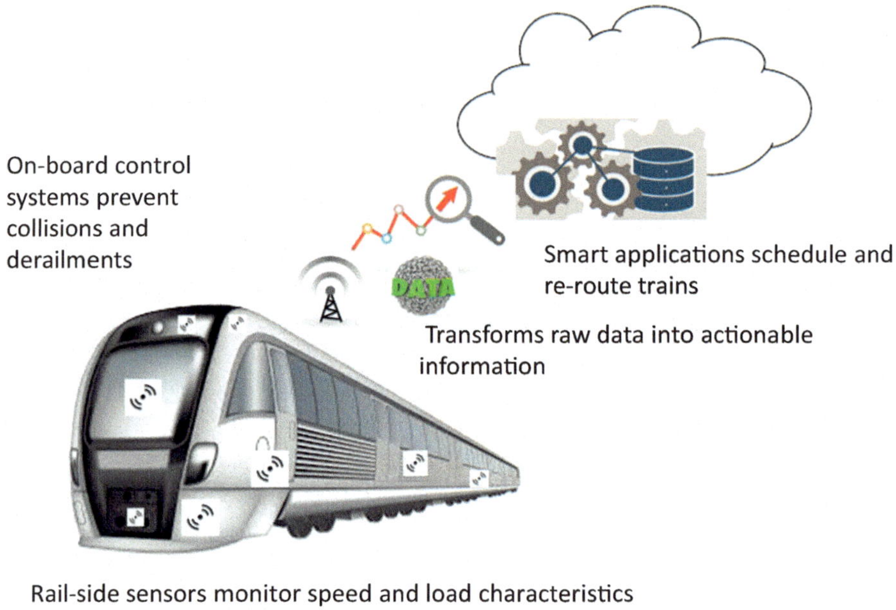

Fig. 4 The intelligent case of Smart train controls

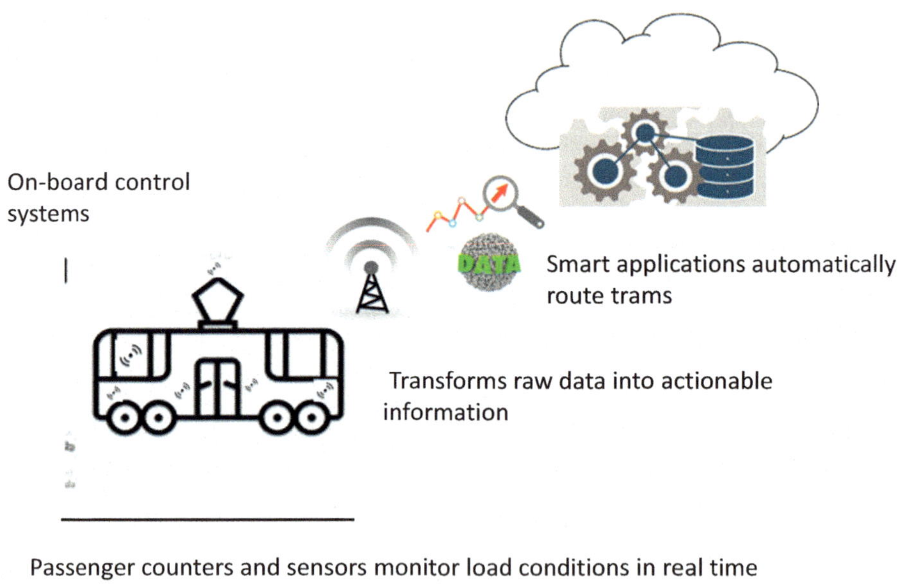

Fig. 5 Fleet telematics based on IoT solutions

such as the Intelligent Transportation System (ITS) [14]. The IoT-based intelligent transportation system (IoT-ITS) helps in automating railways, roadways, airways, and marine, which promotes and develops customers the experience of being served with highest quality assurance and control, which perfectly fits the designing, planning, and developing the Smart city components and architectures. The IoT-ITS will practice several techniques and algorithms to achieve full functioning and implementations of this Smart technology. A case study on intelligent traffic management systems based on IoT modeling and big data transmitting, analyzing, and making decisions was published by Muthuramalingam et al. The publication is entitled "IoT Based Intelligent Transportation System (IoT-ITS) for Global Perspective: A Case Study" [15]. The study concluded that the emerging model, namely big data and IoT, plays a major role in daily activities and applications. Such emergence succeeded to create technologies aiding easiness of preprocessing and analyzing data that are needed for the next cycles of transmitting and making decisions. An intelligent transportation system based on such emerging technology helps in monitoring vehicles' motions and determining the traffic in an area relative to the data received. The system can also substitute the current conventional systems with extreme efficiency and performance for fewer costs and best time management. Future work can be expanded from our view to the human interaction level with such intelligent systems. Enhancing human interaction will increase customer satisfaction and sustain such an amazing solution. Staying with the same ideology of human interaction, in the following part, we will argue the potential use of data transmitted from the social network and media where people interact, communicate, and exchange experience. The analyzed data coming from these websites are empowered with the concept of hashtags.

2.2 *Will the Perceived Data and Information from Social Network endorse Smart Solutions in Transportation Systems?*

To answer this argument, we need first to slightly introduce multimedia social networks and how deep people count on during their life activities. Multimedia social network is a network in which a group of users share and exchange multimedia content, as well as other resources [16]. Social networks have significantly played a role in the distribution of ideas and advertisements. Heidemann et al. [17] provide an overview of online social networks that contributes a better understanding of this worldwide spectacle. Multimedia social networks are classified into three categories [18]:

1. Imagery social networks, which capture social relationships and activities between users through photos, video, or wireless sensors
2. Gaming-driven social networks, which allow users to maximize their payoff by exchanging and sharing their resources for learning and sharing experiences of playing games

3. Interaction-driven social networks, which characterize relationships based on user interaction and online activities such as posting information, searching within the social network, video streaming, and live communications

By tracking the social media, we can analyze community activities and frequency of using resources as well as the value of data and information people are exchanging. We can also derive metadata and aggregate information for better understanding and reasoning about the exchanged content. Multimedia social networks such as Facebook and Twitter have influenced people's lives and impacted their choices as well. Frequently used features and attributes supported by multimedia social networks such as a group of discussions, commenting, interconnecting with a huge number of members, tracking and identifying locations of users, as well as dynamic changing in their status according to their interactions to posts increase the importance of communicating and interacting using these websites and platforms. Furthermore, people do trust the data and information others are giving more than the official sites as most are sharing real-life experiences away from the professional side. These social networks have embedded more tools to enhance the data searching on these platforms; most known is the use of hashtags. So, what are hashtags? Hashtags are metadata tags most likely being used on social networks such as Twitter and other microblogging online activities, allowing to apply for dynamic, user-generated tagging, which increases the possibility for other network users to search for and find [19, 20]. The designed metadata adds specific themes or content to allow informal markup of folk taxonomy discarding the need for any markup-languages. This is a fantastic tool of the hashtag that can be extended to enhance the hashtag performance seeking more accurate search results by adding the semantic taxonomy using the user's previous online profiles, interesting data, experience, suggestion, live streaming data, and accumulated social network information. And this is what we do suggest here as a new development of hashtag tools.

Comprehensively, hashtags can be promoted to be based on IoT solutions that help to reflect real-life data information posted on multimedia social networks and exchanged in terms of the transportation sector. Hashtag such as #trafficJam can be a piece of information alerting local authorities and other drivers that there is a traffic jam in that specific route (location) from where that person posted the hashtag to his/her multimedia social network page or discussion group. Multiple cycles of forwarded such hashtags can notify other interested parties to avoid that jammed route. Other drivers who are with advanced experience and aware of the surrounding environment can suggest other alternative routes to release that congestion. Or previously accumulated data can be brought up for similar situation transpired before and were somehow resolved. Hashtags based on IoT technologies and accessibility to any mobile applications or websites simply produce new e-solutions yielding to more affordable, reliable, and manageable Smart transportation systems.

The number of multimedia social networks is considerable. So, you would assume that complications arise if we want to retrieve the data and analyze given the chance of multioccurrences, frauds, and repeated multiplicity as the action is expected to be dynamic with a flow of information from an acceptable number of social networks. Well, we would like to simplify it for you and inform you that there are lots of researches, studies, and proposed models to integrate the information on social media and manipulate this information as if we are talking about one single platform. Amato et al. presented the "Multimedia Social Network Modeling: A Proposal," which promoted a preliminary work concerning the definition of a novel data model for multimedia social networks (MSNs), such as social networks, that associate information on the users belonging relatively to one or more social communities together with multimedia content that is created and manipulated within an interrelated environment [21].

A hyper graphed structure model allows us to represent the different typical forms of interactions and relationships on a social network in a simple visualized, analyzed, and understood way. The model supports several kinds of applications utilizing the introduction of several ranking functions [21]. The model is taking a step forward to provide a strategy for mapping the proposed model into an object-relational data model for efficiently storing related data. A prototype is shaped to store and integrate information from different online social networks (OSNs) (Facebook, Twitter) and the multimedia sharing systems (such as Flikr and LastFM), to provide some facilities for browsing the entire hypergraph. We do encourage readers and researchers to have further look at the proposed system and promote further models using the data mining concepts that have been functioning well with the concept of the big data set analyzing techniques. Social networks analysis and mining help to rank influential users and track their influence all over the interconnected and integrated networks for learning experiences and anticipated features.

This chapter next discusses recent challenges of deploying the IoT applications in the transportation sector and suggests a further solution that can cope with the technology and be applied to overcome these challenges.

2.3 *Challenges of Internet of Things Deployment*

Most of the challenges are raised because of the unprecedented streams of data presenting challenges for network and data management along with amplified security risks [22]. We are indicating that there is a big data set to consider, especially for real-life environments, such as monitoring a traffic line. Realistically, to be able to address the challenges when we merge IoT technologies into any Smart platform such as transportations systems, we need to discuss possibilities of providing new levels of intelligence, automation, and security consequences. And hence, the possibility of affording and dedicating a reasonable budget to instigate an effective network infrastructure that securely handles vast flows of data and affords framework with manageable platforms to operate and control.

Information technology professionals and researchers are achieving on this front to make plans for more possibilities of IoT emerging toward flexible and reliable Smart transportation systems. They cooperate and incorporate their work in a variety of industries to increase the possibilities of replacing traditional structures with Smart IoT solutions for less and reasonable costs. They all claim that IoT systems will have a significant impact on the economy, the organization's planning and decision-making for more information technology and communication infrastructure budgets and planning strategies [22, 23].

If we consider the Internet of Things (IoT) as the future evolution of the Internet that comprehends machine-to-machine (M2M) learning, then it is rational to merge more of IoT technologies to provide connectivity for everyone and everything. Well-known and widely used IoT applications with intelligent components are embedded into the Internet-connected objects to communicate, exchange information, make decisions, and invoke actions for more Smart services enabling. It is more rational to promote the development trends, the generic architecture of IoT, and deploying its distinguished features with the development of IoT itself. The concept of more artificial expert systems is more desired for treating the environment as smart enough to collect information from the surrounding and analyze the data for learning purposes without humane interference [22–24].

IoT challenges can be presented as the following list which is open for advanced researches and problem- solving models:

The IoT infrastructure must provide a simple, automated process for IoT devices onboard which means that each device should be equipped with sensors, manually provisioning to manage and manipulate data accordingly. Automated onboarding enables the IoT platform to dynamically recognize devices and control them while enforcing decision-making procedures and possibilities.

It is crucial to allocate resources for the IoT systems to operate properly and efficiently. Most of the devices forming the IoT system deliver mission-critical information that requires a specific level of quality of assurance systems such as affording proper bandwidth reservations on a high-performance network infrastructure to ensure service consistency.

There are challenges of providing a secure environment against cyberattack and data loss. Such a challenge that is most disastrous can take a place while the system is running. Many connected devices and sensors in the transportation IoT networks provide a corresponding abundance of potential attack vectors, and security is critical for mitigating risks of cybercrime. Then, security is a major issue and challenge that cannot be easily faced or overcome. Transmitting big data sets is expected with huge stable, sustainable back-end systems that can handle errors, faults, backups, recovery plans storing, and re-storing classified data keys.

As authors of this book, we would like to propose a foggy challenge that has not been yet identified as a challenge of IoT deployment systems, which is the possibility of clouding service discontinuity. In the backbone system of Smart IoT applications and solutions, clouding services play crucial roles, and act as important pillars to ensure the efficiency and consistency of these applications. IoT systems are transmitting a huge amount of data and information sets that need huge physical storages and reliable programming software packages. In addition, the clouding services can reduce the cost of illustrating physical infrastructures and physical layers of networks where the embedded device connects to and communicates over. And clouding service discontinuing will destroy the possibility and potentiality of IoT applications with dreams of globalization. So, this challenge should be also considered for applied solutions for interconnection between IoT devices and cloud devices portable, which can maintain and be replaced without affecting or pausing the progression of these connected devices.

3 Case Study: "Uber," the Internet of Things Car

3.1 Introduction

Most of us, unfortunately, think about the IoT as a founded technology to deploy sensors and wirelessly connected things in our lives with simpler roles and great benefits via applications on our laptops and mobiles, such as Smart homes. The technology of controlling devices from different locations and managing data among these devices have been farther than turning lights on. The Internet of Things is about the insights and competences of such amazing connectivity that provides and promotes our daily activities to better values we can share and enjoy without technological complications [25, 26].

Uber is a successful case that monetized the IoT concept by packaging the concept of e-solutions and e-technologies to grant the "decision-making" choice to customers which in turn upgraded the case to a new level of customer engagement and empowered the Smart economically. Let us fist simplify Uber and then get introduced to how it works.

3.2 What Is Uber?

Uber is basically a smart personal idea of taking the leveraging rapid uptake of smartphones globally and started a private business offering ride services while assembling the user's specifications with the highest priority and customize the whole service framework accordingly to user's specifications, which is the thrilling part of running a Smart business. Uber was started by Travis Kalanick and Garrett Camp in 2015 [27] and is now considered an excellent outline of a "hard to crack" business as the clients need, seek, and being fully satisfied away from complications they used to encounter with the public transportation systems. Uber started as a small business and now has exceeded borders to act as a multinational transportation network company. Uber technologies INC can offer services that comprise a peer-to-peer transportation system based on the Internet of Things technology. Uber tends to identify the company with what Uber can provide in terms of affording services.

It simply affords a comfortable and affordable way of getting to places for the first party, who are passengers, and easy-entry business opportunity for the second party, who are drivers. The relation in between is the potential entity-to-entity for a specific time for specific spot (location), which identifies that the business is based on the privacy and granularity of services being afforded to that passenger. Once the Uber application is launched, "the location sensors associated with both passenger's and the driver's mobile devices (the actual "things" being monitored) board casting their location to a "back end" system that is hosted by uber in the cloud, as the company itself explained the mechanism. The cloud on its turn provides analytics to determine which car/driver combinations are relevant to each service request. It is by default an IoT-based system but with lots of principle modifications to catalyze a swarming marketplace management model relying upon that everyone is holding a connected sensor package wherever they move; the system is easily able to locate them and afford the service [25, 26].

Uber model links information from physical asserts, locations, or activities to an intermediary service that enables transformative and profitable activities based on IoT sensor data that can be used in the form of "information-driven software" as if we are deploying vehicles from other geographical areas to ensure a timely and efficient movement of fragile productions [28, 29]. The consequences are achieving the ultimate outcomes to satisfy the customers and meet their personal goals and facilitating their life activities for less money and less wasted time [30].

Lastly, we would mention that Uber is a technology-based company that affords the service of transportation. It could simply create a global platform where anyone with a car and a smartphone can afford a service of picking up a passenger, deliver a product or at least help that customer and satisfy his/her needs at that time at a specific location.

3.3 How Uber Works?

Uber technologies are ride request corporations using mobile applications and based on emerging concepts of IoT and clouding as well. The application affords three basic types of services [25, 26, 28–30]:

1. Riding with Uber
2. Driving with Uber
3. Rating scores

The Figs. 6 and 7 shows an approximate Uber application:

So we can simply say that Uber application functions in the triple R (RRR: Request-Ride-Rate) as the following diagram RRR indicates.

Once the application is launched, the locations sensors associated with the two parties, the passengers' and the driver's mobile devices, regularly broadcast their location to the back end system that is hosted by Uber company on the cloud. The theme of "the actual things are monitored" makes the scene as if we are directly

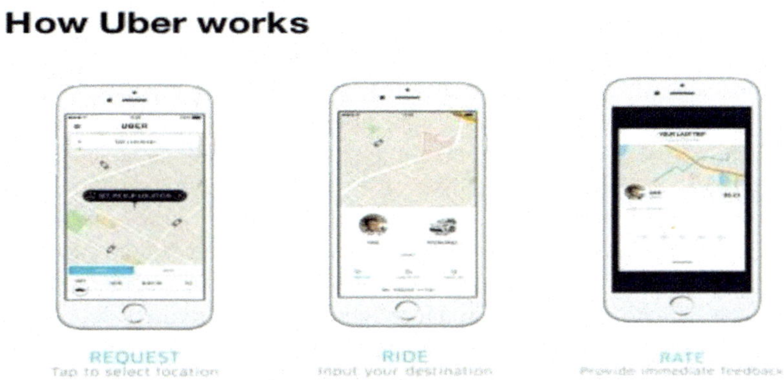

Fig. 6 Uber mobile application. [URL: https://image.slidesharecdn.com/javiercorreosouber-170419170123/95/kurogo-higher-ed-mobile-conference-2017-taking-your-campus-transportation-ondemand-3-638.jpg?cb=1492691393]

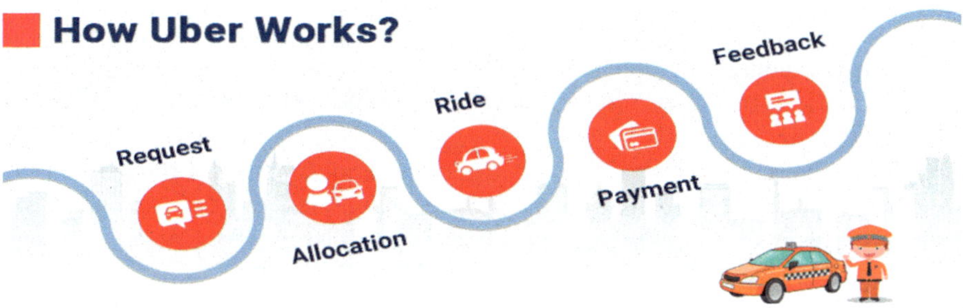

Fig. 7 Uber functions. [URL: https://mobisoftinfotech.com/resources/wp-content/uploads/2018/07/how-uber-works.png]

(peer-to-peer) subpoenaing the closest car (driver) to our location as a passenger, which is not accurate. What exactly happens is that the passenger's device sends a message to Uber's cloud service, declaring your physical address and locating your geographical spot to request the closest car to be sent to that exact location or nearby. Uber's cloud service uses near-real-time analytics to determine which car is the best fir to serve the request. It is based on the hidden object-oriented architecture and emerged artificial intelligence units that all compile the geographical information big data sets and do algorithmic calculations to derive the best choice, which is the closest car to your location accordingly with passenger's request, in other words meeting the customer's requirements [26, 28, 29].

The Uber systems are assumed to manipulate the regularly broadcasting information in the combination of already-stored information such as the middling charge compensated by pick-ups in a certain location at a specific time, and/or the total lifetime value of a particular driver to Uber are prospectively to be pre-computer on a programmed basis in consignment mode [9].

Uber being a Smart business will for sure seek the customer ratings to evaluate the service and being submitted to the quality assurance concept, which is an interconnected cycle of testing, evaluating, and enhancing procedures to meet the customer's requirements. After each expedition, the users of Uber application – the passenger and the driver – must rate the journey and reflect their satisfaction on a scale of 1–5 stars. The rate is taken seriously as riders and drivers with a low rating can be deactivated from the uber scheme. The customer rating is a must and has impacted the process of Uber systems as Uber focuses on building technology to encourage safer driving, for example, and customer's rating is a method to identify problems at Uber's partners' side. Uber Engineering's driving safety team writes code to measure indicators of unsafe driving and help driver-partners stay safe on the road.

The code uses harsh braking and acceleration as indicators of unsafe driving behavior. These crucial factors are highly correlated to unsafe behaviors like tailgating, aggressive driving, and losing focus on the road. To detect such indicators of unsafe driving in the first place, Uber engineered the problem to measure speed. The simplest way of deriving speed from a position is by measuring the difference between two consecutive positions as we assumed that once the passenger's and driver's devices are connected, they are broadcasting the location on a regular basis all over the trip.

So mathematically, if the Uber system knows a location X1 at time t1 and then another location **X2** at **t2**, then the average speed will be:

$$\text{The Average Speed} = X2 - X1/t2 - t1$$

The value will approach the true speed as the frequency of measurements increases. Error is tolerable, especially when we talk about values depend on GPS positional accuracy, which has a fault margin. So, Uber decided to use the Doppler shift principle to get a more accurate measurement of speed depending on the signal's transmitter movement relatively to the receiver of that signal. Doppler shift is often illustrated with fire truck sirens, so the transmitter is the siren and the receiver are your eardrums. The perceived pitch of the siren increases as the truck moves toward you and decreases as the vehicle moves away.

GPS receivers on driver-partner phones work in a similar technique. So, the phone which plays the role of the receiver here is either moving toward or away from a satellite, given that the GPS is a system of 24 active satellites that orbit the earth and the GPS receiver derives its position by determining its distance from at least four satellites as the following Figs. 8 and 9 show:

Using the satellite input and reading data, the Uber system will identify drivers with suspicious behavior that may break rules and agreements of safety and customer satisfaction. Uber stores such incidents for further machine learning and decision-making about the status of dealing and hiring this driver. Uber is not tolerating when it comes to customer safety.

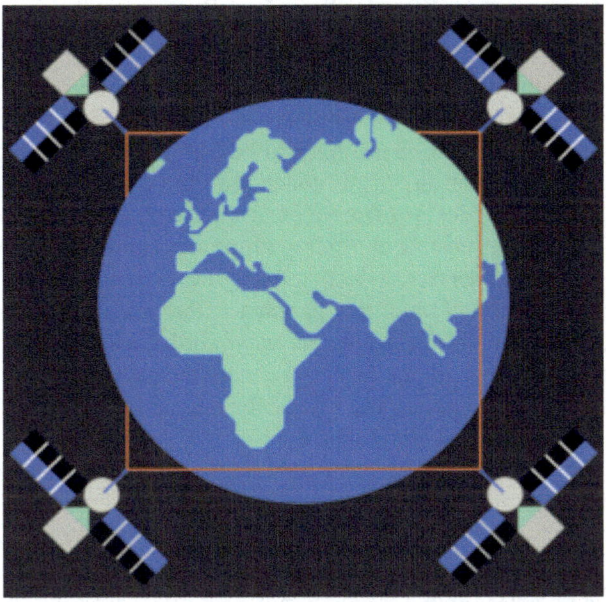

Fig. 8 Satellite transmitting to UBER systems. [URLhttps://1fykyq3mdn5r21tpna3wkdyi-wpengine.netdna-ssl.com/wp-content/uploads/2016/06/Telematics_diagram_02v1-300x300.png]

Fig. 9 Two positions of one car can identify speed at a specific time. [URL: https://1fykyq3mdn5r21tpna3wkdyi-wpengine.netdna-ssl.com/wp-content/uploads/2016/06/Telematics_diagram_01v1.png]

3.4 UBER as an Intelligent Running Business

Talking about the quality assurance in Uber services will be in shortage if we do not mention the smartness of Uber using the social media technologies in terms of customizing the services into icons that can be easily used and memorized. Users of the uber model will get directly in touch with the Smart business model Uber is running by identifying levels of services. Riding with Uber is an experience and the journey starts by requesting the ride with predetermined fare quotes that they will be paying for the service, which reflects the respect of acknowledging the passenger of the price, and upon this parameter, the passenger can pursue or call off the supplication. Withdrawing the entreaty can be with more opportunities to explore and decide further for the level of services the passenger is asking for. Uber uses a dynamic pricing model, and prices for the same route vary based on the supply and demand for rides at the time the ride is requested [25].

While the tip is being broadcasted, passenger can enjoy levels of services that were determined. Similar to the concept of hashtags in social networks that helps to identify specific information and make it easy to follow and track, Uber created its hashtags that specify the service category. UberX is the basic level of service that provides a private ride in a standards car with driver for up to four commuters [28]. UberX and UberXL are codes for cars with child safety seats. These semantic terms come with ontology and taxonomy concepts to hide irrelevant details and arrangements and keep it easier for the user to remember and use. The users have only to choose relating these shortcuts to a specific service they knew before or want to explore and try for the first time. Rider service levels can be limited in certain cities such as:

1. ASSIST provides additional assistance to senior citizens and passengers with a physical disability.
2. UberWAV provides wheelchair-accessible vehicles.
3. BLACK provides a black luxury vehicle with a professional driver.
4. Taxi allows users to summon a taxi using the Uber software application.

Other services are under development, such as UberAIR, which provides short flights using the VTOL aircraft and which was planned to start in 2020 in Dallas and Los Angeles.

At the end of the ride, payment is made based on the rider's pre-preferences, which could be a credit card on file, Google play, apple pay, PayPal, cash, or in India, Airtel mobile wallet. Tips are permitted as the passenger has the choice to provide a token to the driver, which is also billed to the rider's payment method.

Worth to mention is that most of the uber drivers use their cars. Drivers must meet requirements of age, health, car age, and type, have a driver's license, and a smartphone or tablet and must pass a background check. They are considered independent contractors and not employees though this has been the subject of lawsuits in several jurisdictions. Uber application also provides the driver with the option to accept or deny the request and support technically as Uber has invested significantly in mapping technology. Also, the driver's profile on the Uber system is supported by a mechanism called "Real-Time ID Check," which requires some drivers to sporadically take selfies when logging to Uber taking any requests [29].

We come to the question, if Uber can be discussed as a Smart technology provider and if such service helps Smart cities to evolve. To be able to discuss this statement, we need first to conclude the benefits of using Uber and how far the proposed technology has met the requirements and achieved the satisfaction of users. We all agree that Uber has disrupted the taxi industry as it started to afford private ride, but now not limited to it, as we mentioned above. The simplicity of using Uber application and affordability of its application made the service easy to get and utilize. The interface of the application is straightforward and easy to navigate, request, get the ride, and then feedback at the end of that preferable journey. The smartness of utilizing the spreading of smartphones, the Internet of Things, clouding, and the Internet of Mobile things promoted Uber's provision to be a survival business that can compete in the marketplace and other service enablers. Uber makes it very convenient to acquire a ride by simply marking the passenger's location and identifying the nearest driver/s. Not only this, but the convenience is also for the drivers as most of them have private-style vehicles, which means that they are generally more comfortable to serve you and get paid. The convenience in the payment method is also the main factor in playing this game. So, simplicity and convenience of the demand–supply platform make it smart enough to promote Smart cities and be part of their Smart economy.

Uber has disrupted the taxi industry for sure and promoted the concept of sharing economy where services are shared through the use of the Internet and other ICT applications as the platform which ended the "monopoly" of traditional suppliers, manufacturers, and service providers that have lasted for decades and increased the level of competence in the taxi industry. The entrance of Uber into the market had visible impacts and benefits. The following list was suggested by Martin in his online article "7 Strategies Uber Is Using To Disrupt The Taxi Industry" [29]:

1. Raise the awareness of users and encourage them to decline the ongoing prices of taxi licenses and taxi medallions in many cities.
2. Increased the number of passengers and presented a pool of active drivers willing to get paid and cover their living expenses which is a cyclical effect of sorts. People are turning toward the use of this mobile application because of its simplicity with a reasonable time of picking up, and getting served according to their specifications.
3. Creation of a more efficient car-hiring service in the market space, which encouraged and allowed small business using a private car to grow and achieve a higher income for investors (drivers in specific).
4. Decrease the number of taxi operators and companies as they were losing their customers for Uber – for cheaper but with high-quality service.
5. Uber contributed to eliminating the countless transaction costs as simply there is no charge to be paid for just requesting a car through the mobile application. On the contrary, Uber started a strategy of raising your profits by requesting a

service every time, that you can use to cut the amount of money you are paying next trip, which is known as discount codes.
6. Encouraging the general productivity by the growth of employment rates and assets by pushing toward getting additional income for drivers who prefer to work part time.
7. Uber skipped over any licensing systems and encourage the concept of entering a contractual relationship directly with derivers and left a space for them to maintain a degree of freedom without having worries and affairs about the cost associated with employment.
8. Uber affords a flexible pricing model that is not present in the taxi companies whose pricing is controlled by legislation. The price model is transparent as passengers are already aware beforehand of their estimated fair and afford them a service to change their preferences and compare between the proposals they are getting to go on their journey.
9. Uber affords appropriate and frictionless payment methods as we explained before. After each ride, the fare can be automatically charged to the driver's account using different interconnected online systems in case you prefer not to pay with cash.
10. Quality assurance of their services and quality control are implemented through their rating system. Such quality control motivates drivers to serve and maintain the highest quality they can provide starting from the way they drive, obeying the regulation, and respecting the chance of interacting with passengers trusting them for their lives.
11. Uber has developed a strong reputation and reached to customers' loyalty by running a wide base of drivers and affordable services that you can get easily and rely on over your daily routines.

We would also add that Uber has offered a global affiliation and leading platform with a dynamic vacancy to serve dynamic partners according to their request and needs of both drivers and passengers. Uber as well cooperated in many aspects to evolve the cab industry as the Uber application can be used to navigate taxi services if desired. Uber has got interconnected and integrated with running platforms inducting new versions and provisions of a Smart economy that can exert for Smart cities initiatives.

3.5 Challenges Faced by Uber

Uber has proven to be the best example of sharing economy that is based on technology to afford a hassle-free cab service and personal experience everyone can rate and recommend. The huge expansion and transforming to globalization has been the ultimate challenge for Uber so far because of differences in cultural backgrounds and unexpected regulations for each geography where they are willing to run the business. The major transformation is a result of the powerful technological advances and the ability to have stable infrastructure on mobile devices and the Internet facilities in general. So, to be fair enough, technical and infrastructure of the global Internet is not related in specific to Uber and cannot be considered a challenge for the company as it is an international concept with a supported backbone system available and maintained by the strong international economy that cannot be broken easily.

Challenges that can face Uber are more related to the industry itself, by having strong competitors widening the options available for the customers seeking high quality and reliable services. Another challenge can be cloning the Uber system which means that other deceivable service enablers can overwrite the high quality served by Uber and can impact the customers' constancy.

Mahapatra et al. published a study entitled "Challenges faced by Uber Drivers and Customers Satisfaction in Pune City" and focuses on the challenges Uber drivers face providing their services to the customers and the satisfaction derived by the consumers after the ride [30]. The publication also discussed earlier studies, such as the following:

1. Kumar Kishore and Kumar Ramesh argued that the presence of many competitors in the cab industry is crucial for the business to survive and get motivated to develop serving techniques that they can compete with others, such as affording coupons. The study concluded that such affordability intensifies the awareness of customers to be modern, innovative and price sensitive since coupon redemption helps customer retentions [30].
2. Surie Aditi and JyotiKoduganti stated that drivers reflect their satisfaction in terms of earnings. Bonding the stability to the satisfying income they reap, which deviate the perception of security and strategies of income like daily incentives which add to their satisfaction level [30].
3. Hall Jonathan and Krueger Alan investigated the first time the comprehensive analysis of Uber's driver-partners based on survey data and administrative data. The analytical study proved that drivers were attracted to the flexible work hours and incentives provided by the running system. The simplicity of Uber application and running procedure, as well as the convenience and affordability, attracted more customers over all the working years [30].
4. Khalid Farah et al.'s study outlined strategies to improve the situation of the taxi industry as it discussed the level of taxi drivers' attitudes and argued that some important factors were neglected sometimes by the driver such as safety and courtesy. Uber has already overcome such obstacles by including the quality control system and running intelligent units to track the user's behaviors [30].

We do recommend a further reading of the publication "Challenges faced by Uber Drivers and Customers Satisfaction in Pune City," which highlighted more challenges faced by Uber drivers in Pune City as a good practice of Uber application and maintaining all over challenges as the study concluded that most of these challenges are manageable and overall parties are satisfied by the facilities provided by Uber. Customers as well were satisfied highlighting the efficiency of Uber management, and expressed their loyalty.

To end this section, we would clear that the section is not written to advertise Uber technologies. Uber is taken as a case study for Smart transportation that can emerge to promote Smart cities away from traditional transportation systems. Uber is a good example presenting various types of communication and navigation systems that make possible to construct a global Smart system using the concept of driven information system and Smart applications in mobile phones for the profitable Smart economy.

4 Proposed System: Intelligent Transportation Objects

4.1 Online Identity and the Internet Profile

Regardless of whether we like it or not, a huge part of our lives is available on the Internet. Anyone can get some relative information simply using search engines if you mention your name at least. If you are an Internet user, then you, for sure, have what we call Internet identity (IID), also called online identity or Internet persona. Further, if you are a social person, then you have a social identity in the digital world, and parts of your life are available on the Internet, primarily through your name, date of birth, and place of residence and some other media, such as photos, trips, locations, and videos you have recorded for yourself or family. If your work is public, then you are listed on the official sites for where you work. An online persona is an identity that a user of the Internet establishes to represent himself on the Internet. The identity is based on accumulating digital information and store it while communicating with others in social media or building online profiles for recruiting purposes and most of the time we manipulate this information to reflect some of our feelings, interactions, hobbies, places we visit, and the food we eat. This is what most people do when they want to enjoy their time and get some entertainment away from professional character or life. What we post on social media or blogs is traceable. The easiest way to check your Internet profile is to search your first and last name. Search engines can be used to bring up information at any time.

The amazing integrated social network structures lead you to find a lot about yourself and others with whom you were sharing moments and interacting within the digital world. Facebook, Twitter, Instagram, and Snap chat are in fact websites or applications meant to connect friends, family members, groups of similar interests, and, most importantly, satisfying something you need or you seek information about without the hassle of formal routines and paying the unacceptable amount of money. Most of the small businesses have started using these platforms and could survive as the cost of advertising and commercial stuff go to zero if we compare it to the importance of commencing that business [31].

Athens's availability of the Internet and the huge promotion in information technology and communication have decreased the cost on users for which comes to their benefits at the end. We have witnessed an incredible number of home startup started using only a page on the social network to advertise and compete in e-commerce. Unbelievably, the innovation came with social networks and media presented a pool of ideas and interconnected people to interact and exchange experiences that notably impacted the others' choices and decisions. Utilizing such a phenomenon is a smart act, especially with the technology accelerating the promotion of smartphones and the decreasing requirements and cost of being online on the Internet.

To be able to enjoy all these Internet facilities and amenities, an online identity helps a lot. The identity may be determined by the relationship to a certain social group you are following, interested in their post of your Internet contributions as in the case of researchers and academics. Such information increases your credential on the Internet that helps others to track you online and frame your online identity for recruiting, for example, or consulting when you have a valuable experience of what you do and have the valuable background of knowledge or practice.

You will be impressed if you try to look for information about yourself on the web using any search engine. No limits or a frame of retrieving information, especially if your work is public and you are listed on the official sites for where you work. Whatever you post on social media or blogs is traceable.

Elaboration about conceptualization of the online identity is a must to present the sociocultural approach to its notion while considering that the online identity is founded on experiences of participation in gaining knowledge, learning, training information to make the most acceptable decisions. Lots of previous studies have made attempts to operationalize the notion of identity to justify the claim about its potentiality as an analytical tool for investigating learning [32]. Identity learning activities form a set of personifying and endorsable autobiographies of that related personality this identity presents in the digital world. Moreover, these related and interconnected stories create digital products of conjoint storytelling that are valuable to be exchanged and trained smartly to come with a higher quality of next experience. For that and more of communicating purposes on the social networks in specific, the online identity conceptualizing should have a theme of learner identity with identity functions designed for the learning context where it mediates the personality's process of attribution of meaning and sense-making concerning any learning objectives [33]. Learning objectives interest takes the simplest form of identity notion from simplicity, such as being a character willing to connect to other notions to complexity such as attitudes, conceptions, feelings, and impacting other's decisions for quality purposes. More specifically, basic forms of online identity do not exist anymore; nevertheless, the definition should include operational and behavioral functions that support the identity with analytical tools and decision-making on the web. The intelligence concept we are referring to here is supported by learning objects that we can bring into to actuate the online identity to insatiate communication and web interacting. Learning objects are procedural mechanisms to personalize the learning experience any identity can practice on the web based on self-motivation and knowledge acquiring as the next section discusses and suggests that the identity conceptualization can enable a holistic view of learning and knowledge attainment based on self-interest and own experiences for meaning and sense-making purposes.

Expanding the concept of conceptualizing the online identity with Smart (intelligent) learning objects will make it mandatory to investigate the possibility of integrating this concept with multiagent systems. A multiagent system (MAS or self-organized system) is a computerized system composed of multiple interacting intelligent agents that are dedicated to solving problems with intelligent, methodic, functional procedural approaches, algorithmic search, or reinforcement learning [34]. What makes the multiagent systems are more needed to utilize the learning objects is the need for practical architecture with an applied approach to retrieve, search and composition of the relative learning objects which at the base, built using open systematic, distance learning aspects to activate the online learning systems. Emerging multiagent systems permits for more interoperability, reuse, self-origination system, and information recovery. Multiagent systems also permit models to organize the learning objects' content and analyze the potential impact on the knowledge appropriation process using the quality concept [35]. Valentina et al. present a multiagent system for expert evaluation of learning objects from repository which verifies and validates the quality of learning object content by investigating the level of significance of the educational resources of the teaching-learning process, associated with the educational goal and other characteristics which were not an easy task at all [35]. A group of expert systems had been deployed to reckon the characters of that leaning object. Valentina et al. [35] cross-examined different criteria such as teaching, interoperability, scalability, and re-usability. The next section discusses this potentiality and gives more details, and the following figure presents the validation through a developed multiagent system. Figure 10 presents an example of a multiagent system architecture diagram.

4.2 Intelligent Learning Objects

With the advent of the Internet, people communicate and do businesses in innovative ways that emergence the Internet itself to be poised to bring paradigm shifts in the way people learn and accumulate their knowledge. Thus, and as expected, a key change should be revolving in the way educational materials are created, designed, developed, and delivered to those who wish to learn and interact via any learning network. This instructional technology paradigm shift is called "learning objects" and currently leads other paradigm shifts as the technology of choice for next generations of instructional design, development, and delivery [33]. So, learning objects as a topic is not new at all, but it has been neglected in the research field even though learning objects is a major technological innovation and can result in entire paradigm shifts because of its potentiality for reusability, generativity, adaptability, and scalability. Then, if we define the learning objects according to An Agent-Based Approach of Learning Objects, Ricardo Silveira, Eduardo Gomes and Rosa Vicari Universidade Federal de Pelotas, - UFPEL, Campus Universitdrio learning content

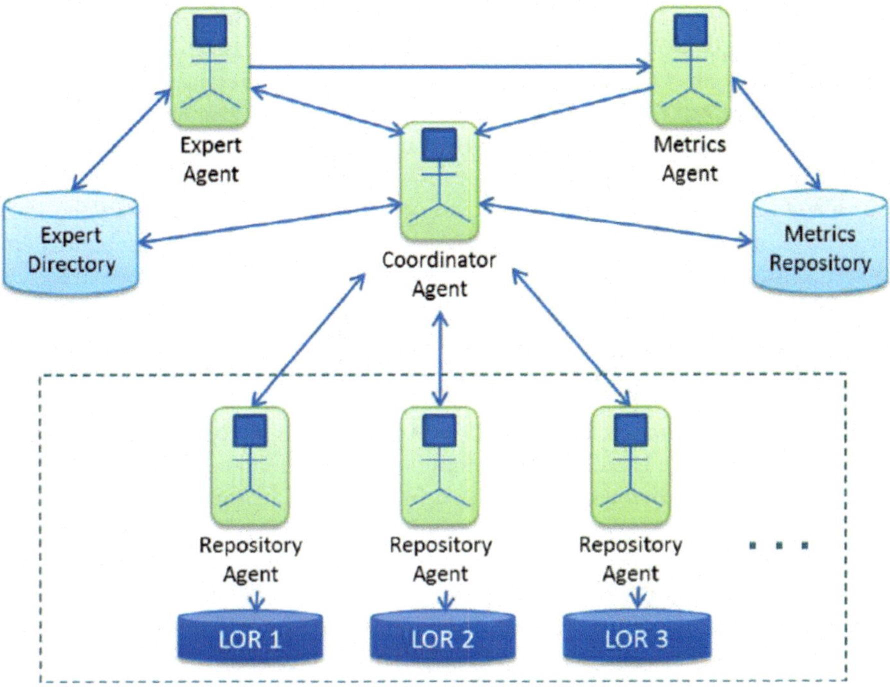

Fig. 10 Multiagent system architecture diagram. (Source: https://www.researchgate.net/figure/Multi-agent-System-Architecture-Diagram_fig 1_282726998, Ref. [35])

which can be used several times in different courses or different situations. The learning technology standards committee had possibly extended "Learning Object (LO)" as a term from Polsani [36] when the term of the title CedMA working group called ("Learning Architecture API's AND Learning Objects"). The term LO is manipulated to describe small instructional components, as the IEEE learning technology standards committee[1] (LTSC)'s working definition states:

> "Learning Objects are defined here as any entity, digital or non-digital, which can be used, re-used, or referenced during technology-supported learning. Examples of technology-supported learning include computer-based training systems, interactive learning environments, intelligent computer-aided instruction systems, distance learning systems, and collaborative learning environments. Examples of learning objects include multimedia content, instructional content, learning objectives, instructional software, and software tools and persons, organizations, and or events referenced during technology supported learning [(LOM 2000)", & IEEE 1484.12.1-2002, 15 July 2002, Draft standards for learning objects metadata].

Any education system based on multiagent architectures is highly motivated to improve the efficacy and reusability of learning content and material to design dynamic learning environments for real-life learning based on an object-oriented

[1] https://adlnet.gov/working-groups/learning-technology-standards-committee/

design paradigm [32]. Therefore, education professionals have been working so hard on intelligent learning environments to instruct the use of artificial intelligence through architectures based on agent societies and deploy the concept to create intelligent platforms producing huge metadata specifications toward the construction of more interactive and adaptable systems to the end specifications and requirements.

This chapter proposes the customization of learning objects to improve the efficiency and reusability of Smart transportation systems. It discusses the integration between these learning objects and multiagent systems to expand the Smart transportation technologies and services. The metadata specification should be highlighted to meet the real requirements and anticipate the needs of Smart transportation users while the employment of the reusable learning objects creates transportation environments that are quick, flexible, and sustainable when we discuss the Smart economy. The first idea of renaming the learning object to be a transport-object to refer the customization of learning object concept into an object-oriented structure that is built, given definition and value, and being used and reused for a transportation theme but still with a full-frame of learning objects behavior. So, we have decided to give this new customization a T-Learning object referring to a learning object being used for Smart transportation technologies and services, not education. Both systems have a common base of manipulating gained knowledge and information systems, so the emerging concept of a multiagent systems approach will work perfectly with the learning objects. To produce intelligent data units and intelligent agents with entity-based software that work in a continuous and autonomous technique in particular situations that are able to cope with Smart technologies, interact with other components, and interfere in a specific environment without the intervention of a human or guidance [37].

To utilize the features of learning objects and build the Transportation-learning object, we need to track efforts that focus on the definition of standardization, an organization such as IMS global learning consortium (2004), CanCore (2004), ARIADNE (2004) and IEEE (2004), which have defined indexing standards called metadata structures (data about data) contain information and guidelines explaining exactly what the learning object is about and how to manipulate the concept to search, identify, and retrieve any educational material according to user's demand. Downes (2002) clarifies that to be able to use the concept of the learning object, lots of efforts should be paid which is also applied when we need to customize this learning object to act as a T-learning object in a real-life application. The first obstacle we might encounter is defining a Smart transportation environment where these T-learning objects function. Identifying the Smart system will for sure involve behavioral and functional areas to enable the useful features of these objects and make it possible to locate and arrange them in their proper order according to the designed and specified mission. Discussing the real fact that we are customizing information and knowledge content should bring into consideration the high quality of information systems permitting data streaming, installation, configuration, and appropriate viewing to deliver the appropriate context according to the user's specification and need. This means that the system should be robustly built on Smart learning objects with intelligent features for easier handling and achieving the ultimate performance of a Smart transportation system suitable for Smart cities and adorable for Smart citizens. The conversance of artificial intelligence (AI)-based learning objects will overcome obstacles of instructional design problems which developers have claimed before and avoided the risk of time-consuming when involving human control and interaction despite all metadata specifications as it is not sufficient for pedagogical decisions without artificial intelligence Units. Silveira et al. [8] propose the multiagent system approach to designing intelligent tutoring systems in the educational system based on intelligent learning objects (ILO) that result in faster, more versatile, and low-costs systems. The learning objects with agent characteristics such as autonomy, knowledge about itself and its environment, sociality, and goals can be easily customized to play the same role but with Smart transportation systems.

The importance and benefits of deploying learning objects into the education system are uncounted. Being small sharable "knowledge packages" that include all related learning materials needed to cover a specific objective of the e-learning course and give the chance later to re-use, and customize for other purposes is a huge benefit and basic definition of the purpose behind using these objects. Learning object functions with autonomous learning units that can be used in multiple learning environments as long as the same learning objectives need to be covered. Learning objects notably has a significant impact on both learning expansion, development time, and budget. Such abilities and features are perfect for online education considering learning objects are SCORM files that can be in practice incorporated with systems without any human interference or add any additional cost.

Before jumping to how we can deploy the concept of learning objects in our system, it is beneficial to know the main visual components of learning objects in any instructional design for e-learning as an example and practice field. Components can be as:

1. Title: The way we choose the name and the title of learning objects taken into consideration its prospectives will help to grip learners' attention and match perspectives they want to learn and acknowledge.
2. Subtitles: It adds additional metadata to provide further information about the content.
3. Learning objectives to be covered: Learning objectives are expected to be related to the rest of the e-learning course.
4. Overview, which means that learners can overview all details and requirements of joining such course which give a chance of cascading in case of being unable to meet the requirements.
5. Specify the content to be presented to learners for acquiring the necessary information they need to master to proceed with their learning experience and later to practice the gained knowledge when they are employed.
6. Additional metadata to help retrieval processes in case of future use or customization.

Bear in mind that thinking about learning objects greatly facilitates both the instructional design and development process of the system they were embodied in for a huge variety of learning objectives and purposes. Learning objects have been proven to have professional and great benefits in terms of Smart technologies and rapid transformation such as we need in Smart cities.

4.3 Developing an Intelligent Transport-Learning Object (IT-Learning Object)

Silveira et al. [8] propose an agent-based approach to produce more Intelligent Learning Objects (ILO) according to the FIPA agent architecture reference model ad LOM/IEEE 1484 learning object specifications. The agent will communicate and exchange messages using a high-level communication language called agent communication language (ACL) based on speech act theory. SCORM – The complete current learning object learning model – is being used in real-life learning environments and lays a platform to perform the communication by calling methods (functions) and passing parameters, according to the object-oriented programming paradigm. Deploying the ACL language is to enable communication among these agents and ending up with more powerful data manipulation layers and afford semantic communication using the formal protocol and formal content-language (CL) based on logic formalization to express the messages content, facts, rules, mental states, not only the variable values.

Basing IT-learning objects on multiagent systems increases the potentiality of autolearning capability as it is the main feature in expert systems. Assuming the Internet of Things platforming the interaction of these agents, the Smart transportation system will promote a simple trip to be a real-life environment for shareholders, users, and the relevant economy. The Internet of Things is a revolutionary technology that proffers to connect a plethora of digital devices endowed with several sensing, actuation, and computing capabilities with Internet, which is the most suitable manifold service in the context of component technologies of a Smart city [6]. The tempting services of the Internet of Things (IoT) are more appealing if joined with big data concept as it is supported with analytics that enables the intelligent transportation-learning object and multiagent system to emerge in well-framed taxonomy podium to best bring forth a generic overview of the IoT paradigm and present it into the new revolution of the Internet of Mobile things affording services to improve quality of human life.

Emerging the Internet of Mobile things (IoM) is crucial indeed as personalized access to the Internet and multimedia resources has been already become of great importance in transportation systems by having the 4th generation of mobiles and 5th mobile generation launched already. Even though daily technologies are detached at the infrastructure level, they are merging and integrating into a compatible framework to deliver multifaceted and multimedia channels for content that makes traveling more reliable for travelers. Including the IoM will assist to communicate Smart user's specifications to the internal infrastructure of the relevant Smart transportation system. The perception will enable a Smart transportation system to virtually plan the transportation contour and customize it to the user's demand. The designed virtual plan is a set of methods and standards that enable easy use, recombination, and transfer of the content between individuals, institutions, and other interested parties globally, if needed. Conferring more details, goals of such emerging can be as follows [33]:

1. Personalization

   The intelligent Smart system will innovate a unique IT-learning experience relevant to the real-life environment specification defined and requested by the Smart user. IoM applications will characterize profiles of each user and match the requirements with the right content combinations of one or combined IT-learning objects. The obligation has a multicycle to develop and enhance analytic practice.

2. Contextualization

   This term means the process of providing relevant context for source string which provides a meaningful description for translators to ensure the correctness and quality of translation. In programming, the process discusses the context related to software application intended to serve a particular purpose through a full procedure of understanding the purpose, providing required instructions, manuals, and graphical user interface (GUI) dialogues. For our mutual system design, context awareness is an important feature and should be precisely highlighted since the design focus is on conceptual modeling, understandability, and clarity as precedence for reasoning and analyzing.

3. Adaptivity and flexibility

   Content can be automatically adjusted to be retrieved by the largest base of interested devices while any changes at any time will be dynamically adapted.

4. Time limitless

   Requesting and delivering the content will be with no time constraints on both cases synchronously and asynchronously.

5. Location-free constraints

   Transporting the content anywhere with the IOM applications.

Incorporating all these techniques is mainly to increase and diversify user's access to any transportation opportunities and investigates specific ways in which digital resources are designed and stored into different systems and organizations and automatically transported to the users.

Smart traffic and parking management can transform the way that drivers get from point A to point B. Not only do IoT-enabled traffic and parking solutions make driving more efficient and convenient for a city's residents, but they can also make transportation safer for everyone in the city. If you want to connect and monitor transportation assets with ease, partner with Telit when creating end-to-end IoT solutions that are scalable and secure.

4.4 Mutual Scenario

Mark is a student who lives in London, England. The following is a map taken from the Google map of London city. Mark wants to arrange for a trip and share riding using a regular service provider such as Uber, but not limited to it. Mark can arrange the trip geographically and view all places on Google Maps, as shown in Figs. 11, 12, and 13. Then once he is determined, he will use a separate mobile application to do RRR actions as explained before – request and ride, and finally, rate the trip. No personal specification in terms of what Mark wants to have except what already the service provider can provide, which is the level of services as we also detailed above. You can request a large car with safety and also you have some luxurious means on board. The following is shown in Fig. 11.

Next, Figs. 12, 13, and 14 display many features already supported by Google Maps.

If we want to re-design the trip using the concept of intelligent transport-learning objects, we would assume that each trip is ITOL, and once it is created, it is given an active ID so other interested parties can track, follow dynamic posts, reviews, and can recommend as well. Such a trip is active once the creator specifies the start, destination, and approve the offer from the driver; then it is shown on the map for others to join and interact. All details such as additional luxurious stuff as requesting a large car, but most importantly, the features, data, and information we have embedded to this trip to increase user's satisfaction. Data and information can be retrieved from any online identity to represent the user as an object with associated preferences, interests, and requirements as traveling is not only to reposition your self or a product, it is more about entrainment and sharing. The base of interacting and exchange experience is crucial for all social networks such as Facebook and Twitter.

So, Mark lives two building blocks from Joe, but they are not aware of that. Analyzing their Internet identity – the semantic analysis of information, profile, interest, and reactions to some posts on social media – we can conclude that they share around 40% of interests. Add to this, they both like to have a hot cup of espresso with a chocolate chips cookie on their way to the university. They can buy from a close café but time is not helping when they are in a hurry or late and sometimes, they cannot handle the long queues of waiting. They are not aware that nearby their living location there is Julia, Julia which runs a private business from her house selling hot espresso and warm and fresh cookies she prepares according to customer's requests, but selling to raise more money with what she is good at and is well known on social media. Most of the social media have expanded their service further than just posting and commenting, they support voice communication, live broadcasting, and also location tracker. A fourth person can be added to this scenario, who is Rick. Rick is a graduate student with a car, and he uses this car to give a ride to people from time to time to cover some of his living expenses. Every morning, he drives down the hill close to Mark and Joe's locations

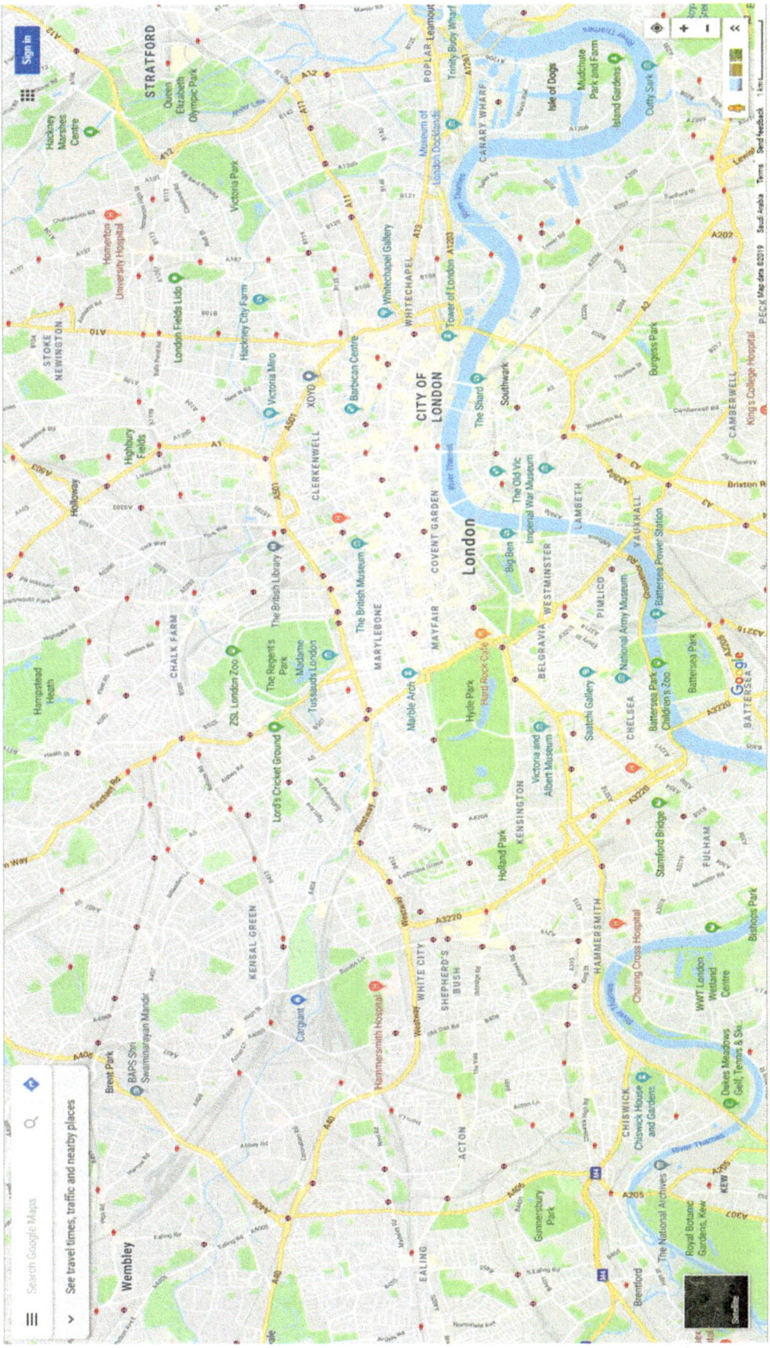

Fig. 11 Google map of London city

Fig. 12 A close-up Google Map of London

passing Julia's small café. Rick has studied engineering like Mark is studying now, and loves reading as Joe does.

If Mark and Joe can be online and use "ME-Online" application, they will be aware of Julia and Rick's activities and will most likely arrange the trip to be with Mark and drop by Julia to pick up their coffee and warm cookies that they have ordered by the application and specified the time they will drop by to pick it up. Rick is also a nice personality, can discuss a few things with Mark and Joe as they have common things instead of just being the driver. Mark and Joe are communicating more often now as they have a common interest. The whole trip starting from location x to z is recommended with the same persons to anyone willing to share such an exciting trip. Photos, reviews, and recommendations are added to the trip ID to pass the experience for others with the same requirements and needs.

Others can request the same trip ID with details or can customize it as their personal needs such as buying a book from Jack who operates a bookstore on the route of the specified trip. Accommodation of trips can build a huge Smart trans-objects with easy access and remarkable memories and feedback.

Such a scenario is almost as having a prearranged Smart digital deal and has high potentiality and various building blocks for successful Smart city strategies have been examined and articulated framing the Smart transportation into interconnected objects with learning and experimenting abilities and strategies have been examined and articulated.

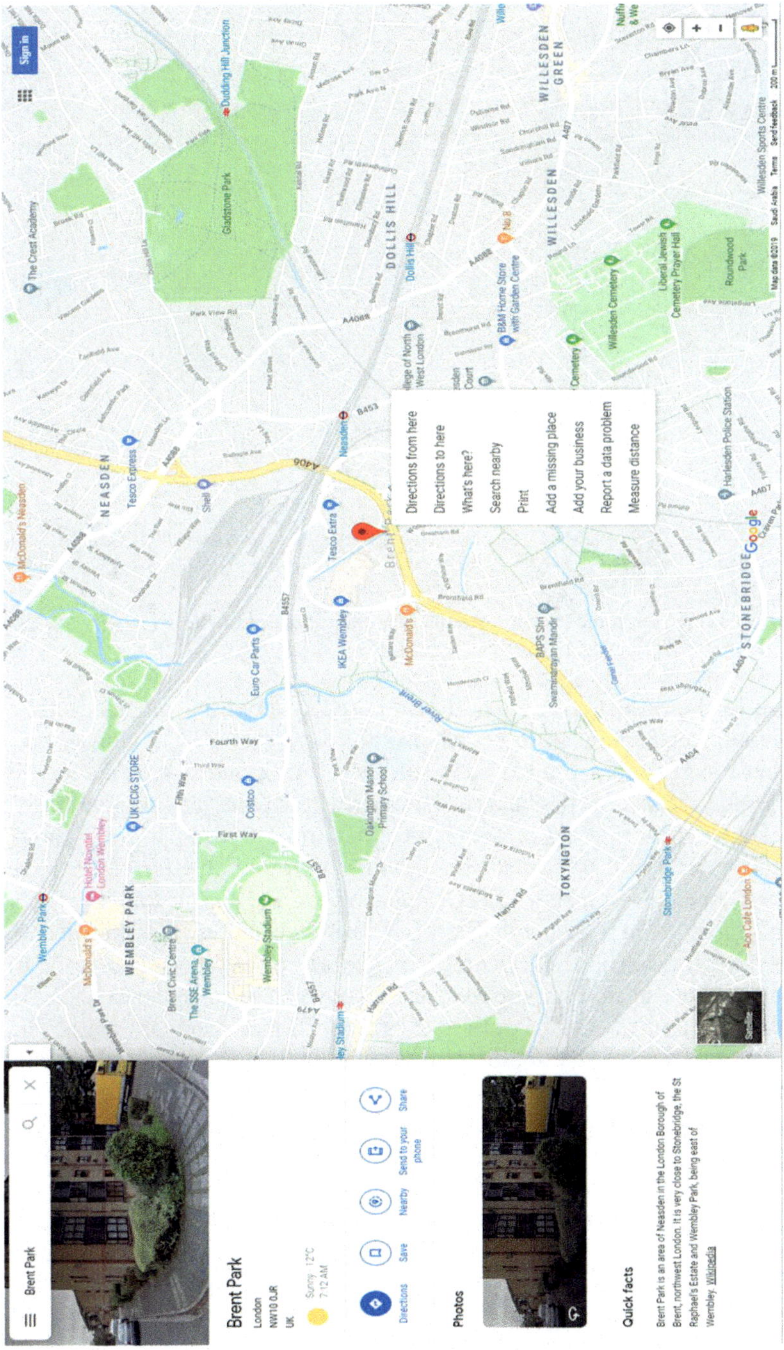

Fig. 13 May features when you decide to arrange for a trip

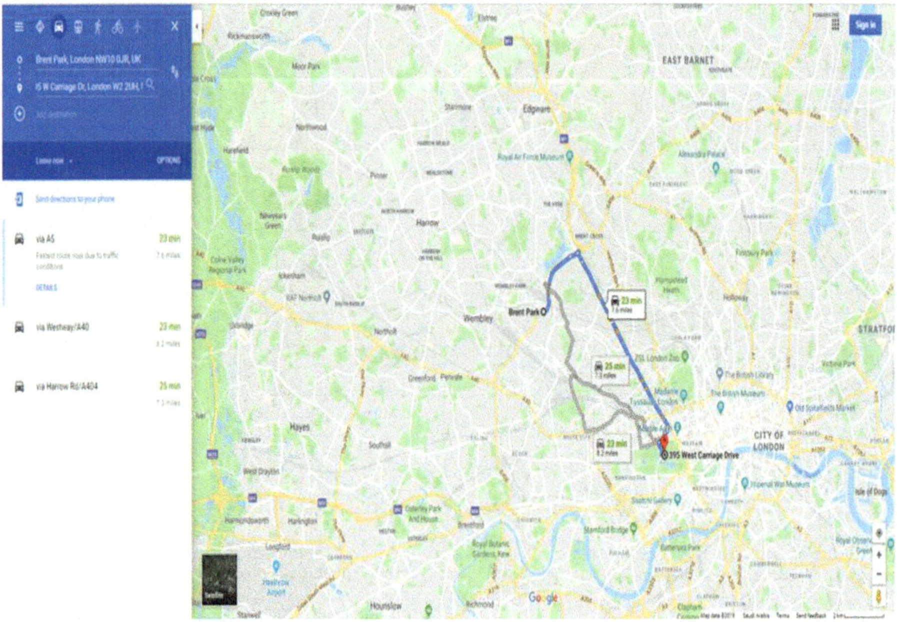

Fig. 14 Specifying your route showing destinations and estimated times

Figure 15 shows such use case of the proposed system using Google Maps.

With more details, Fig. 16 shows a close-up conceptual view of simple trip.

We have suggested a pseudo code for such a trip, and it is open for further argument and discussion to enhance the system functions. Please note that to be able to later review the trip or adopt for other trip purposes, the trip should be completed; else, it will be cascaded and will not be available for others to reuse. The system can support a built-in ITOL the user can use and customize. If the user refers that the mission is accomplished, then the system stores all related details and associates it with the trip ID. Dynamic storages will be afforded all over the trip to temporarily save and storing purposes, and a copy can be later stored once the trip is over and the user (creator) agrees to save it for further share and reuse. The use of hashtags will promote the solution to identify the right and expected information and contributions from others according to the customer needs and requirements to fully enjoy the trip and make benefits. Also, with no doubts, such system is an IoT-based as well as the deep usage of Internet of Mobile things as the system is depending on locating positions and visualizing using the concept of GIS to be able to communicate the information and display it for others to get involved without any concerns of irrelative internal details. Figure 17 suggests steps to launch and accomplish a mission while Fig. 18 presents also a suggestion on how to manage this mission.

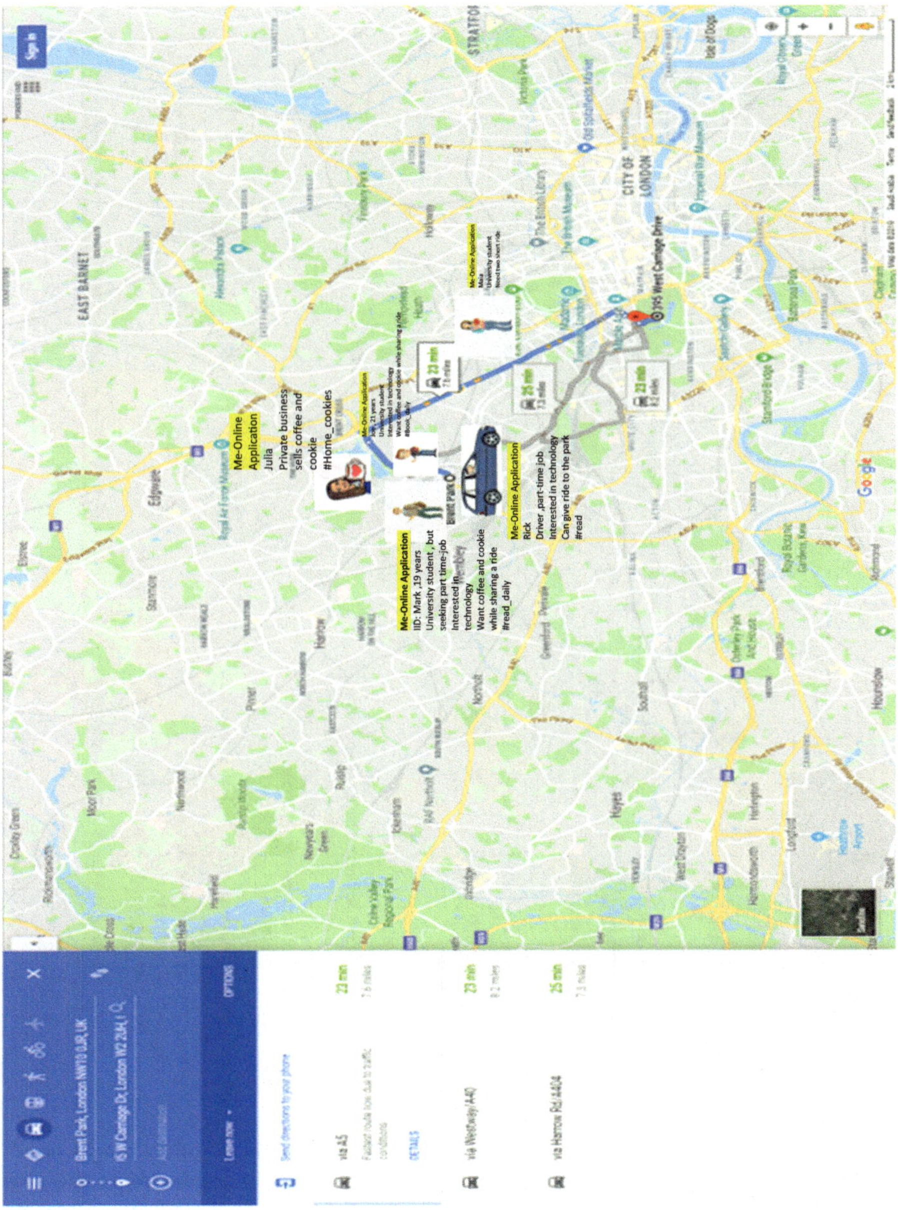

Fig. 15 Google Map ME-ONLINE application

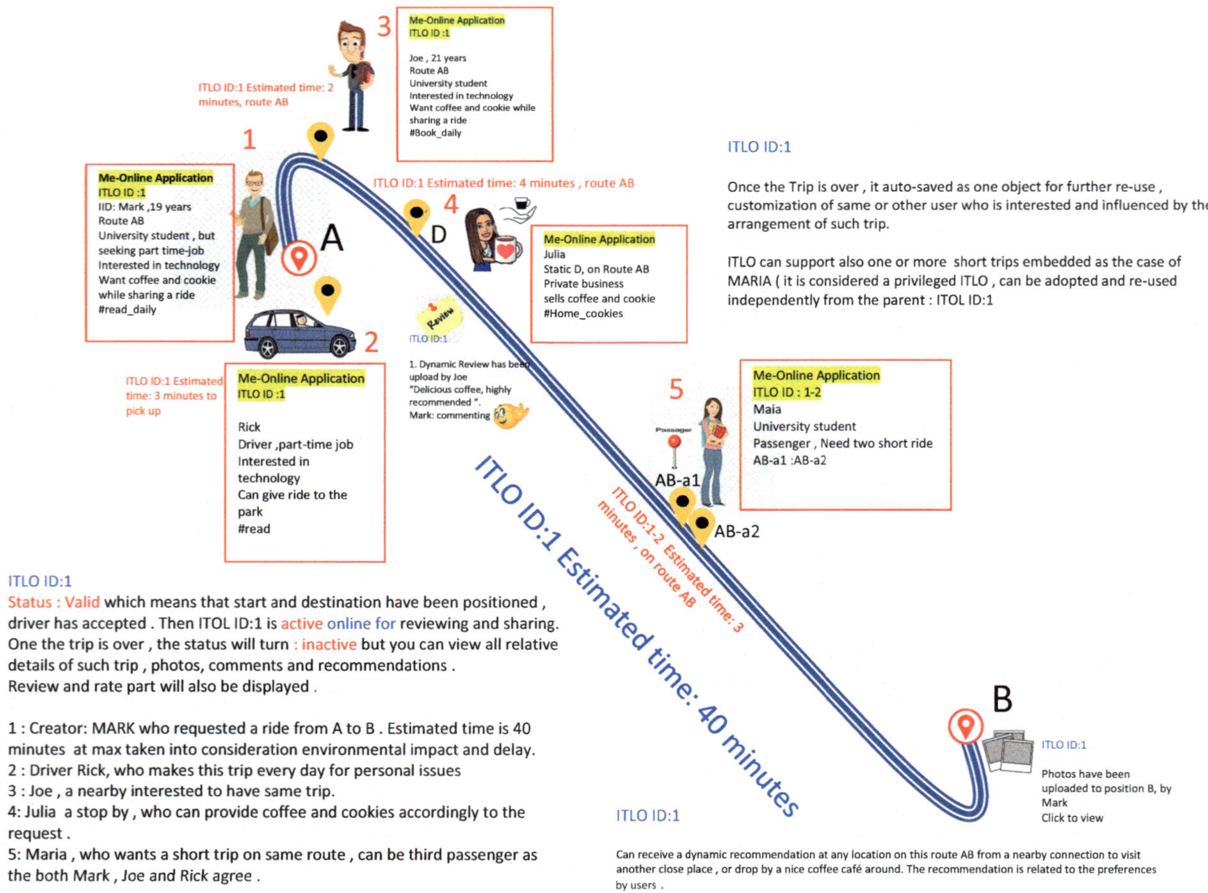

Fig. 16 Use case of ME-ONLINE mobile application

```
If ITOL True
  // means completed
      Then
         Store (ITOLD, ID);
      Else
         {
           Cascade (ITOLD , ID);
           Delete (ITOLD , ID);
         };
```

Fig. 17 If statement to accomplish the mission

```
1  Launch MeONline application
2   Retrieve pre-preference data
3  Create composite Journey
4 Accept an offer from the driver
5 Set value of Status to Online
6  While Online
7 DO
8 Create hashtags
9 Create Composite journey
10  Input N    number of transport-object
11  For int x=1    to   int = N-1
12       Do
14      Add destination node
15   Input   Char satisfy,
16    if satisfy = Y then
17   Accept offers
18     Dynamic Rate   // Other features can be suggested, like commenting, posting photos and so on.
19   else
20   Cascade
21    If retry = yes
22    Then go to 6
23     Else cascade
24    The End
```

Fig. 18 Suggested Pseudocode

```
If  ITOL True
// means completed
Then
Store  (ITOLD, ID);
Else
{
Cascade (ITOLD , ID);
Delete (ITOLD , ID);
};
```

4.5 Pseudocode

```
1   Launch MeONline application
2    Retrieve pre-preference data
3  Create composite Journey
4 Accept an offer from the driver
5 Set value of Status to Online
6  While Online
7 DO
8 Create hashtags
9 Create Composite journey
10  Input N     number of transport-object
11  For int x=1    to   int = N-1
12       Do
14      Add destination node
15   Input   Char satisfy,
16    if satisfy = Y then
17   Accept offers
18     Dynamic Rate    // Other features can be suggested,
like commenting, posting photos and so on.
19   else
20   Cascade
21    If retry = yes
22    Then go to 6
23     Else cascade
24    The End
```

5 Conclusion and Further Suggestions

The Internet of Things (IoT) is a novel cutting-edge technology that profits and proffers to connect a plethora of digital devices enabled with several sensing, actuation, and computing proficiencies with the Internet. The appealing IoT services are worthy of further studies and development considerations as technology promotes and solutions for smarter and beneficial outcomes. Emerging Internet of Things along with big data concepts, Internet of Mobile things, and embedding artificial units are a smart combination for Smart transportations systems for a better life and continuous cycles of development and enhancing procedures that anticipate risks in advance as we discussed through this chapter.

References

1. Dale Janssen – Co-founder Janalta Interactive In. https://www.techopedia.com/definition/31494/smart-city
2. A. Chang, R. Kalawsky, in *European Transport Sector Interventions for Smart City*, 2018. https://doi.org/10.1109/PESA.2017.8277778
3. https://www.toronto.ca/city-government/accountability-operations-customer-service/long-term-vision-plans-and-strategies/smart-cityto/
4. L. Jones, E. Carabine, *Exploring Political and Socio-Economic Drivers of Strategy Transformational Climate Policy: Early Insights from the Design of Ethiopia's Climate Resilient Green Economy*, 2013. https://www.odi.org/sites/odi.org.uk/files/odi-assets/publications-opinion-files/8617.pdf
5. https://digitalindia.gov.in/content/good-governance
6. S.P. Mohanty, Everything you wanted to know about smart cities: The internet of things is the backbone. IEEE Consum. Electron. Mag. **5**, 60–70 (2016)
7. V. Giordano, I. Onyeji, G. Fulli, M. S. Jimenez, C. Filiou, *Guidelines for Best Analysis of Smart Metering Deployment*, 2012 https://ses.jrc.ec.europa.eu/sites/ses/files/documents/guidelines_for_cost_benefit_analysis_of_smart_metering_deployment.pdf
8. R. Silveira, E. Gomes, M.V. Rosa, *Intelligent Learning Objects* (International Federation for Information Processing Digital Library; Information and Communication Technologies and Real-Life Learning, 2004). https://doi.org/10.1007/0-387-25997-X_12
9. P. Gawade, A. Meeankshi, IOT based smart public transport system. Int. J. Sci. Res. Publ. **7**(7), 390 (2017) ISSN 2250-3153, http://www.ijsrp.org/research-paper-0717/ijsrp-p6745.pdf
10. Online: https://www.redhat.com/cms/managed-files/iot-transportation-technology-overview-201608-en.pdf
11. Online: https://www.al-enterprise.com/-/media/assets/internet/documents/iot-for-transportation-solutionbrief-en.pdf
12. J. Sherly, D. Somasundareswari, *Internet of Things Based Smart Transportation Systems*, 2015. https://pdfs.semanticscholar.org/7bfe/05e20ce3845bd940d596223591fabc0490c9.pdf
13. S. Saarika, K. Sandhya, T. Sudha, Smart transportation system using IoT, in *2017 International Conference On Smart Technologies For Smart Nation (SmartTechCon)*, (Bangalore, 2017), pp. 1104–1107. https://doi.org/10.1109/SmartTechCon.2017.8358540
14. https://www.redhat.com/cms/managed-files/iot-transportation-technology-overview-201608-en.pdf
15. S. Muthuramalingam, A. Bharathi, S. Kumar, N. Gayathri, R. Sathiyaraj, B. Balamurugan, IoT based intelligent transportation system (IoT-ITS) for global perspective: A case study, in *Based Intelligent Transportation*, (Springer, Cham, 2019), pp. 279–300. https://doi.org/10.1007/978-3-030-04203-5_13
16. D. Kanellopoulos, University of Patras, Greece, Multimedia Social Networks, Copyright © 2015, IGI Global. Copying or distributing in print or electronic forms without written permission of IGI Global is prohibited. https://doi.org/10.4018/978-1-4666-5888-2.ch662. http://www.irma-international.org/viewtitle/113137/
17. J. Heidemann, M. Klier, F. Probst, Online social networks: A survey of a global phenomenon. Comput. Netw. **56**, 3866–3878 (2012). https://doi.org/10.1016/j.comnet.2012.08.009
18. T. Wen (Stella), Y. Yu, D. Vogel, R. Kwok, The impact of online social networking on learning: A social integration perspective. IJNVO **8**, 264–280 (2011). https://doi.org/10.1504/IJNVO.2011.039999

19. University of Illinois at Urbana-Champaign, *Library Trends [e journal]* (Johns Hopkins University Press, Baltimore, 2005)
20. https://www.hashtags.org/featured/what-characters-can-a-hashtag-include/
21. F. Amato, V. Moscato, A. Picariello, G. Sperlí, Multimedia social network modeling: a proposal, in *2016 IEEE Tenth International Conference on Semantic Computing (ICSC)*, (Laguna Hills, CA, 2016), pp. 448–453. https://doi.org/10.1109/ICSC.2016.20
22. E. Ahmed, I. Yaqoob, A. Gani, M. Imran, M. Guizani, Internet-of-things-based smart environments: State of the art, taxonomy, and open research challenges. IEEE Wirel. Commun. **23**(5), 10–16 (2016). https://doi.org/10.1109/MWC.2016.7721736
23. Y. Chen, Challenges and opportunities of internet of things, in *17th Asia and South Pacific Design Automation Conference*, (Sydney, NSW, 2012), pp. 383–388. https://doi.org/10.1109/ASPDAC.2012.6164978
24. D.G. Korzun, S.I. Balandin, A.V. Gurtov, Deployment of smart spaces in internet of things: Overview of the design challenges, in *Internet of Things, Smart Spaces, and Next Generation Networking. ruSMART 2013, NEW2AN 2013. Lecture Notes in Computer Science*, ed. by S. Balandin, S. Andreev, Y. Koucheryavy, vol. 8121, (Springer, Berlin/Heidelberg, 2013)
25. W.N.A.A.B. Wan, A.F.M.F. Mohamad, N.S. Shahib, A. Azmi, M. Kamal, S. Bahari, D.A. Abdullah, A framework of customer's intention to use Uber service in tourism destination. Int. Acad. Res. J. Bus. Technol. **2**, 102–106 (2016)
26. https://www.smart-industry.net/iot-car-the-internet-of-things-is-already-here-and-its-called-uber/. June 7, 2016
27. A. Scott, Co-founding Uber made Calgary-born Garrett Camp a billionaire. Canadian Business (November 19, 2015).
28. https://www.smart-industry.net/iot-car-the-internet-of-things-is-already-here-and-its-called-uber/
29. https://www.cleverism.com/strategies-uber-is-using-to-disrupt-the-taxi-industry/
30. S. Mahapatra, P. Telukoti, Challenges faced by the Uber drivers and consumers satisfaction in Pune City. Glob. J. Res. Anal. **7**(2), e36106 (2018)
31. A. Zareie, A. Sheikhahmadi, M. Jalili, Influential node ranking in social networks based on neighborhood diversity. Futur. Gener. Comput. Syst. **94**, 120–129 (2019). https://doi.org/10.1016/j.future.2018.11.023. Copyright © 2019 Elsevier B.V. https://www.sciencedirect.com/science/article/pii/S0167739X18319009
32. A. Sfard, A. Prusak, Telling identities: In search of an analytic tool for investigating learning as a culturally shaped activity. Educ. Res. **34**(4), 14–22 (2005). https://doi.org/10.3102/0013189X034004014
33. C. Coll, L. Falsafi, Learner identity. An educational and analytical tool La identidad de aprendiz. Una herramienta educativa y analítica. Rev. Educ. **353**, 211–233 (2010)
34. Y. Kubera, P. Mathieu, S. Picault, Everything can be Agent! (PDF), in *Proceedings of the Ninth International Joint Conference on Autonomous Agents and Multi-Agent Syllstems (AAMAS'2010)*, (1548, 2010), p. 1547
35. T.M. Valentina, N. Duque, D. Ovalle, Multi-agent system for expert evaluation of learning objects from repository. Commun. Comput. Inform. Sci. **524**, 320–330 (2015). https://doi.org/10.1007/978-3-319-19033-4_27
36. P. Polsani, Use and abuse of reusable learning objects. J. Digit. Inf. **3** (2003).
37. T. Raffaillac, S. Huot, Polyphony: *Programming Interfaces and Interactions with the Entity-Component-System Model*. EICS 2019 – 11th ACM SIGCHI Symposium on Engineering Interactive Computing Systems (Valencia, Spain, June 2019). ff10.1145/3331150ff. ffhal-02147180f

Part III
Policy, Cyber Security and Data Management of CAV

Formal Modeling: A Step Forward to Cyber Secure Connected Car Systems

Branka Stojanović, Katharina Hofer-Schmitz, Kai Nahrgang, Heribert Vallant, and Christian Derler

1 Introduction

The paradigm *communicate with anyone anywhere at anytime* nowadays spans to cyber-physical systems in general and impacts many fields, including different types of industry (e.g., transportation, manufacturing, IT, etc.), health, and mobility [1]. An increase in connectivity demands, including a built-in connectivity, is reflected in a great deal by vehicle manufacturing industry. The established path and goal for automotive industry include connected cars and autonomous driving [2].

Existing and potential services in the connected cars industry should increase the road safety, bring more comfort to all passengers, and add more efficiency in traffic flows. Different sensors and services inside connected cars communicate and synchronize in order to enhance drivers' experience and make processes smoother. In addition to in-vehicle communication, connected cars communicate and interact with their environment, including other vehicles, roadside users, and external infrastructure and devices and even share processing efforts between other entities.

An increase in vehicles' connectivity demands influences in a great deal a rise of security issues. The ***security-by-design*** frameworks, including threat modeling and formal methods, have potential to respond to these challenges.

This chapter covers two main objectives – (i) a comprehensive overview of the *connected cars' communication architecture* and most important communication protocols under the V2X umbrella, with a special focus on the security perspective and (ii) *security-by-design* frameworks application within this domain, *threat modeling* state of the art methodologies and the ability to adapt those for the

B. Stojanović (✉) · K. Hofer-Schmitz · K. Nahrgang · H. Vallant · C. Derler
Joanneum Research Forschungsgesellschaft mbH, Graz, Austria
e-mail: Branka.Stojanovic@joanneum.at; Katharina.Hofer-Schmitz@joanneum.at; Kai.Nahrgang@joanneum.at; Heribert.Vallant@joanneum.at; Christian.Derler@joanneum.at

U. Z. A. Hamid, F. Al-Turjman (eds.), *Towards Connected and Autonomous Vehicle Highways*, EAI/Springer Innovations in Communication and Computing,
https://doi.org/10.1007/978-3-030-66042-0_6

automotive industry and *formal verification* tools and their applications in V2X protocols space. Additionally, it discusses challenges and future research directions, as research and development path within this industry. In Fig. 1, an illustration of a security-by-design procedure for in-vehicle communication is presented as research guideline. It includes threat modeling and formal verification based on inputs from previous steps and standards/specifications and their redesign according to findings.

The chapter is structured as follows: In Sect. 2, a basic introduction into the communication architecture and the protocols in use for connected cars is given. Section 3 focuses on threat modeling and describes different modeling methodologies and their possible applications, benefits, and limitations to model connected cars. Section 4 focuses on formal methods and describes used tools and considered protocols. In Sect. 5, key points are summarized, and open challenges for future work are stated.

2 Connected Cars Communication Architecture and Protocols

This section focuses on the communication architecture and most important automotive communication protocols from the security perspective.

The term vehicle-to-everything, commonly known by abbreviation V2X, encompasses all types of communications in the automotive domain, involving different types of communication entities, like vehicles, infrastructure units, motorcycles, cycles, pedestrians, etc. The heterogeneous connected car network consists of two main subnetworks [3] – intra-vehicle network, which covers a communication between in-vehicle devices, and inter-vehicle network, including the communication between the vehicle and surrounding.

2.1 Connected Car Network

Intelligent transportation system usually refers to the connected car system. It encompasses diverse entities and technologies, like vehicles, infrastructure units, and roadside users, and then data processing, communication and sensor technologies, etc. The heterogeneous network of such a system that connects different types of entities using different types of communication technologies consists of two main subnetworks [4]:

- *Intra-vehicle* network – covers a communication between in-vehicle devices, including controlling units, sensors, and actuators.
- *Inter-vehicle* network – commonly refers to the communication between vehicles; in this paper this term will be extended to all communication types among the vehicle and surrounding devices: on-board unit in-vehicle and external

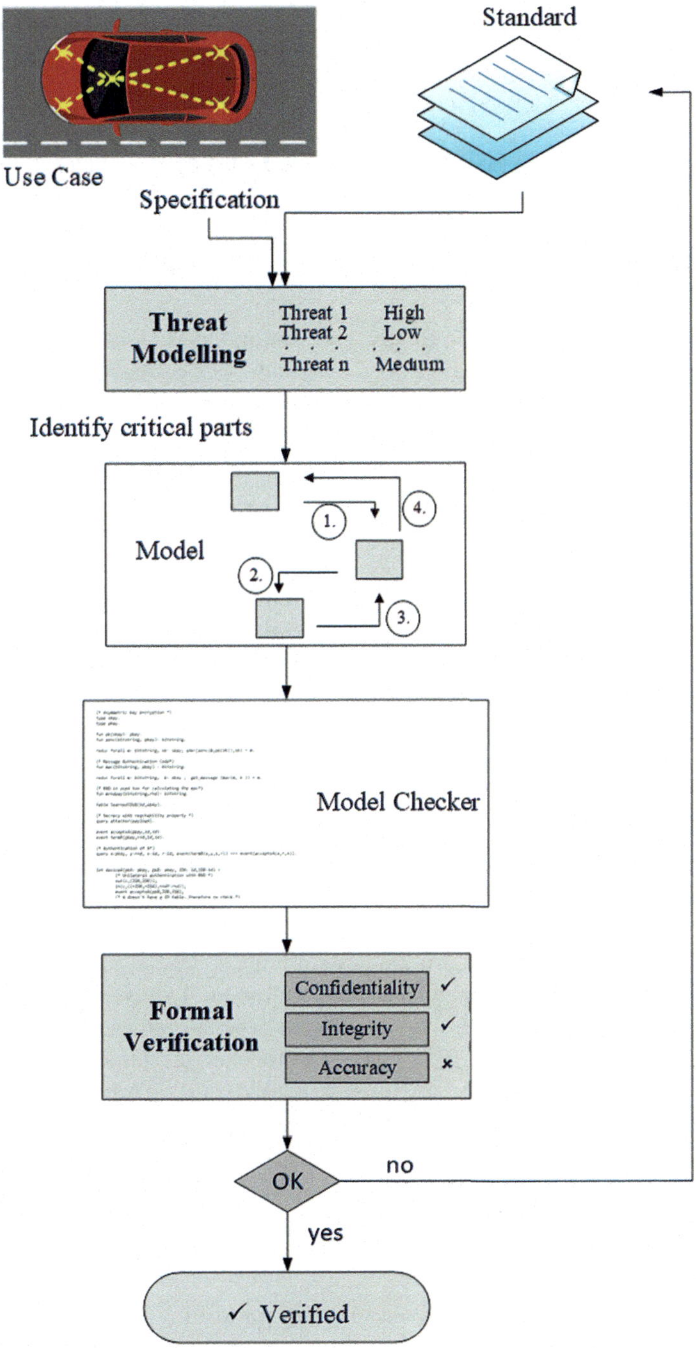

Fig. 1 Illustration of security-by-design framework within connected cars industry

entities, like roadside users (pedestrian, motorcyclists, etc.), infrastructure units, and central processing units (central/cloud server).

V2X on the other hand supports a unified connectivity platform for all connected end points and allows road entities to transmit information such as their current speed, position, and direction to the neighboring entities. It includes both *intra-* and *inter*-vehicle networks and can be categorized in different types of communication (Fig. 2):

- *In-vehicle* communication – represents the communication between entities in *intra*-vehicle subnetwork;
- *Vehicle-to-vehicle (V2V)* – covers the communication between vehicles, for example, the vehicle can broadcast the message of a pedestrian crossing the road to other vehicles, or the vehicle learns of another vehicle ahead braking suddenly and communicates this alert with other vehicles.
- *Vehicle-to-infrastructure (V2I)* – represents the communication between road entities and infrastructure units, for example, the vehicle can communicate with the traffic lights to know the speed at which he can drive to get green at the next traffic light, etc.
- *Vehicle-to-grid (V2G)* – supports the communication between vehicles and the electric grid, for example, plug-in electric vehicles communicate with the power grid to sell services on a return basis either by returning electricity to the grid or by throttling their charging rate;
- *Vehicle-to-pedestrian (V2P)* – provides the connection between the vehicle and vulnerable road users (VRU), including pedestrians, cyclists, and motorized two-wheeler operators; a typical *V2P* crash prevention system involves periodic exchange of safety messages among vehicles and VRUs [5].

This book chapter focuses on the security and safety perspectives of the most important automotive communication protocols, including in-vehicle communication and V2V and V2I protocols, because they address safety applications that are crucial for a rapid, robust, and timely performance, where any delay in message delivery could lead to a potentially fatal collision. Safety applications include various warnings (e.g., red light violations, curve speeds, reduced speed/work zones, emergency electronic brake lights, forward collisions, etc.) that are sent from their place of occurrence, picked by the closest vehicle, and then further propagated to the surrounding vehicles.

2.2 Intra-vehicle Communication

The interaction between various sensors and controlling units inside the vehicle requires an information exchange using specific communication protocols [2, 6].

Intra-vehicle communication usually involves LIN (Local Interconnect Network), CAN (Controller Area Network), FlexRay, MOST (Media Oriented System

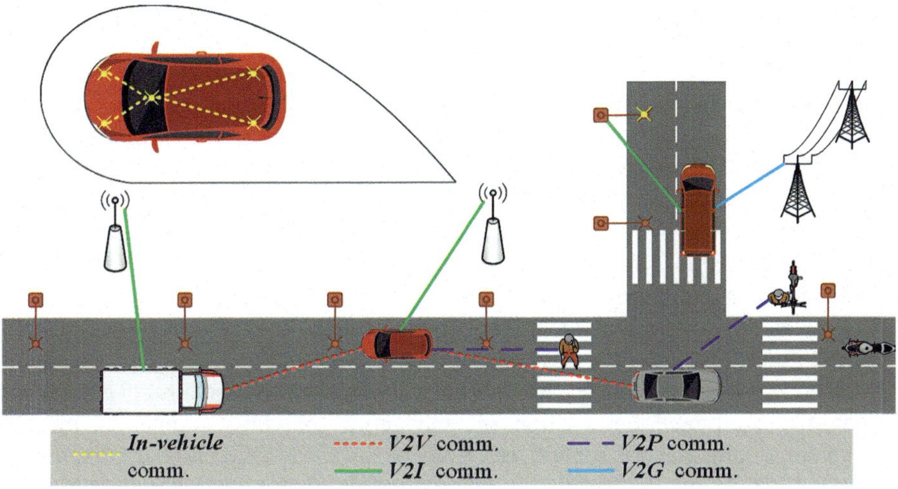

Fig. 2 Connected vehicle communication illustration

Table 1 *Intra*-vehicle protocols

Protocol	Data rate	Medium	Standard	Alliance	Year
CAN	1 Mbps	Twisted pair	ISO 11898	ISO	1991
MOST	150 Mbps	Optical fiber	Proprietary	MOST Coop. consortium	2001
LVDS	655 Mbps	Twisted pair	TIA/EIA-644	TIA	2001
LIN	19.2 kbps	Single wire	ISO 17987	LIN consortium	2002
FlexRay	20 Mbps	Twisted pair/optical fiber	ISO 17458	FlexRay consortium	2005
Automotive Ethernet	10 Mbps	Single twisted pair	IEEE802.3cg-2019	OPEN alliance	2019
	<10 Gbps	Single twisted pair	IEEE P802.3ch	OPEN alliance	tba

Transport), LVSD (Low-voltage differential signaling), or Automotive Ethernet. An overview of intra-vehicle protocols is presented in Table 1.

2.2.1 CAN: Controller Area Network

CAN [7] is a robust automotive-specific bus standard. It defines the functionality of the first two layers of the Open Systems Interconnection (OSI) network model – Layer-1 and Layer-2. CAN's design allows communication between different devices inside vehicles, including microcontrollers, or ECUs (electronic control units) [6]. CAN was first developed and released in 1986 by Robert Bosch GmbH. I1991 is the year of production of the first vehicle featuring this protocol [8].

CAN standard ISO 11898 was released in 1993 by the ISO – International Organization for Standardization. It was later restructured into two parts, with a third part released afterward. The most recent versions of those parts of ISO 11898 standard are as follows: (i) ISO 11898-1:2015[1] covering the data link layer and physical signaling; (ii) ISO 11898-2:2016[2] covering CAN, high-speed medium access units; (iii) ISO 11898-3:2006[3] covering CAN, the low-speed, fault-tolerant, medium-dependent interface.

Typical applications include the communication between ECUs controlling engine, power transmission, gearbox, antilock braking/ABS, electric power steering, etc. Beside passenger vehicles, it is used in trucks and buses, agricultural equipment, electronic equipment for aviation and navigation, building automation, medical equipment, industrial automation, etc. CAN bus is used in the on-board diagnostics (OBD)-II [9] vehicle diagnostics standard, as one of five supported protocols. CAN nodes are connected through a twisted pair bus, and data rates supported are up to 1 Mbps.

CAN is a low-level protocol and contains no direct support for security features. The implementations do not contain an encryption standard, and it leaves networks using CAN protocol open to cyber attacks, like man-in-the-middle frame interception and inserting messages on the bus. Security mechanisms are customized and usually implemented on the application and manufacturer level.

2.2.2 MOST: Media Oriented Systems Transport

MOST [10] is an automotive-specific high-speed multimedia network technology. It defines the physical and the data link layer as well as other layers of the ISO/OSI model of data communication. It was first introduced in 2001 by the MOST Cooperation consortium and has been implemented in ten vehicle models in the same year. The technology is nowadays used in almost every car brand, including Audi, General Motors, BMW, Hyundai, Honda, Lancia, Jaguar, Porsche, Mercedes-Benz, Land Rover, Toyota, Saab, Volkswagen, SKODA, Volvo, and SEAT. It is used to transport data signals, video, and audio inside vehicles. MOST nodes are connected via plastic optical fiber (POF) (MOST25, MOST150) or electrical conductor (MOST50, MOST150) physical layers, and it supports data rates up to 150 Mbps (MOST150).

[1]https://www.iso.org/standard/63648.html
[2]https://www.iso.org/standard/67244.html
[3]https://www.iso.org/standard/36055.html

The MOST protocol is secured by an automatically generated CRC sum (4 bytes) and ACK/NAK mechanism with automatic retry. There is no automatic retransmission in case of an error, and it has to be handled by the higher layers [10].

2.2.3 LVDS: Low-Voltage Differential Signaling

LVDS [11] is a technical standard that specifies high-speed signaling, using a differential, serial communication protocol. It specifies only the physical layer, while different data communication standards and applications that are built on top of it specify a data link layer of the ISO/OSI model.

The LVDS standard was defined in 2001, as ANSI/TIA/EIA-644-A standard.[4] It is used for high-speed video, graphics, video camera data transfers, and general-purpose computer buses. Although LVDS was not specifically developed for the automotive industry, its high-speed bandwidth of 655 Mbps over twisted-pair copper cable made it the top choice for automotive camera manufacturers. Besides automotive infotainment system, it is used in LCD-TVs, industrial cameras and machine vision, notebooks, tablets, etc.

LVDS protocol, as Layer-1 protocol, does not define any security mechanisms.

2.2.4 LIN: Local Interconnect Network

LIN [12] is an automotive-specific bus standard, defined as a cheaper alternative to CAN for less important components of the in-vehicle network [6], like the seats and steering wheel adjustment. It is a broadcast master-slave serial network protocol, which supports a data rate up to 19.2 kbps, via a single wire. Similar to CAN, it specifies the first two layers of OSI model.

The first fully implemented version of LIN protocol was specified in 2002, by the LIN Consortium, founded by five car manufacturers – BMW, Volkswagen Group, Audi, Volvo Cars, and Mercedes-Benz. It is standardized in the ISO 17987 series, where ISO/AWI 17987-8[5] is the standard defined for LIN over DC power line (DC-LIN).

LIN supports only error detection and checksums and faces similar risk exposures as CAN.

---

[4]https://www.ti.com/lit/an/slla038b/slla038b.pdf

[5]https://www.iso.org/standard/71044.html

2.2.5 FlexRay

FlexRay is an automotive-specific bus standard. Its advantages over CAN are higher reliability and speed, while disadvantage is additional cost overhead. Similar to the previously described bus standards, it specifies the first two layers of the OSI model – the physical layer and the data link layer.

The FlexRay Consortium developed it in 2009, mainly for high-performance onboard automotive computing applications, including drive electronics and safety (e.g., proximity control, active suspension, drive-by-wire, etc.). It comes with a bandwidth from up to 10 Mbps and uses unshielded cable pairs. The consortium, which later disbanded, included BMW, Volkswagen, Daimler, and General Motors (GM). FlexRay is specified in ISO 17458-1[6], 17458-2[7], 17458-3[8], 17458-4[9], and 17458-5[10] standards.

FlexRay, like other previously described bus protocols, was engineered in the absence of any security concerns. Therefore, it is highly vulnerable to adversarial attacks [13].

2.2.6 IEEE 802.3: Automotive Ethernet

Ethernet standard, commonly utilized as communication bus, is introduced to automotive industry as automotive Ethernet. The driving force for Ethernet usage in the automotive industry was primarily the high bandwidth. Additionally, the usage of UTP (unshielded twisted single-pair) cabling, its size, flexibility, and cost, also contributed to Ethernet applicability in vehicles. UTP cabling reduces network complexity and cabling costs and also contributes to free space and less weight of cars [6].

There are several revisions to the IEEE 802.3 standard that were made to fully meet the automotive requirements:

- IEEE 802.3bw[11]: 100BASE-T1 – 100 Mbps Ethernet over a single twisted pair for automotive applications, released 2015, superseded;
- IEEE 802.3bp[12]: 1000BASE-T1 – 1 Gbps Ethernet over a single twisted pair, automotive and industrial environments, released 2016, superseded;

[6] https://www.iso.org/standard/59804.html
[7] https://www.iso.org/standard/59806.html
[8] https://www.iso.org/standard/59807.html
[9] https://www.iso.org/standard/59808.html
[10] https://www.iso.org/standard/59809.html
[11] https://standards.ieee.org/standard/802_3bw-2015.html
[12] https://standards.ieee.org/standard/802_3bp-2016.html

- IEEE 802.3bv[13]: 1000BASE-RHx – 1000 Mbps Ethernet over plastic optical fiber (POF), intended for home, industrial, and automotive use, released 2017, superseded;
- IEEE 802.3cg[14]: 10BASE-T1 – 10 Mbps Ethernet over a single twisted pair, intended for automotive and industrial applications, released 2019, active;
- IEEE P802.3ch[15]: IEEE draft standard for multi-Gig automotive Ethernet (2.5, 5, 10 Gbps) over 15 m, release date tba, active.

Comparing to previously described bus standards, Automotive Ethernet, with high bandwidth gives leeway to better authentication or encryption mechanisms (e.g., Media Access Control (MAC)), and due to point-to-point characteristics of the Ethernet, a stricter separation into and within functional domains can be achieved, using Virtual Local Area Network (VLAN), Quality of Service (QoS), and firewall concepts [14].

2.3 Inter-*vehicle Communication*

The interaction between vehicles and surrounding devices, including other vehicles and road side users, usually includes discussions about two types of protocols – WiFi based, often referred to as IEEE 802.11p from the name of the first standard designed to this scope, and cellular technologies including LTE-V2X and recently 5G, as part of the fourth generation of Third Generation Partnership Project (3GPP) standards and under the broader umbrella of the C-V2X (Cellular-V2X) [1]. Inter-vehicle protocols overview is presented in Table 2.

2.3.1 IEEE 802.11p

IEEE 802.11p[16] is the name of the first WiFi-based standard designed for V2X communication, released in 2010. Later, IEEE 802.11p was included in the IEEE 802.11-2012, which is afterward superseded by the IEEE 802.11-2016[17]. IEEE 802.11p defines the layer-1 (PHY) and layer-2 (MAC) layer protocols. A number of other standards have been defined above IEEE 802.11p standard, creating two different pillars – one in the USA, known as DSRC (dedicated short-range communication) and WAVE (wireless access in vehicular environments), and one in Europe, known as ETSI-ITS-G5 [1].

[13]https://standards.ieee.org/standard/802_3bv-2017.html

[14]https://standards.ieee.org/standard/802_3cg-2019.html

[15]https://standards.ieee.org/project/802_3ch.html

[16]https://standards.ieee.org/standard/802_11p-2010.html

[17]https://standards.ieee.org/standard/802_11-2016.html

Table 2 *Inter*-vehicle protocols

Protocol	Standard	Description	Status	Alliance	Year
IEEE 802.11p	IEEE 802.11p	Amendment 6: Wireless Access in Vehicular Environments	Supers.	IEEE	2010
IEEE 802.11p	IEEE 802.11	802.11-2016 – includes IEEE 802.11p functionalities	Active	IEEE	2016
C-V2X	3GPP Release 14	Mission Critical (MC) enhancements, LTE support for V2X services, IoT, voice and multimedia-related items, location and positioning items, etc.	Frozen	3GPP	2016
C-V2X	3GPP Release 15	New Radio (5G), 5G Phase 1, massive IoT, V2X Phase 2, MC networking with legacy systems, LTE improvements, etc.	Frozen	3GPP	2018
IEEE 802.11p	P802.11bd	Amendment: Enhancements for Next Generation V2X	Draft	IEEE	2018
C-V2X	3GPP Release 16	5G Phase 1, industrial IoT, V2X Phase 3, etc.	Frozen	3GPP	2019
IEEE 802.11p	IEEE 1609.12	1609.12-2019 – IEEE Standard for Wireless Access in Vehicular Environments (WAVE) Identifiers	Active	IEEE	2019
IEEE 802.11p	ETSI EN 302 663	ITS-G5 Access layer specification for Intelligent Transport Systems operating in the 5 GHz frequency bands	Active	ETSI	2020
C-V2X	ETSI EN 303 613	LTE-V2X Access layer specification for Intelligent Transport Systems operating in the 5 GHz frequency band	Active	ETSI	2020
C-V2X	3GPP Release 17	NR enhancements, enhanced V2X, unmanned aerial systems, etc.	Sched.	3GPP	2021

In the USA, IEEE 1609 standards define protocols below the application layer as wireless access in vehicular environments (WAVE), with IEEE 1609.12-2019[18] as the active version. In Europe, IEEE 802.11p was adopted by ETSI under the

[18]https://standards.ieee.org/standard/1609_12-2019.html

Cooperative Intelligent Transport Systems (C-ITS[19]) as ITS-G5[20], together with a large number of other documents dealing with all layers above it, dedicated to automotive ITS and Road Transport and Traffic Telematics (RTTT).

In May 2018, IEEE announced a new study group focused on the evolution of 802.11 technology for next-generation V2X communications. Their work resulted in publishing the amendment IEEE 802.11bd[21] later the same year. The ability to communicate for relative vehicle speeds of 250 kmph is a key feature of 802.11p. It operates in the licensed ITS band of 5.9 GHz with 10 MHz channel. IEEE 802.11p typically supports the range of 150–300 m. Its data rate is typically 6–27 Mbps, and it uses mesh network topology.

2.3.2 Cellular V2X

In parallel with IEEE 802.11p development, cellular technologies have been evaluated as long-range alternative. In 2016, 3GPP created the so-called C-V2X within Long-Term Evolution (LTE) Release 14[22]. It included a short-range interface that can be used also outside the cellular coverage and that poses an alternative to IEEE 802.11p [1].

In general, LTE is a wireless broadband communication standard designed for data terminals and mobile devices. It is based on the 2G/2.5G GSM/EDGE and 3G UMTS/HSPA technologies. LTE is specified in the 3GPP Release 8 and 9 document series, where Release 9 defines minor enhancements. It is also known as 4G LTE, Advance 4G, and 3.95G, since it does not meet the technical criteria of a 4G wireless service (defined in the 3GPP Rel. 8 and 9). In the beginning of 2020, ETSI published LTE-V2X[23] standard – LTE-V2X Access layer specification for Intelligent Transport Systems operating in the 5 GHz frequency band.

LTE-V2X advantages include easy implementation – it can use existing cellular infrastructure. It supports relative speeds of up to 500 kmph [15]. It provides rates of 300 Mbps for downlink and 75 Mbps for uplink. A transmission range depends on application mode and can be up to 100 km in the radio network, while in Direct C-V2X applications, it is greater than 450 m. It provides a longer reaction time for driver, than in 802.11p communications [16].

[19]https://www.etsi.org/technologies/automotive-intelligent-transport

[20]https://www.etsi.org/deliver/etsi_en/302600_302699/302663/01.03.01_60/en_302663v010301p.pdf

[21]https://standards.ieee.org/project/802_11bd.html

[22]https://www.3gpp.org/release-14

[23]https://www.etsi.org/deliver/etsi_en/303600_303699/303613/01.01.01_30/en_303613v010101v.pdf

In 2018, 3GPP published the Release 15[24] that describes 5G NR (5G New Radio), including vehicle-to-everything communications (V2X) Phase 2. 5G is the successor of GSM (2G), UMTS (3G), and LTE and LTE Advanced Pro (4G). The International Telecommunication Union Radiocommunication Sector (ITU-R) has lists following main uses for 5G:

- eMBB – Enhanced Mobile Broadband: an enhancement of 4G LTE mobile broadband services that includes more capacity, higher throughput, and faster connections;
- URLLC – Ultra-Reliable Low-Latency Communications: includes support for applications that requires uninterrupted and robust data exchange, like mission critical applications (deployment expected after 2021);
- mMTC – Massive Machine Type Communications: connects a large number of low-power and low-cost devices in a wide area; it should have increased battery lifetime and high scalability (deployment expected after 2021).

The three key frequency ranges for 5G spectrum, necessary to deliver widespread coverage and support all use cases, are:

- <1 GHz, which supports IoT services and provides widespread coverage across urban, suburban, and rural areas
- 1–6 GHz, expected to operate within the 3.3–3.8 GHz and to provide a good mixture of coverage and capacity benefits
- >6 GHz, expected to operate in 26 GHz and/or 28 GHz band, needed to meet the ultrahigh broadband speeds envisioned for 5G.

The targeted air latency in 5G is 1–4 ms. 5G should operate with throughput up to 10 Gbps, a hundred times faster throughput than 4G (LTE) speed of 100 Mbps. 5G Phase 2 is announced in 3GPP Release 16[25], with final submission planned for June 2020. It includes V2X Phase 3, with platooning, extended sensors, automated driving, and remote driving as main key points. More 5G system enhancements are set to follow in Release 17[26]. It is scheduled for delivery in 2021. Enhanced V2X services are announced in this release.

3 Threat Modeling

The increasing connectivity demands of various handheld devices, Internet of Things (IoT) and infrastructure assets together with the built-in automotive components, result in new threats from cyber space that are striking directly without any warning time. Therefore, theoretical modeling about the security status of a complex

[24]https://www.3gpp.org/release-15
[25]https://www.3gpp.org/release-16
[26]https://www.3gpp.org/release-17

system is becoming increasingly important. A theoretical modeling approach is threat modeling, which has the goal to identify potential threats and vulnerabilities based on the architecture of the given IT system. Conceptually different methodologies are used ranging from secure and agile application development to operative and business-driven concepts. Threat modeling is especially useful when applied during the design phase, as it delivers a semiformal security assessment which identifies security issues and the most likely attack vectors.

This section describes different threat modeling methodologies and their possible applications and limitations to model the domain of connected cars.

Threat modeling is a process for identifying security issues for various IT systems. By using different methodologies, threat modeling can be used for plenty of scenarios and is not limited in any way of creating new methodologies or even adapting existing ones for new purposes. Hence, with regard to connected cars, the research community has already adapted well-established methods and achieved great results.

3.1 Threat Modeling Overview

Different threat modeling techniques have been developed addressing not only different aspects, like data and data flow, application and assets, and risk based or impact oriented but also different application areas like the software engineering or system architectures overall. However, more general threat modeling is split into two approaches [17]:

- Application Threat Modeling
- Operational Threat Modeling

The former is focusing on identifying threats of applications or IT architectures, which are represented using process-flow diagrams (PFD). These threats can then be addressed by software developers, software testers, as well as system architects and cyber security experts to work on mitigation. The latter are created from an attacker's point of view using data-flow diagrams (DFD). Operational threat models provide a visualization of the infrastructure's big picture in order to give a better view on the full attack surface. The result is usual used within (Sec)DevOps life cycles. Regardless of the approach and the application field, threat modeling is usually performed in four steps [18]:

1. Model system
2. Find threats
3. Address threats
4. Validate

While the first point is usually done using different software tools, the second point usually differs from the used threat modeling approach, which will be discussed in more detail in the following subsections. The third point then focuses on addressing the found threats by coming up not only with mitigation strategies but also, depending on the used approach, with a risk assessment. In the last step, a validation of the work done in point one to three should be performed.

The following subsections discuss several techniques of threat modeling, outlining the different aspects they address, in order to understand the different approaches.

3.1.1 ATT&CK

Adversarial Tactics, Techniques, and Common Knowledge (ATT&CK) was created by MITRE and is used as a knowledge base and model for cyber adversary behavior. ATT&CK reflects the life-cycle phases of an adversary attack and the corresponding platforms. It can be used during various scenarios like red teaming or to improve defenses against network intrusion attacks. It started for Windows systems only, but now includes also Linux, macOS, cloud platforms, and mobile devices [19]. As stated in [20], the behavior model consists of three core models:

- Tactics, denoting short-term, tactical adversary goals during an attack (the columns);
- Techniques, describing the means by which adversaries achieve tactical goals (the individual cells);
- Documented adversary usage of techniques and other metadata (linked to techniques).

To illustrate an example, Table 3 [21] shows the ATT&CK Cloud Matrix. Since this technique has already been adapted for various platforms and systems, it could also be possible to adapt it for connected cars.

3.1.2 Attack Trees

Attack trees are a rather old but a still valid and valuable approach to discover threats in various environments. The concept was invented by *Bruce Schneier* [22] for modeling threats against computer systems. Attack Trees are not limited to computer systems, but in the information technology, they are a formal and methodical way to describe security threats based on possible attacks. *Shostak* describes in [18] three ways of using them: (1) use an attack tree created by someone else for your purposes; (2) create a tree specifically for your project; and (3) create a tree for general use, with the indent others will use it. Now, if you want to use a tree created by someone else, a modeled system is necessary first. Once this is done, the attack tree can be applied for each node of the model to see if the threat might

Table 3 ATT&CK cloud matrix

Initial access	Persistence	Privilege escalation	Defense evasion	Credential access	Discovery	Lateral movement	Collection	Exfiltration	Impact
Drive-by compromise	Account manipulation	Valid accounts	Application access token	Account manipulation	Account discovery	Application access token	Data from cloud storage object	Transfer data to cloud account	Resource hijacking
Exploit public-facing application	Create account		Redundant access	Brute force	Cloud service dashboard	Internal spearphishing	Data from information repositories		
Spearphishing link	Implant container image		Revert cloud instance	Cloud instance metadata API	Cloud service discovery	Web session cookie	Data from local system		
Trusted relationship	Office application startup		Unused/ Unsupported cloud regions	Credentials in files	Network service scanning		Data staged		
Valid accounts	Redundant access		Valid accounts	Steal application access token	Network share discovery		Email collection		
	Valid accounts		Web session cookie	Steal web session cookie	Permission groups discovery				
					Remote system discovery				
					System information discovery				
					System network connections discovery				

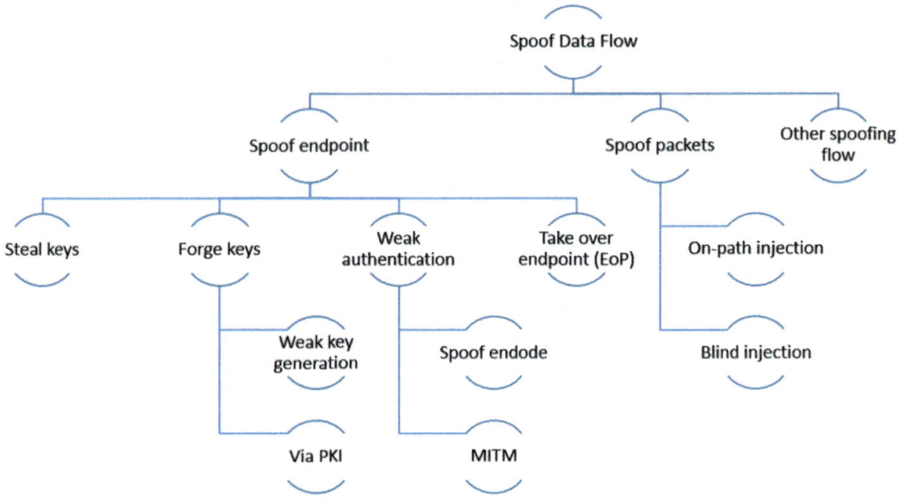

Fig. 3 Attack tree: spoofing of data flow

be applicable or not. To illustrate the approach, an example attack tree for spoofing data flow is given in Fig. 3.

The illustration shows that the root node is most properly the goal of the attack but might also represent a component of the system. If it represents a component, the subnotes should indicate what could get wrong. If the root node is the goal of the attack, the next steps show how to achieve it. When adding multiple subnotes, it is important to decide if the relationship between the nodes represents AND or OR. However, most of the time, the attack trees are using OR relationships. Once the goal of the tree as well as every single step how to achieve it is drawn, you should consider to prune the tree. This way, each node will once again be evaluated if it is duplicative or maybe even already mitigated by your system. Lastly, an attack tree should not exceed a single page in order to keep it clearly represented. If that is not the case, a tree might need to be split up into several trees.

To sum up, in order to create an attack tree, six steps need to be followed:

1. Decide on representation (AND or OR)
2. Create the root node (goal or components)
3. Create subnotes
4. Consider completeness
5. Prune the tree
6. Check the representation

As there are plenty of general attack trees which already can be applied to various projects, this approach can also be applied on automotive systems. Also, as this section discussed, there is always the possibility to create new, specifically for automotive vehicles, attack trees.

3.1.3 STRIDE

STRIDE was originally introduced by Microsofts' cyber security professionals Loren Kohnfelder and Praerit Garg as part of Microsoft's Security Development Lifecycle (SDL) concept. STRIDE uses data flow diagrams to describe the communication between processes and data stores in order to generate threats that are divided into the following six categories [23]:

- **S**poofing identity: A user or service illegally accesses and uses other authentication information to gain illegitimate access to a system or data.
- **T**ampering with data: Data tampering occurs when data is maliciously modified. This includes data at rest, data in use, as well as data in transit.
- **R**epudiation: This means that an entity may plausibly deny an action that it has taken. Countering these threats usually requires a combination of authentication, authorization, and logging, ideally in a cryptographically secured way.
- **I**nformation disclosure: Refers to any information exposed to unauthorized users.
- **D**enial of service (DoS): DoS attacks deny services availability to valid users.
- **E**levation of privilege: These threats occur when unprivileged users gain privileged access and, thus, are able to compromise an entire system.

The Microsoft Threat Modeling Tool[27] offers different templates for various scenarios and also gives the possibility to create new templates. STRIDE therefore has already been adapted also for the automotive domain [24–27] and hence will be discussed in Sect. 3.2.1.

3.1.4 TARA

Threat Agent Risk Assessment (TARA), developed by Intel, is a methodology that not only identifies threats but also shows which of them are most likely to occur [28]. This is achieved by focusing on threat agents, their motivations and methods. Threat agents represent attackers with certain capabilities of skills and resources. These properties of a threat agent are then mapped to methods that can occur, which might lead to possible threats. TARA is also taking a step further, by considering acceptable levels of corporate risks. This means that although an attack is likely to occur, the impact might be too little, and TARA might not identify this threat has a high-priority item.

TARA consists of three components:

- Threat agent library (TAL)[2]: The library identifies 22 threat agent archetypes including from internal employees to different kinds of external criminals.
- Common exposure library (CEL): The CEL includes common security vulnerabilities and exposures and maps them to known mitigations.

[27] https://www.microsoft.com/en-us/download/details.aspx?id=49168

- Methods and objective library (MOL): The MOL lists a set of methods on how threat agents usually plan to achieve their objectives.

When mapping these three components together, it becomes clear how the methodology works. As an example, Table 4 shows a sample from the MOL library. In Sect. 3.2.2, it is discussed how the approach and its components are adapted for the automotive industry.

3.2 Examples of Threat Modeling in the Automotive Industry

Since the previous section gave an overview about the different threat modeling methodologies, this section focuses on how to adapt threat modeling approaches for the automotive industry. Therefore, two examples on how popular approaches were already adapted in related works are given.

3.2.1 STRIDE for the Automotive Industry

The adaption of STRIDE is done using the template feature of the Microsoft Threat Modeling Tool. Here, a new template with regard to the automotive industry has been created by the NCC Group [26, 27]. The authors claim to provide following features:

- Processes and Data Stores related to connected cars;
- External Interactors tailored to an automotive system;
- Data Flows including over the air (OTA) and CAN bus;
- Trust Boundaries including vehicle-to-vehicle (V2V) networks;
- Known threats to components of connected cars, following the STRIDE categories.

Based on this template, a sample architecture of a connected car has been created and can be seen in Fig. 4. Here, a driver using a Human Machine Interface (HMI)/In-Vehicle Infotainment (IVI) and various sensors and cameras of the connected car is illustrated. These sensors and cameras are gathering information from the environment entity and are then passed to the Sensor Fusion Electronic Control Unit (ECU). The ECU sends the data to the gateway, which stores the data to the respective database. Also, a firmware update server and the respective data storage are drawn.

Figure 4 shows a simple example of a connected car threat model using an automotive industry template. A sample of the generated threats can be seen in Fig. 5, which shows newly added threats like tricking the sensor fusion ECU in into triggering an emergency stop, which, for example, would affect the safety of the vehicle, the passengers, and most probably also outside traffic participants.

Table 4 Sample from MOL library

Agent name	Attacker					Objective		Method									Impact				
	Access	Trust				Motivation	Goal	Acts					Limits								
		None	Partial trust	Employee	Administrator			Copy, Expose	Deny, withhold, ransom	Destroy, delete, render unavailable	Damage, alter	Take, remove	Code of conduct	Legal	Crimes against property	Crimes against people	Loss of financial assets	Business operation impact	Loss of competitive advantage, market share	Legal or regulatory exposure	Degradation of reputation, image, or brand
Employee error	Internal		X	X	X	Accidental/Mistake	No malicious intent, accidental	X		X	X		X				X	X	X	X	X
Reckless employee	Internal		X	X	X	Accidental/Mistake	No malicious intent, accidental	X		X	X			X			X	X	X	X	X
Information partner	Internal		X			Accidental/Mistake	No malicious intent, accidental	X		X	X						X	X	X	X	X
Competitor	External	X				Personal gain (Financial)	Obtain business or technical advantage	X							X				X		
Radical activist	External	X				Social/Moral gain	Change public opinion or corporate policy	X	X	X	X	X				X		X			X
Data miner	External	X				Personal gain (Financial)	Obtain business or technical advantage	X							X				X		
Vandal	External	X				Personal gain (Emotional)	Personal recognition or satisfaction			X	X				X			X			X
Disgruntled employee	Internal		X	X	X	Personal gain (Emotional)	Damage or destroy organization		X	X	X				X			X	X		X

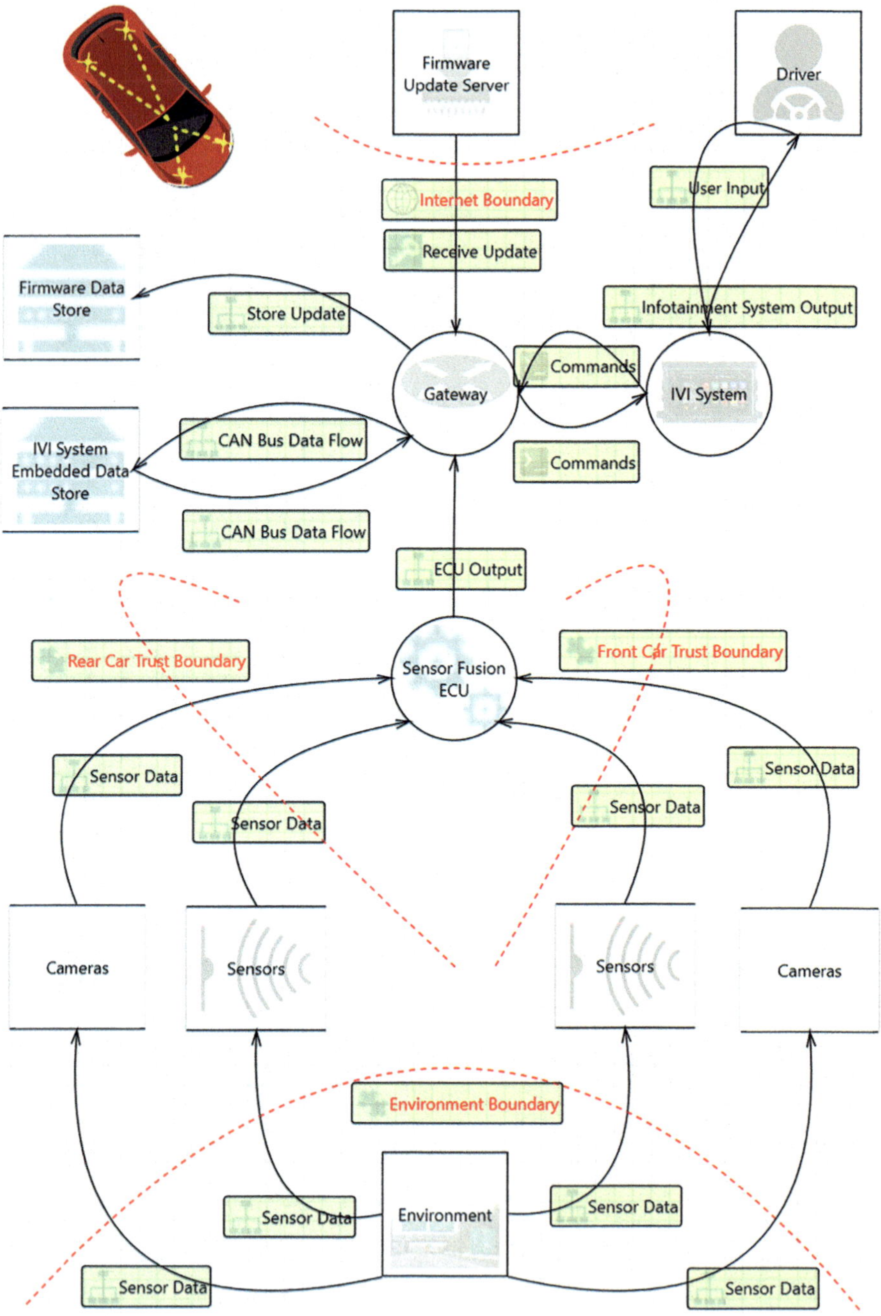

Fig. 4 Connected car: sample threat model

Title	Category	Interaction	Priority	Description	Attack method	Recommendation
Flood IVI System With Invalid Data	Denial of Service	Commands	Medium	DoS on IVI System by flooding with invalid data.	Either physically by clipping onto the sensor wires and inject valid data or with external input e.g. a bright torch.	Rely on additional sensors in the event of one is unavailable.
Take the IVI System Offline	Denial of Service	Commands	Medium	DoS on IVI System.	Perform an network attack and case resource exhaustion.	Have a number of IVI System delivery servers across a broad geographic radius, in the event of one server failing the system should continue unhindered.
Pretend to Be the Sensors in Order to Exploit the Sensor Fusion ECU	Elevation of Privilege	Sensor Data	Medium	Elevation of privileges in order to exploit the Sensor Fusion ECU.	If data from the server is not sufficiently validated an attacker could pretend to be the Sensors in order to deliver a malicious update to the Sensor Fusion ECU.	Ensure that connections to the Sensors are authenticated and encrypted and access should be limited to only the required files. All firmware should be encrypted and signed to prevent modification.
Manipulate Sensor Fusion ECU Data in Order to Exploit a Software Parsing Vulnerability	Elevation of Privilege	Sensor Data	Medium	Elevation of privileges in order to manipulate Sensor Fusion ECU data.	Manipulate the camera stream by clipping onto the sensor wires and injecting malformed sensor data.	All video data should be treated as unsafe. The software handling the data should follow the SDLC.

Fig. 5 Connected car: sample of generated threats

Although most of the threats are created specifically for the automotive template, all of them are still categorized in the STRIDE categories.

3.2.2 TARA for the Automotive Industry

In order to adapt TARA for the automotive industry, *Karahasanovic et al.* in [29] extended the TAL and MOL with industry-specific threat agents, methods, and objectives. The adapted version of the TAL includes a total of 19 threat agents for the automotive industry, which all have 9 different attributes. When used by security experts during the first two steps of TARA, the library helps to determine which threat agents present the greatest risk to the system. Table 5 illustrates the adapted TAL, showing the 19 threat agents and their attributes.

Next, the Common Exposure Library is extended with all interfaces of a modern vehicle. Beside the interfaces, it also shows the impact level, the type of access, as well as the impact potential. Figure 6 illustrates the adapted library, which however is not complete as the properties might differ from various car manufactures.

Lastly, for the methods and objectives library (MOL), the "Acts" and "Limits" sections were removed, and the "Method" section was newly introduced, containing the values "Theft of PII and business data," "Denial of Service," "Intentional Manipulation," "Unauthorized Physical Access," and "Unpredictable Action." These methods conclude most of the cyber attacks which threaten the connected vehicles. Furthermore, the attribute "Impact" has new impact levels: "reputation damage," "privacy violation," "loss of financial assets/car," and "traffic accidents and injured passengers." The updated MOL can be seen in Table 6.

Table 5 Adapted TAL

		Non-hostile intent			Hostile intent															
Threat agent attributes		Reckless employee	Untrained employee	Outward sympa-thizer	Information partner	Hackti-vist	Compet-itor	Cyber vandal	Data miner	Online social hacker	Script kiddies	Government cyber warrior	Organized crime	Radical activist	Sensatio-nalist	Cyber terror-ist	Cyber crimi-nal	Govern-ment spy	Internal spy	Disgr-untled empl-oyee
Access	Internal	X	X	X	X													X	X	X
	External					X	X	X	X	X	X	X	X	X	X	X	X			
Outcome	Acquisition/t							X		X	X		X				X		X	
	Business advantage						X		X									X		
	Material damage	X	X					X				X	X	X		X	X			X
	Harm to the passengers											X	X			X		X		
	Reputation damage	X	X	X	X	X		X			X	X		X	X					X
	Technical advantage						X		X									X	X	
	15 min of fame										X				X					
Resources	Individual	X	X	X													X			X
	Club										X									
	Contest							X												
	Team								X	X										
	Organization				X	X	X						X	X		X			X	
	Government											X						X		

Table 5 (continued)

Threat agent attributes		Reckless employee	Untrained employee	Outward sympa-thizer	Information partner	Hackti-vist	Compet-itor	Cyber vandal	Data miner	Online social hacker	Script kiddies	Government cyber warrior	Organized crime	Radical activist	Sensatio-nalist	Cyber terror-ist	Cyber crimi-nal	Govern-ment spy	Internal spy	Disgr-untled empl-oyee
		Non-hostile intent			Hostile intent															
	None																			
Skills	Minimal		X								X				X		X			
	Operational				X			X		X										X
	Adept	X		X		X	X		X			X	X	X		X		X	X	
Visibility	Overt		X											X	X					
	Covert	X		X		X		X					X			X	X			
	Clandestine				X		X		X	X								X	X	
	"Don't care"										X	X								X
Limits	Code of conduct		X		X															
	Legal	X																		
	Extra-legal – minor			X		X	X	X	X	X	X			X	X		X		X	
	Extra-legal – major											X	X			X		X		X

Table 5 (continued)

Threat agent attributes		Non-hostile intent			Hostile intent															
		Reckless employee	Untrained employee	Outward sympa-thizer	Information partner	Hackti-vist	Compet-itor	Cyber vandal	Data miner	Online social hacker	Script kiddies	Government cyber warrior	Organized crime	Radical activist	Sensatio-nalist	Cyber terror-ist	Cyber crimi-nal	Govern-ment spy	Internal spy	Disgr-untled empl-oyee
Objective	Copy					X	X		X	X								X	X	
	Deny											X								
	Injure											X	X	X		X		X		
	Destroy											X				X				X
	Damage			X								X				X				X
	Take												X			X				
	All above/Don't care	X	X		X			X			X			X	X		X			
Motivation	Accidental	X	X		X															
	Coercion																			
	Disgruntleme																			X
	Dominance							X				X								
	Ideology					X								X		X		X		
	Notoriety														X					
	Organization: gain						X		X				X							
	Personal financial gain									X							X		X	
	Personal satisfaction			X							X									
	Unpredictabl																			

Level	Exposures	TYPE OF ACCESS		IMPACT POTENTIAL		
		Physical access	Wireless access	Safety	Data Privacy	Car-jacking
HIGH	OBD II port	X		X		
	Wi-Fi		X	X		
	Cellular connection (3G/4G)		X	X		
	Over-the-air update		X	X		
	Infotainment System		X	X		
	Smart-phone	X		X		
MEDIUM	Bluetooth		X	X		
	Remote Link Type App		X	X		
	KeyFobs and Immobilizers		X			X
	USB	X		X		
	ADAS System		X	X		
	DSRC-based receiver (V2X)		X	X		
LOW	DAB Radio		X	X		
	TPMS		X		X	
	GPS		X		X	
	eCall		X	X		
	EV Charging port	X		X		
	CD/DVD player	X		X		

Fig. 6 Adapted CEL

4 Formal Modeling and Verification

Another possibility to identify possible attacks and to minimize the attack vectors at an early stage is the use of formal verification methods. By using a diverse set of mathematical and logical methods, security guarantees with respect to a given model developed from, e.g., a protocol specification, an implementation or (parts of) a system can be obtained.

In general, there are two types of formal verification tools, *model checkers* and *theorem provers*. Model checkers are usually more restricted to a certain problem domain and the verification of properties in that field. Based on a given model and its specification, the dependent state space is automatically and exhaustively checked. Theorem provers are useable for a wider field of potential problem settings. However, they often need human expertise as the formulation of algebraic constrains or theorems to guide a proof of correctness [30, 31].

For *intra-* and *inter-*vehicle protocols, a wide variety of tools for formal verification are applied. Approaches include:

- techniques to prove functional correctness
- detection of possible attacks

(continued)

Table 6 Adapted MOL

Agent name	Attacker					Objective		Method					Impact				
	Access	Trust				Motivation	Goal										
		None	Partial trust	Employee	Administrator			Theft of PII and business data	Denial of service	Intentional manipulation	Unauthorized physical access	Unpredictable action	Reputation damage	Privacy violated	Loss of financial assets/car	Traffic accidents	Injured passengers
Competitor	External	X				Organizational gain	Technical advantage	X					X				
Car thief	External	X				Personal financial gain	Acquisition/Theft				X		X		X		
Cyber terrorist	External	X				Ideology	Physical harm; Damage			X						X	X
Cyber vandal	External	X				Dominance	Personal satisfaction	X	X	X			X	X	X	X	
Data miner	External	X				Organizational gain	Technical advantage	X					X	X			
Disgruntled employee	Internal		X	X	X	Disgruntlement	Reputation damage	X		X			X		X		
Government cyber warrior	External	X				Dominance	Physical harm; Damage	X	X	X						X	X
Government spy	Internal		X	X	X	Ideology	Technical advantage	X	X	X	X			X		X	X
Hacktivist	External	X				Ideology	Reputation damage	X					X	X			
Information partner	Internal		X			Organizational gain	Business advantage					X	X	X			
Internal spy	Internal		X	X	X	Personal financial gain	Acquisition/Theft	X					X	X	X		
Online social hacker	External	X				Personal financial gain	Acquisition/Theft	X							X		
Organized crime	External	X				Organizational gain	Acquisition/Theft	X	X	X	X			X	X	X	X
Outward sympathizer	Internal		X	X	X	Personal satisfaction	No malicious intent		X	X			X	X		X	
Radical activist	External	X				Ideology	Material damage	X	X	X			X	X		X	
Reckless employee	Internal		X	X	X	Accidental/Mistake	No malicious intent					X	X	X			
Script kiddies	External	X				Personal satisfaction	"15 Minutes of Fame"	X	X	X			X	X	X		
Sensationalist	External	X				Notoriety	"15 Minutes of Fame"	X					X	X			
Untrained employee	Internal		X	X	X	Accidental/Mistake	No malicious intent					X	X	X			

- considerations of the performance or worst-case scenarios

The focus of most of the publications is different depending on the protocols:

- *intra-vehicle protocols:* publications mainly focus on proving functional correctness and investigation of performance
- *inter-vehicle protocols:* especially for 5G, the focus is on security properties as secrecy and authentication.

Enhanced protocols are mainly considered for CAN, which do not provide authentication by default, and for 5G, where most of the work focus on 5G-AKA.

4.1 Formal Verification Tools Overview

Commonly used tools in literature are the security protocol model checkers `AVISPA`, `ProVerif`, `Scyther`, and `Tamarin`. In the class of probabilistic/statistical model checkers, the tools `UPPAAL` and `PRISM` are widely used (see also [32]). For those tools, a short description shall be given.

The push-button tool *AVISPA*[28] stands for Automated Validation of Internet Security Protocols and Applications. The tool suite contains different verification techniques as On-the-Fly model checker (OFMC), Constraint-Logic-based Attacker Searcher (CL-AtSE), SAT-based model checking (SAT = satisfiability problems in propositional logic), and tree automate-based automatic approximation [33] for the security's protocols analysis, applicable on the same protocol specification. Some of the techniques can deal with unbounded verification. Furthermore attack finding and visualization is supported.

The command-line tool *ProVerif*[29] was developed for the automatic analysis of the security of cryptographic protocols. It can handle an unbounded number of runs of a protocol. The analysis of (weak) secrecy properties can be performed via reachability properties, authentication properties by using correspondence assertions. Additionally, ProVerif can prove observational equivalence, which can, e.g., be used for proving strong secrecy or real or random secrecy. In case a proof fails, the tool assists in the reconstruction of an attack.

The tool *Scyther*[30] has a similar goal as ProVerif. Proofs can be obtained based on a symbolic representation of sets of protocol runs with the backward search

[28]http://www.avispa-project.org/

[29]https://prosecco.gforge.inria.fr/personal/bblanche/proverif/

[30]https://people.cispa.io/cas.cremers/scyther/

algorithm. Moreover, it can be used for attack finding and visualization, can handle an unbounded number of sessions, and additionally has a GUI.

Another similar tool is *Tamarin*,[31] which is both, a model checker and a theorem prover. Tamarin uses a generalization of Scyther's backward search, which makes it capable of handling protocols with non-monotonic mutable global states and complex control flows such as loops. Tamarin enables attack finding and visualization. It can deal with models such as the eCK model for key exchange protocols and equational theories as Diffie-Hellman. Moreover, it can handle bilinear pairings as well as user-specified subterm-convergent theories.

The toolbox *UPPAAL*[32] focuses on system's modeled as a collection of non-deterministic processes with finite control structures and real-value clocks, where the communication is performed via shared variables or through channels. Therefore, suitable application areas of the tool are, e.g., real-time controllers and communication protocols including critical timing aspects. The toolbox has three main parts, a description language, a simulator used for validation, and a model-checker based on timed automata theory.

The probabilistic model checker *PRISM*[33] developed at the University of Birmingham is intended for formal modeling and the analysis of systems that exhibit random or stochastic behavior. The tool can handle several probabilistic models as probabilistic automata and probabilistic timed automata, discrete-time and continuous-time Markov chains, as well as Markov decision processes. The underlying probabilistic verification techniques include quantitative abstraction refinement and symmetric reduction. Furthermore, the generation of optimal adversaries/strategies is supported.

4.2 Examples of Formal Modeling and Applications in Connected Cars

In this section, an overview of different intra- and inter-vehicle protocols where formal methods are applied is given. It extends our previous work in [2], provides more details, and addresses a wider range of protocols.

4.2.1 Intra-vehicle Protocols Formal Verification

For intra-vehicle protocols, formal methods are applied to the CAN, Automotive Ethernet, and FlexRay.

[31] https://tamarin-prover.github.io/manual/tex/tamarin-manual.pdf

[32] http://www.uppaal.org/

[33] https://www.prismmodelchecker.org

Table 7 *Intra*-vehicle protocols

Protocol	Tool(s)	Model	Properties of interest	Reference
CAN	SHVT	Four car components: fieldbus, telemetric ECU, backend-server, terminals representing clients	Replay messages, unlocking someone else's car, downgrading	[34]
CAN	UPPAAL	Focus on arbitration and transmission process and the fault confinement mechanism	11 properties out of the categories: safety, liveness, invariant	[35]
CAN/ Ethernet	CPA	Multiplexing strategies at gateways	Buffering, triggering, and mapping with focus on worst-case and end-to-end latency and load.	[36]
FlexRay	Isabelle/ HOL	FlexRay bus guardian component	Correct relay and integrity	[37]
FlexRay	CPN	AUTOSAR FlexRay transport protocol	Deadlock-free for selected configurations	[38]

Most of the work there are applications of formal method to specifications/standards in order to check selected properties. For CAN there are also enriched schemes/protocols checked. Additional security for the low-level protocol CAN is considered in [39], where an authentication protocol for CAN is presented. Furthermore, a clock synchronization service for CAN is proposed. An overview about different approaches is given in Table 7. Details on the existing approaches are given below.

Guergens et al. propose in [34] an abstract vehicle communication system model providing telemetric functions and onboard communication. It considers four car components, namely, the fieldbus, the telemetric ECU (electronic control unit), a backend server, and also terminals representing clients. As main attack points, the interface GSM/GPRS for a remote attacker and the fieldbus interface for a local attacker is considered. For formal modeling under the Dolev-Yao attacker model [40], the authors use Asynchronous Product Automata (APA), an operational description concept for cooperating systems. As tool, the Simple Homomorphism Verification tool (SHVT), providing components for the complete cycle from formal specification to exhaustive analysis and verification and supports APA, is used. The authors consider a real-world example, which is – with support from SHVT – analyzed for three different scenarios which differ by the foreknowledge of the attacker. There, especially *replay messages, unlocking someone else's car*, and *downgrading of security mechanisms* are taken into account. Additionally, a formal model of a fieldbus is given.

Pan et al. consider in [35] a formal verification with UPPAAL of the *CAN* bus protocol with a focus on the arbitration process, the transmission process, and the fault confinement mechanism. The authors formalize 11 properties, which can be divided into three categories, namely, *safety, liveness*, and *invariant*. The formal verification with UPPAAL shows that the main security issues of the CAN bus system are *deadlock, starvation, data inconsistency*, and *the fault confinement mechanism*. The authors state that the detected problems can at least be partly solved in the application layer.

Bruni et al. give in [39] a formal analysis of *MaCAN*. MaCAN is an authentication protocol developed in order to enable authentication in the CAN bus. By using ProVerif, the authors detected two flaws. The first one leads to unavailability during key establishment. The second one allows a re-using of authenticated signals for different purposes. The authors state that some aspects of MaCAN had to be adjusted (e.g., the usage of timestamps for ensuring message's freshness). However, it is stated by the authors that they could not express the freshness of timestamps in ProVerif, since ProVerif abstracts away the state information. Furthermore, the presence of an attack in their own implementation of the protocol is experimentally verified.

Rodriguez-Navas et al. apply in [41] model checking on a proposed clock synchronization service for the Controller Area Network (*CAN*) for highly synchronized clocks even in the occurrence of faults in the system. For modeling and verification, the tool UPPAAL is used. The model is based on timed automata, and a novel technique for drifting clocks is proposed. The author's solution achieves the desired precision event in case of the presence of various node and channel faults. Furthermore, their results indicate that inconsistent channel faults pose a big threat to clock precision. However, it is possible to reduce their negative impact by using a suitable resynchronization period.

Thiele et al. focus in [36] on an analysis of timing impact, which is introduced by various *CAN/Ethernet* multiplexing strategies at gateways. The authors state that the timing determinism of critical control and streaming data is crucial in the automotive network design. In particular, three different aspects of multiplexing are considered: buffering, triggering, and mapping. By using Compositional Performance Analysis CPA [42], the authors model and analyze three different multiplexing scenarios. In the evaluation, the authors focus on the effect of multiplexing on the design metrics worst-case and end-to-end latency and load. Furthermore, their analysis allows to capture and quantify differences between different multiplexing strategies.

Zhang considers in [37] the *FlexRay* bus guardian component. The bus guardian component helps to protect the communication channel against faulty behavior of communication controllers in FlexRay. The author uses Isabelle/HOL, a theorem prover for higher-order logic for specifying and verifying. The focus in the paper is on two properties of the bus guardian, namely, the *correct relay* and the *integrity*. In order to verify the properties, the correctness of the FlexRay clock synchronization is assumed.

Gordon and Choosang give in [38] a formal analysis of the AUTOSAR *FlexRay* Transport Protocol by using Colored Petri Nets, a mathematical modeling language. The authors prove that the FlexRay Transport Protocol is *deadlock-free* for certain configurations in case of delivering a single-protocol data unit from the sender to the receiver. Furthermore, it captures the desired service language. Moreover, it is stated by the authors that their results indicate the absence of

Table 8 *Inter*-vehicle protocols

Protocol	Tool(s)	Model	Properties of interest	Reference
IEEE 802.11p	PRISM	MAC protocol abstracted into four modules	Collision avoidance mechanism	[48]
5G	Tamarin	5G AKA	Confidentiality, authentication, privacy	[49]
5G	Tamarin	5G AKA, modelled all four parties involved in the protocol	Fine-grained analysis	[50]
5G	Scyther	5G-EAP-TLS, mutual authentication between subscribers and home network	Secrecy of SUPI and session key, non-injective synchronization of events, non-injective agreement on data	[51]
5G	ProVerif	5G-EAP-TLS, severing network and home network are considered as single entity	Authentication and secrecy statements	[52]

functional errors in the protocol specification and that the protocol is likely error-free.

4.2.2 Inter-vehicle Protocols Formal Verification

Formal verification is considered for the MAC of IEEE 802.11p and a wide variety of approaches for 5G (see Table 8). For 5G especially, 5G-AKA and 5G-EAP-TLS are considered. Furthermore, there are several approaches to formally verify enhanced versions of 5G in general – not focusing on the automotive domain explicitly – as [43–47]. Details of the approaches are given below. A review of formal verification method approaches considering 5G is also given in [32].

Zou et al. in [48] consider the Media Access Control (MAC) of *IEEE 802.11p*. The MAC abilities are essential in order to reach requirements as high-speed data transmission and self-organization of networks. In the MAC protocol of 802.11p, the collision avoidance mechanism is used. That means in a first step, a node needs to listen to a channel. Then, two cases can be distinguished: The channel is free (for a specific time period), and then data packets can be sent directly. Otherwise, the node has enter the backoff procedure and wait. For modeling probabilistic timed automaton (PTA) is used, since it fully takes the characteristics of the MAC into account due to the non-deterministic existence and the support of continuous time and probabilistic choice. As tool `PRISM` is used. For modeling, the MAC protocol is abstracted into four modules, a destination node, two sending nodes, and a transmission channel, which are sufficient to cover any transfer case in 802.11p. As performance measures, two different types, the probabilistic and the expected reachability, are considered. With the probabilistic reachability, the successful completion of the data transmission process in 802.11p can be verified. Moreover, the probability of reaching the max backoff counter of any station is much less for 802.11p than for the 802.11 standard. Therefore, the data can be transmitted forward under a high speed. Furthermore, by using expected reachability, it is shown that a collision event in 802.11p is less likely than in 802.11. The authors also point out an approximately four times higher average transmission speed for 802.11p compared to the one of 802.11 standard.

Basin et al. provide in [49] an extensive formal analysis of the Authenticated Key Exchange protocol used in *5G* (5G AKA). This protocol and especially its security guarantees are important for ensuring the security of the users' calls, text messages, and mobile data. The contribution of the authors is very broad. First, the authors formalize the standard, targeting a wide range of properties – *confidentiality, authentication*, and *privacy* – and fine-grained versions of them. Second, the authors create a formal model, which is then evaluated by using the `Tamarin` tool. The formal, systematic security evaluation shows that some critical requirements are underspecified (especially for authentication) or even missing. It is pointed out that without further assumptions, some properties are violated, as the agreement properties on the session keys. Furthermore, the authors criticize the standard's choice of implicit authentication as well as the absence of key confirmation. The authors explicitly state that this introduces weaknesses if the protocol is not used in the way it is intended for. Moreover, the authors detect a likely realistic privacy attack, due to the fact that 5G AKA does not provide unlinkability against an active attacker. Additionally, the authors suggest a fix for the security issue.

Cremers and Dehnel-Wild also study in [50] *5G* AKA, performing a fine-grained formal analysis with the `Tamarin` tool. The authors state several challenges which complicated their work: first, the complexity of the specification documentation; second, the complexity of the protocol involving all four parties which are defined in the protocol specification and third, the informal nature of the security requirements, forcing the modeler to make complex assumptions on the basis of possible use cases. All four parties are modeled by the authors. Furthermore, possible assumptions on the channels connecting these four parties have been modeled precisely. The proposed formal model from 5G AKA standard enables a detailed view of the interactions between several security-critical components. The results show that 5G AKAs security is based on unstated assumptions on the inner workings of underlying channels. This results in an attack which exploits a potential race condition. However, even for the honest case, solving the race condition does not necessarily prevent the attack. It is stated that in practice, the standard can be implemented "correctly" in an insecure manner. Moreover, the authors propose a possible fix based on their findings.

Zhang et al. focus in [51] on the *5G*-EAP-TLS protocol, which is defined in 5G networks for subscriber authentication in limited use cases as private networks or IoT environments. One main security goal is to ensure mutual authentication between subscribers and their home network. The authors state to provide the first 5G-EAP-TLS formal protocol model and perform a security analysis with the use of the `Scyther` model checker. In their model two roles are considered, user equipment (UE) and network (NW). The last is a composition of the home network and the server network. The authors check four security related properties: the *secrecy* of the Subscription Permanent Identifier (SUPI) and the one of the session key; for UE, the secrecy hold for SUPI and the session key; and for NW, only the secrecy of the SUPI can be verified. The other two security-related properties, the *non-injective synchronization* of events and the *non-injective agreement* on data, are falsified with Scyther, for both UE and NW.

Zhang et al. focus in [52] on the *5G*-EAP-TLS protocol. There, the protocol is modeled in applied pi calculus, while `ProVerif` is used for the security analysis. The authors extend their previous work in [51] by using a more expressive formal language which is capable of modeling the protocol's behavior more precisely. Moreover, a more fine-grained formal model is provided, i.e., the severing network and the home network are considered to be a single entity. The authors check three secrecy and two authentication statements. For the authentication, it is falsified with ProVerif that the subscriber and the home network agree after successful termination on the identification of each other. Furthermore, ProVerif falsifies that after successful termination, both parties agree on the pre-master key. The secrecy statements can be successfully verified with ProVerif. Those statements include: The adversary must not be able to obtain the SUPI of an honest subscriber, nor the pre-master key, nor the session key. The authors propose a provable fix, showing that their revised version fulfills the stated security properties.

Koutsos considers in [43] the privacy of *5G*-AKA. Although asymmetric randomized encryption is used in order to reach a better privacy than for 3G or 4G, only the IMSI-catcher attacker can prevented. Other known privacy attacks as the Failure Message Attack and Encrypted IMSI Replay Attack still hold. In a second step, the 5G-AKA protocol is modified for the prevention of those attacks. The security proof is performed by `Bana-Comon` indistinguishability logic and shows the absence of those privacy attacks.

Braeken et al. propose in [44] based on the detected security issues in *5G*-AKA in [49] a new version. There a non-monotonic logic – also known as `RUBIN` – is used to successfully verify the proposed scheme. The reason for choosing RUBIN was that this method is quite close to the actual protocol's implementation.

Sharma et al. proposes in [45] an enhanced handover AKA protocol for being used in *5G* communication networks in order to overcome security vulnerabilities as false base-station attack, key compromise, DoS attacks, and high authentication complexity. The authors use `AVISPA` to show that their proposed protocol is not vulnerable to the stated attacks.

Han et al. in [46] suggest the employment of Mobile Edge Computing (MEC) servers into the traditional authentication architecture for re-authentication. Furthermore, instead of using one-way hash functions and permanent names for authentication, the use of existing Extensible Authentication Protocol-Authentication and Key Agreement (EAP-AKA) protocol pseudonyms is proposed. In their security analysis, the authors use `AVISPA` and especially consider *mutual authentication, confidentiality*, and *anonymity*. All those security attributes can be verified.

Cao et al. propose in [47] a group-based handover authentication and re-authentication protocol for massive machine type communication (mMTC) in *5G* wireless networks. By using BAN logic and the model checkers `AVISPA` and `SPAN`, the authors verify that their proposed protocol is secure against various malicious attacks.

5 Conclusions and Key Points

Connected cars services, which increase road safety, and contribute to traffic flows' efficiency and passengers' comfort, require complex communication infrastructure behind. V2X, the most important technology in connected cars communication, includes two main subnetworks – *intra*-vehicle network, including a collection of in-vehicle controlling and processing units and sensors, and *inter*-vehicle network, including the communication between the vehicle and surrounding.

Intra-vehicle communication usually involves bus protocols and media-oriented protocols. The most common bus protocols are LIN, CAN, and FlexRay. Widely used media protocols are MOST and LVSD. Nowadays, Automotive Ethernet, due to increased bandwidth and cheap components, is taking over both purposes.

Inter-vehicle communication includes two types of protocols, categorized by the used technology. The first type is a WiFi-based protocol, often referred to as IEEE802.11p from the name of the first standard designed to this scope. The IEEE802.11p protocol is now superseded and became part of WiFi protocol IEEE802.11. Two other initiatives based on IEEE802.11p are ETSI ITS-G5 in Europe and IEEE 1609 in the USA. In parallel to WiFi-based protocol, cellular technologies also offer solutions for V2X communication. Cellular solutions are known as C-V2X (Cellular-V2X) and include LTE-V2X and recently 5G, with additional V2X functionalities announced for future releases, including platooning, extended sensors, automated driving, and remote driving.

Because of significant growth and advancements in V2X technology, security issues related to them are on the rise. The ***security-by-design*** frameworks, including threat modeling and formal methods, have the potential and means to answer these challenges.

The threat modeling section discussed state-of-the-art methodologies and the ability to adapt those for the automotive industry. It was shown that various methodologies already exist for plenty of scenarios by either using more general approaches or even adapting those general approaches for more specific settings. For the latter, two explicit, for the automotive domain adapted, methodologies were discussed. Upcoming challenges will therefore not only include enhancing those methodology in the research but more likely to consider this research into the development process of the automotive domain.

Another security-by-design framework – formal verification and its applications in automotive industry were also discussed in this chapter. It was shown that formal verification approaches are clearly different depending on the type of the protocol. While for *intra*-vehicle protocols the focus of the approaches are mainly *functional correctness* and the *investigation of performance*, for *inter*-vehicle protocols – especially for 5G – the focus is clearly *security properties* as secrecy and authentication. However, none of the approaches focus on implementations of the corresponding protocols. So far applying formal verification tools to verify those implementations, with different purposes as checking for implementation errors, but also to check if the implementation follows the standard/specification and does not pose additional security issues, is still an open issue. As stated in [53] and the references therein, for several implementations for widely used implementations of different application layer protocols, several security issues have been detected, opened by the implementation since they do not follow the corresponding standard.

Further research might consider – as stated in [2] – forced protocol downgrading, which might arise due to the unavailability of the technology. Further research also might deal with [32] a combination of tools for better overall results in case of restrictions of the model checker, model checking for different versions of a protocol, and an in-depth analysis in order to provide a very broad verification for the connected vehicle by a suitable combination of different in-depth verifications of pieces in the protocol and some an overall analysis.

Acknowledgments This work was funded by the Austrian Federal Ministry for Climate Action, Environment, Energy, Mobility, Innovation and Technology (BMK).

References

1. A. Bazzi, G. Cecchini, M. Menarini, B.M. Masini, A. Zanella, Survey and perspectives of vehicular wi-fi versus sidelink cellular-V2X in the 5G era. Future Internet **11**(6), 122 (2019)
2. B. Stojanović, K. Hofer-Schmitz, Formal methods for connected vehicle protocols, in *2019 27th Telecommunications Forum (TELFOR)* (IEEE, 2019), pp. 1–4
3. E. Hamida, H. Noura, W. Znaidi, Security of cooperative intelligent transport systems: standards, threats analysis and cryptographic countermeasures. Electronics **4**(3), 380–423 (2015)
4. A. Alnasser, H. Sun, J. Jiang, Cyber security challenges and solutions for V2X communications: a survey. Comput. Netw. **151**, 380–423 (2019)
5. P. Sewalkar, J. Seitz, Vehicle-to-pedestrian communication for vulnerable road users: survey, design considerations, and challenges. Sensors **19**(2), 358 (2019)
6. S. Tuohy, M. Glavin, C. Hughes, E. Jones, M. Trivedi, L. Kilmartin, Intra-vehicle networks: a review. IEEE Trans. Intell. Transp. Syst. **16**(2), 534–545 (2014)
7. C. Specification, Version 2.0, Robert Bosch GmbH
8. R.I. Davis, A. Burns, R.J. Bril, J.J. Lukkien, Controller Area Network (CAN) schedulability analysis: refuted, revisited and revised. Real-Time Syst. **35**(3), 239–272 (2007)
9. A.X.A. Sim, B. Sitohang, OBD-II standard car engine diagnostic software development, in *2014 International Conference on Data and Software Engineering (ICODSE)* (IEEE, 2014), pp. 1–5
10. I.A. Grzemba, *MOST: The Automotive Multimedia Network* (Franzis Verlag, Munchen, 2012)
11. S.B. Huq, J. Goldie, An overview of LVDS technology. Natl. Semicond. Appl. Note **971**, 1–6 (1998)

12. M. Ruff, Evolution of local interconnect network (LIN) solutions, in *2003 IEEE 58th Vehicular Technology Conference. VTC 2003-Fall (IEEE Cat. No. 03CH37484)*, Vol. 5 (IEEE, 2003), pp. 3382–3389
13. P.-S. Murvay, B. Groza, Practical security exploits of the FlexRay in-vehicle communication protocol, in *International Conference on Risks and Security of Internet and Systems* (Springer, 2018), pp. 172–187
14. C. Corbett, E. Schoch, F. Kargl, F. Preussner, Automotive ethernet: security opportunity or challenge? Sicherheit 2016-Sicherheit, Schutz und Zuverlässigkeit
15. A. Masmoudi, K. Mnif, F. Zarai, A survey on radio resource allocation for V2X communication. Wirel. Commun. Mob. Comput. **2019**, 1–12 (2019)
16. F. Arena, G. Pau, An overview of vehicular communications. Future Internet **11**(2), 27 (2019)
17. ThreatModeler, Application Threat Modeling vs Operational Threat Modeling (2016). https://threatmodeler.com/operational-application-threat-modeling/
18. A. Shostack, *Threat Modeling* (Wiley, Indianapolis, 2014)
19. B. Strom, ATT&CK 101 (2018). https://www.mitre.org/capabilities/cybersecurity/overview/cybersecurity-blog/attck-1011
20. B.E. Strom, A. Applebaum, D.P. Miller, K.C. Nickels, A.G. Pennington, C.B. Thomas, *MITRE ATT&CK: Design and Philosophy* (MITRE, 2018)
21. MITRE, Cloud Matrix (2019). https://attack.mitre.org/matrices/enterprise/cloud/
22. B. Schneier, Attack trees. Dr. Dobb's J. **24**(12), 21–29 (1999)
23. M. Howard, D. LeBlanc, *Writing Secure Code* (Microsoft Press, Redmond, 2014)
24. S. Marksteiner, H. Vallant, K. Nahrgang, Cyber security requirements engineering for low-voltage distribution smart grid architectures using threat modeling. J. Inf. Secur. Appl. **49**, 102389 (2019)
25. R. Ankele, S. Marksteiner, K. Nahrgang, H. Vallant, Requirements and recommendations for IoT/IIoT models to automate security assurance through threat modelling, security analysis and penetration testing, in *Proceedings of the 14th International Conference on Availability, Reliability and Security* (2019), pp. 1–8
26. C. Corradini, The Automotive Threat Modeling Template (2016). https://www.nccgroup.trust/uk/about-us/newsroom-and-events/blogs/2016/july/the-automotive-threat-modeling-template/
27. NCCGroup, The Automotive Threat Modeling Template (2017). https://github.com/nccgroup/The_Automotive_Threat_Modeling_Template
28. Intel, Prioritizing Information Risk Security with Threat Agent Risk Assessment, Technical report, Intel (2009)
29. A. Karahasanovic, K. Pierre, M. Almgren, Adapting threat modeling methods for the automotive industry, in *15th ESCAR Conference*
30. D. Basin, C. Cremers, C. Meadows, Model checking security protocols, in *Handbook of Model Checking* (Springer, Cham, 2018), pp. 727–762
31. K. Keerthi, I. Roy, A. Hazra, C. Rebeiro, Formal verification for security in IoT devices, in *Security and Fault Tolerance in Internet of Things* (Springer, 2019), pp. 179–200
32. K. Hofer-Schmitz, B. Stojanović, Towards formal verification of IoT protocols: a review. Comput. Netw. **174**, 107233 (2020). `doi:https://doi.org/10.1016/j.comnet.2020.107233`. http://www.sciencedirect.com/science/article/pii/S1389128619317116
33. L. Viganò, Automated security protocol analysis with the AVISPA tool. Electron. Notes Theor. Comput. Sci. **155**, 61–86 (2006), in *Proceedings of the 21st Annual Conference on Mathematical Foundations of Programming Semantics (MFPS XXI)*. `doi:https://doi.org/10.1016/j.entcs.2005.11.052`. http://www.sciencedirect.com/science/article/pii/S1571066106001897
34. S. Gürgens, N. Lahr, D. Zelle, On formal security analysis of automotive systems, in *15th Embedded Security in Cars (escar)* (2017). http://sit.sit.fraunhofer.de/smv/publications/download/guergensESCAReu2017.pdf

35. C. Pan, J. Guo, L. Zhu, J. Shi, H. Zhu, X. Zhou, Modeling and verification of CAN bus with application layer using UPPAAL. Electr. Notes Theor. Comput. Sci. **309**, 31–49 (2014)
36. D. Thiele, J. Schlatow, P. Axer, R. Ernst, Formal timing analysis of CAN-to-Ethernet gateway strategies in automotive networks. Real-Time Syst **52**(1), 88–112 (2016)
37. B. Zhang, On the formal verification of the FlexRay communication protocol, in S. Merz, T. Nipkow (eds.), *Automatic Verification of Critical Systems, Automatic Verification of Critical Systems (AVoCS 2006)* (Nancy, 2006), pp. 184–189. https://hal.inria.fr/inria-00091667
38. S. Gordon, S. Choosang, Verification of the FlexRay transport protocol for AUTOSAR in-vehicle communications. Int. J. Veh. Technol. **2010**, 1–23 (2010)
39. A. Bruni, M. Sojka, F. Nielson, H.R. Nielson, Formal security analysis of the MaCAN protocol, in *International Conference on Integrated Formal Methods* (Springer, 2014), pp. 241–255
40. D. Dolev, A. Yao, On the security of public key protocols. IEEE Trans. Inf. Theory **29**(2), 198–208 (1983). doi:10.1109/TIT.1983.1056650
41. G. Rodriguez-Navas, J. Proenza, H. Hansson, An UPPAAL model for formal verification of master/slave clock synchronization over the controller area network, in *Proceedings of the 6th IEEE International Workshop on Factory Communication Systems*, Torino (IEEE Computer Society Press, Los Alamitos, 2006)
42. C. Tofts, Compositional performance analysis, in E. Brinksma (ed.), *Tools and Algorithms for the Construction and Analysis of Systems* (Springer, Berlin/Heidelberg, 1997), pp. 290–305
43. A. Koutsos, The 5G-AKA authentication protocol privacy, in *2019 IEEE European Symposium on Security and Privacy (EuroS P)* (2019), pp. 464–479. doi:10.1109/EuroSP.2019.00041
44. A. Braeken, M. Liyanage, P. Kumar, J. Murphy, Novel 5G authentication protocol to improve the resistance against active attacks and malicious serving networks. IEEE Access **7**, 64040–64052 (2019). doi:10.1109/ACCESS.2019.2914941
45. A. Sharma, I. Sharma, A. Jain, A construction of security enhanced and efficient handover AKA protocol in 5G communication network, in *2019 10th International Conference on Computing, Communication and Networking Technologies (ICCCNT)* (2019), pp. 1–6. doi:10.1109/ICCCNT45670.2019.8944569
46. K. Han, M. Ma, X. Li, Z. Feng, J. Hao, An efficient handover authentication mechanism for 5G wireless network, in *2019 IEEE Wireless Communications and Networking Conference (WCNC)* (2019), pp. 1–8. doi:10.1109/WCNC.2019.8885915.
47. J. Cao, M. Ma, H. Li, Y. Fu, X. Liu, EGHR: efficient group-based handover authentication protocols for mMTC in 5G wireless networks. J. Netw. Comput. Appl. **102**, 1–16 (2018). doi:https://doi.org/10.1016/j.jnca.2017.11.009. http://www.sciencedirect.com/science/article/pii/S1084804517303776
48. C. Zhou, Y. Wang, M. Cao, J. Shi, Y. Liu, Formal analysis of MAC in IEEE 802.11 p with probabilistic model checking, in *2015 International Symposium on Theoretical Aspects of Software Engineering* (IEEE, 2015), pp. 55–62
49. D. Basin, J. Dreier, L. Hirschi, S. Radomirovic, R. Sasse, V. Stettler, A formal analysis of 5G authentication, in *Proceedings of the 2018 ACM SIGSAC Conference on Computer and Communications Security* (ACM, 2018), pp. 1383–1396
50. C. Cremers, M. Dehnel-Wild, Component-based formal analysis of 5G-AKA: channel assumptions and session confusion, in *Network and Distributed Systems Security (NDSS) Symposium 2019* (Internet Society, San Diego, 2019)
51. J. Zhang, Q. Wang, L. Yang, T. Feng, Formal verification of 5G-EAP-TLS authentication protocol, in *2019 IEEE Fourth International Conference on Data Science in Cyberspace (DSC)* (2019), pp. 503–509. doi:10.1109/DSC.2019.00082
52. J. Zhang, L. Yang, W. Cao, Q. Wang, Formal analysis of 5G EAP-TLS authentication protocol using proverif. IEEE Access **8**, 23674–23688 (2020). doi:10.1109/ACCESS.2020.2969474
53. K. Hofer-Schmitz, B. Stojanović, Towards formal methods of IoT application layer protocols, in *2019 12th CMI Conference on Cybersecurity and Privacy (CMI)* (IEEE, 2019), pp. 1–6

Towards Autonomous Vehicles in Smart Cities: Risks and Risk Governance

Araz Taeihagh and Hazel Si Min Lim

1 Introduction

In transportation, smart and sustainable mobility solutions can be utilised to improve transportation outcomes [1], one such solution being autonomous vehicles (AVs) that offer improved levels of safety, congestion, traffic efficiency, as well as opportunities for mobility for the elderly and disabled, and reduced environmental impact [2–5]. Many of these benefits stem from AVs' ability to exchange data with other vehicles and infrastructure and, through algorithmic decision-making, optimise and execute driving tasks more efficiently and accurately than the typical human driver, without the latter's human errors [6–9]. With continued support from governments and further technological advancements, AVs are expected to occupy 25% of the global market by 2040 [10].

However, scholars have cautioned against rushing to develop smart mobility solutions such as AVs without preparation for the management of their unintended consequences [11, 12]. Firstly, many existing technological issues in the AV system lead to inaccurate and unpredictable outcomes and create new safety hazards [13, 14]. Safety risks can also arise from the limitations of existing approaches to incorporating ethics into AV design [15, 16], and the incentives for AV stakeholders to design AVs in ways that maximise profit but promote risky driving behaviour, which conflict with the public interest [17, 18]. Secondly, AVs are susceptible to algorithmic biases that could perpetuate discriminatory driving outcomes [19, 20], and the AVs' displacement of professional drivers could widen skill gaps in the workforce and economic inequalities [21–23]. AVs' external connectivity can

A. Taeihagh (✉) · H. S. M. Lim
Lee Kuan Yew School of Public Policy, National University of Singapore, Singapore
e-mail: spparaz@nus.edu.sg; araz.taeihagh@new.oxon.org

U. Z. A. Hamid, F. Al-Turjman (eds.), *Towards Connected and Autonomous Vehicle Highways*, EAI/Springer Innovations in Communication and Computing,
https://doi.org/10.1007/978-3-030-66042-0_7

expose their users to potential privacy infringements and cyberattacks that can undermine safety, the security of critical infrastructure, and public trust in AVs [24].

These new societal risks from AVs demonstrate how an overemphasis on technological solutions alone for economic development could risk neglecting social and environmental considerations, and thus hinder AVs' true 'smartness' [25, 26]. Scholars highlight that smart city concepts heavily emphasise economic imperatives, prompting calls for appropriate governance strategies to ensure that technological solutions such as AVs are aligned with environmental and social sustainability [12, 27, 28]. However, limited research exists to assess these issues and the governance of AVs. Most scholarly works examine the effects of AVs (e.g [29, 30].), such as AV-related privacy and cybersecurity risks [31–34], and do not explicitly analyse government strategies more broadly. To fill these research gaps, this article aims to address the following questions:

- Why are AVs important for smart cities?
- What are the major risks arising from AVs and their implications for smart cities?
- What are the emerging government responses to address these risks?

2 Background to AVs

AVs rely on both software and hardware algorithms to perceive their environment, make and execute decisions. At the hardware level, the AV obtains data about its environment through embedded sensors [35], exchanges data with other vehicles and infrastructure through vehicle-to-vehicle and vehicle-to-infrastructure communication technologies, and executes its physical movements through its actuators (e.g. steering wheel, brakes and accelerator pedals) [14, 36]. In the AV software, the perception component comprises algorithms that process information obtained from sensors and communication networks; the decision-making component comprises mission planning algorithms that decide on 'high-level objectives' (e.g. deciding which route to take), behavioural planning algorithms that produce 'local objectives' (e.g. overtaking and merging), and 'motion planner' (or local planning) that 'generates appropriate paths and/or sets of actions to achieve local objectives' (e.g. reaching a target destination) [14, p. 16]. These decisions are executed in the control component, where algorithms compute the inputs, such as the steering angle and vehicle speed, to be implemented by the AV's actuators [36, 37].

The Society of Automotive Engineers (SAE) defines five levels of vehicle automation: at levels 1 and 2 (assisted automation and partial automation, respectively), the human driver performs the driving operations but the vehicle is aided with advanced driver assistance systems, such as rear-view video systems and automatic emergency braking [38], while at levels 3–5, the automated driving system carries out all of the dynamic driving tasks such as steering, acceleration and monitoring the environment, but the human driver is required to resume control occasionally at level 3 (conditional automation) [26]. Only vehicles at level 5 are expected to operate under all environmental conditions [29]. Throughout this chapter, we focus on vehicles classified under the SAE's levels 4 and 5 of autonomy, which we will refer to as 'AVs'.

3 AVs in Smart Cities

First gaining popularity in the 1990s, the idea of a 'sustainable city' is widely understood as the strong interrelationship between economic growth, environmental protection and social equity [12, 39]. Since 2010, interest has significantly shifted towards the concept of smart cities amidst the acceleration of competition among cities to utilise technological solutions to realise desired objectives of increased efficiency and economic growth in urban development [39, 40]. These smart technologies and connected devices, such as sensors, actuators and wearables, are implemented in the cities' infrastructure that stores and transmits data with other smart devices, that is, the 'Internet of Things' (IoT). Smart devices and infrastructure can therefore synchronise their actions across multiple smart applications, such as community development, grid distribution and transportation [12, 41, 42]. Smart mobility aims to integrate IoT and intelligent transportation systems, such as connected road infrastructure and smart vehicular technologies, into the transportation system, to enhance sustainability in transportation [43–46]. AVs' benefits of increased accuracy, efficiency and timeliness of decision-making relative to human drivers hinge on the performance of on-board sensors that shape its situational awareness [42, 46], its autonomous nature and high computational power, and on its external connectivity that facilitates platooning, which also increases spatial efficiency and road capacity [7, 29, 47–49]. The use of V2V and V2X communications enables AVs to cooperate with other vehicles and comply with traffic rules more efficiently than conventional vehicles (e.g. negotiating with other vehicles for lane-change manoeuvres, assigning priority to other vehicles to cross intersections without the need for traffic lights, [50]), and maintaining a safe distance from other vehicles [51]. Data retrieved from other connected infrastructure that is processed by machine-learning (ML) algorithms enable AVs to continuously adapt their decision-making to environmental changes, and thus tailor on-demand ride-sharing transportation services to changing consumer demands more efficiently than existing ride-sharing services [7, 31, 52]. This can potentially increase transport accessibility for new users, such as the elderly and disabled, strengthening social inclusivity and mental well-being [53–55]. Thus, the 'smartness' and sustainability of AVs hinges upon the performance of algorithms in the AV system and the coordination of connected devices throughout the AV ecosystem [56, 57].

However, AVs introduce new risks that can undermine their promise as a smart and sustainable transportation solution. Firstly, limitations in the AV's software and hardware can undermine the safety of and public trust in AVs, and are more difficult to correct in ML systems, as their internal logics are not explicitly programmed [58]. AV manufacturers and other stakeholders [12] can also design AV algorithms to maximise profit rather than for public safety. Secondly, AVs can potentially exacerbate social inequalities through the displacement of lower-skilled workers from driving occupations and through discrimination arising from algorithmic biases and inappropriate ethical rules, exposing some individuals to greater safety risks than others [16, 59, 60]. Data privacy infringements and cyberattacks on AV systems through various channels can further undermine safety and security. Governing these risks is key to ensuring consumer acceptance of AVs and the realisation of their benefits for smart cities [61]. In the following sections, we examine these risks and the steps taken to govern them.

4 AV Risks and Risk Governance for Smart Cities

In this chapter, we focus on technological risks, defined as potentially negative social, physical and economic consequences related to citizens' concerns about the adoption of novel technologies [62–64]. To amplify societal gains from the deployment of AVs, it is important for governments to introduce new measures and to manage risks associated with this technology.

4.1 Safety

While AVs can eliminate human error, various sources of machine error still exist and will increase as the technology grows in complexity. This section first discusses the technical issues in the AV system's perception, decision-making and control components, and existing AV safety verification and testing methods, before discussing the ethical issues in AV design that potentially create safety risks. We then explore the government responses adopted to address these issues.

4.1.1 Technical Issues

New safety risks can arise from inaccuracies in the AV's perception component due to potential inaccuracies and manipulation of sensors, and from decision-making algorithms' failure to anticipate unexpected obstacles and interpret unforeseen situations. Studies have shown that inaccuracies in the Global Position System (GPS) [65] and obstruction of the GPS signal by external objects [66] can compromise the accuracy of Global Navigation Satellite sensors (GNSS). Costly Light Detection and Ranging (LiDAR) sensors could yield inaccurate estimates of the surrounding environment, particularly of 'non-grounded objects' [67]. LiDAR data, subsequently processed by perception algorithms, could prevent the latter from accurately detecting unexpected movements of objects. Alternatively, less costly visual sensors remain inaccurate in adverse weather conditions [68, 69]. ML-based perception systems can also be manipulated by modifying camera images such as road signs [69–71]. Furthermore, decision-making algorithms face challenges in anticipating unpredictable movements [14, 72] and interpreting traffic rules [13], which can result in erroneous decisions and unsafe driving behaviour. Failure to accurately model human–vehicle interactions, such as determining when the AV user will be prepared to regain control of the vehicle [73], and understanding the intent of other road users [72], can create mismatched expectations that can lead to accidents [13, 74].

Control algorithms are responsible for modelling and ensuring that the AV's motion is aligned with the path determined by its decision-making algorithms [75, 76], but they remain prone to inaccuracies, particularly amidst unexpected road conditions, and correcting them remains computationally costly. Studies have also highlighted the difficulties of validating AV safety due to the limitations of extensive road testing and current safety specification standards, and new issues posed by the unpredictable behaviour of ML systems. In addition, existing standards for systems safety requirements were designed for traditional systems where requirements are 'known' and 'unambiguously specified' (e.g. ISO Standard 26,262), but this is incompatible with adaptive systems in AVs that learn from new data in real-time rather than just relying on clearly defined requirements [26, 77]. Furthermore, regulators have yet to define legitimate methods of determining the safety of AVs [22]. Assessing AV safety remains challenging as AV systems' non-deterministic algorithms produce non-repeatable and probabilistic outputs that are highly sensitive to minor environmental changes, which implies potential differences in behaviour during testing and deployment [77]. Other challenges that AV developers need to overcome are the difficulties and high costs of acquiring manually labelled data to correct accidental correlations in the AVs' training data that lead to erroneous predictions (overfitting) [9, 77]. Erroneous corner-case behaviours that have already led to fatal accidents in AV trials, such as that of Tesla and Google [78], are also difficult to detect, due to algorithmic opacity. Simulated data on corner cases can also have biases and may not cover all kinds of driving scenario [77].

4.1.2 Ethical Issues

Studies highlight the need for ethical rules to guide how AVs allocate risks among multiple persons during accidents [4, 79] and routine driving scenarios, in ways that comply with legal and ethical standards [16, 80]. However, it is unclear how to arrive at these rules, and the choice of rules programmed into the AV can influence the AV's driving decisions and its interactions with other road users in unexpected ways that create new safety risks. Thought experiments can be used to formulate ethical rules for AVs through the creation of ethical dilemmas that reveal individuals' ethical and decision-making preferences [81, 82]. However, trolley problems may be unreliable as unrealistic assumptions can be made, such as assuming that outcomes are certain and passengers can choose how harm is distributed [17, 82], and as inconsistencies among participants can render the insights derived from these experiments questionable [83].

Various technical approaches are also proposed to programme ethical theories into AVs, referred to as a 'top-down' approach [84, 85], but each theory has limitations that can result in unsafe driving behaviour. Programming AVs based on utilitarian ethics would imply minimising the total quantity of harm to all parties, which requires AV algorithms to compute all possible outcomes, actions and their consequences [16, 60, 86]. These computations could be prone to inaccuracies and delays, and dependent on the programming of decision-making criteria and cost function representing the harms being minimised [15, 86]. Alternatively, deontology ensures that actions maintain respect for all humans [87], but this requires rules to be explicitly programmed and strictly adhered to, which may not be possible at all times and may not cover all scenarios [16, 88]. Combining different ethical theories can potentially address both limitations and broaden the AV's perspective on a decision [81]. Alternatively, a bottom-up approach allows algorithms to learn from past driving experiences and human judgements to construct its own ethical rules [84, 85, 89], but this requires specifying high-level goals [90], and the system can modify pre-programmed ethical rules to yield unethical behaviour [81]; also, the opacity of complex decision-making logic hinders the detection of errors [17, 89].

Furthermore, economic incentives can motivate AV stakeholders to programme AV algorithms to maximise profits at the expense of safety.[1] Following profit-seeking objectives, manufacturers can differentiate AVs' algorithmic decision-making criteria and configured safety optima, which shape the AV's assumptions about other road users' behaviour [17]. But heterogeneity between different AVs can create mismatched expectations among road users and generate unexpected dynamics [91] that increase the risk of collisions. This necessitates greater standardisation among AV developers, but it is unclear how data and software code can be shared without compromising intellectual property rights [17, 91]. Secondly, there is a lack of examination into potential misalignments of different profit incentives that motivate different AV stakeholders. Profit incentives could vary from the number of vehicles sold (AV manufacturers), average km/mi travelled per trip or the number of trips completed (ride-sharing and transportation network companies), to the value derived from transportation and behaviour data (data aggregators) [92], which could create negative externalities and undermine safety [93].

4.1.3 Governance Responses to Safety Risks

Several governments have adopted strategies to address the ethical issues in artificial intelligence (AI) through voluntary guidelines, advisory committees and expanding

[1]Incentives can also motivate manufacturers and programmers to design AVs that prioritise the safety of some road users over others and, consequently, perpetuate discrimination.

research. Singapore released voluntary guidelines for AI that emphasise human dignity and human centricity, and recently created a Fairness, Ethics, Accountability and Transparency Committee to issue AI guidelines and codes of conduct that address bias and other ethical issues [94]. On the other hand, China intends to develop laws and ethical norms for AI governance [95]. Similarly, the EU's newly established expert group on AI released AI ethics guidelines in April 2019, emphasising the need for AI to be human-centric and outlining a framework in which AI can be lawful, ethical and robust. It presents four ethical principles for AI that include respect for human autonomy, prevention of harm, fairness and explicability, which address issues around safety, discrimination and algorithmic opacity [96]. The group highlights the need to adapt these guidelines to 'a particular AI-application' and the need for additional sector-specific guidelines.

The United States and United Kingdom have released voluntary AV safety guidelines to address safety risks. The US government displays a market-oriented approach [97], relying on voluntary standards to 'bring cost-effective innovation to the market more quickly' and adapt flexibly to rapid innovation, such as by recommending that AV developers, testers and operators provide a Voluntary Safety Assessment specifying safety strategies and design redundancies for addressing AV malfunctions [98, 99]. Its latest AV guidance prioritises safety as one of its key principles and outlines the National Highway Traffic Safety Administration (NHTSA)'s primary role in enforcing performance standards for all vehicles, developing safety programmes and regulating safety defects in all states, along with the role of other agencies in conducting research and education [99]. The Self Drive Act, passed in 2017, also stipulates that the Department of Transportation (DOT) should require each manufacturer to 'show how it is addressing safety' in AVs with updates 'every five years thereafter' [100]. Similarly, the UK's Department for Transport (DfT) published voluntary AV testing guidelines in 2015 that legalise AV testing without requiring prior approval or a surety bond [101], and the updated guidance in 2019 includes more details on safety scenarios and the need to access AV data [102]. This laid-back approach reflects both countries' intention to avoid excessively restricting innovation [103, 104].

The Chinese national government released its first regulations for AV testing in April 2018, which enable local governments to evaluate local conditions and arrange road tests for AVs, and also require a manual override function and a specification of how certain areas can be reconstructed to improve AV testing [105, 106]. The Chinese government identifies AVs as a key sector in which to advance its leadership in AI by 2025 and in which to compete with the US' core AI industries, creating a 'friendly policy environment' for AV development [107, 108].

At both the EU and national level, European governments have not established permanent regulations but are working to harmonise regulations with EU members. The 1968 Vienna Convention was amended in 2016 to legalise AVs, and AV testing rules have also been implemented, such as confining AV tests to 'private streets' and 'pre-defined routes' [109]. The EU released guidelines to coordinate 'national ad hoc assessments' of AVs, clarify manufacturers' expectations from regulators, 'harmonise regulation with international partners' and promote discussions on amendments to national laws for AVs [110]. Though not specific to AVs, the AI ethics guidelines address several technical issues in AI that are applicable to AVs. For instance, it recommends best practices to test and validate AI systems that address their unpredictability, such as to 'carefully monitor' the underlying model at all stages to ensure robustness, and conducting adversarial testing to find vulnerabilities [96].

Singapore has begun amending regulations to control safety risks in AV testing. The Singapore Road Traffic Act that was amended in February 2017 now recognises that a motor vehicle need not have a human driver [111], and the Minister for Transportation can set standards for AV designs, create new rules and acquire the data from AV trials. A 5-year regulatory sandbox was created to ensure that innovation is not stifled, and the government intends to enact further legislation in the future. Meanwhile, AVs must pass safety assessments, robust plans for accident mitigation must be developed before road testing, and the default requirement for a human driver can be waived once the AV demonstrates sufficient competency to the Land Transport Authority (LTA). After displaying higher competencies, AVs can trial on increasingly complex roads [64, 112].

4.2 Socio-economic Equity

The societal impacts of AVs could reach beyond road safety, through the decisions they make on the safety of different groups of individuals, as well as their impacts on the redistribution of jobs in incumbent industries. In this section, we examine how AVs can undermine socio-economic equity through these channels and explore the steps taken by governments to address these risks.

4.2.1 Sources of Risk to Socio-economic Equity

There are concerns that biased algorithms in AVs can potentially make discriminatory driving decisions, resulting in disparate safety outcomes for different groups of individuals. Biased computer systems can yield discrimination, which can be defined as the 'application of different rules to people of different groups' and/or results that 'differ for different groups' [19, 113]. Statistical bias exists where the input data misrepresent different groups in the population [114], such as using driving data from only one country to train an AV for deployment in another context, which would cause the AV to learn localised patterns and make inaccurate classifications [115–117]. Algorithms can also be biased when data categories representing individual characteristics, such as age and gender, are 'penalised' to meet the algorithm's pre-defined preferences (e.g. minimising the total quantity of harm) [59]. This could expose individuals associated with these characteristics to more safety risks, for example, if more safety risks were allocated to 'more protected' road users (e.g. those wearing a helmet), because they would suffer fewer injuries compared to others [118]. In addition, the designers themselves could be unintentionally biased in their construction of the algorithms' models and parameters [113, 119].

Bias can also be introduced intentionally by AV stakeholders to achieve profit-maximising objectives at the expense of the safety of some groups of road users. As the customer seeks to ensure their own safety, the manufacturer can programme AV algorithms to disproportionately allocate safety risks away from the AV passenger towards third parties to maximise AV sales and profits [60], which may be construed as unethical, even if AVs reduce the overall number of fatalities [120]. Furthermore, under product liability frameworks that determine liability damages based on the amount of income 'lost to dependents' [17], manufacturers could programme the AV to take more safety precautions in more 'affluent' districts to minimise liability claims during accidents [17]. Studies also highlight potential discrimination along income levels when both public- and privately-owned AVs are deployed. It has been shown that consumers do prefer riding in AVs that prioritise passenger safety [15, 121], so AV manufacturers could programme their algorithms to prioritise the safety of private AV purchasers to maximise profits [83, 118]. This could consequently transfer safety risks from private AV users, who benefit from their use, to users of public AV services who cannot afford to purchase AVs. If other group characteristics associated with private ownership, such as income levels, correlate with lower allocated risks relative to that in publicly-owned AVs, individuals who can afford to purchase AVs (and therefore are likely to have higher average incomes) are more shielded from safety risks, leading to discriminatory safety outcomes [26].

Potential methods to detect and correct bias have been proposed for autonomous systems. This includes modifying algorithms' outputs to offset the effects of bias between protected and unprotected groups, ensuring that potentially disadvantaged groups are represented in the data [122], establishing agreed-upon ethical standards such as fairness to evaluate bias [117], and creating auditing procedures for algorithms [116, 123]. Nevertheless, detecting bias can be hindered by complex and opaque decision-making processes in ML systems [115, 116], the unpredictability of their decisions [113], and the inadequacy of existing legal frameworks to hold AV developers and designers accountable [124].

Other risks to socio-economic equity can arise from the displacement of professional drivers and mechanics [125]. Truck and bus drivers are particularly at risk due to the massive cost savings from eliminating labour [126, 127], which make up the majority of the gains from adopting AVs in the trucking and delivery industries. Furthermore, the displaced workers could spill over to other low-skilled occupations, creating downward pressure on their wages, which would exacerbate inequality [125].

4.2.2 Governance Responses to Socio-economic Equity

Actions taken thus far to tackle algorithmic bias and discrimination are not specific to AVs and range from releasing voluntary AI guidelines, improving the design and testing of algorithms in AI systems, and the EU's General Data Protection Regulation (GDPR) [26]. Singapore's Personal Data Protection Commission released Asia's first Model AI Governance framework in January 2019, followed by the latest version, released in January 2020. The first framework provides guidelines that emphasise the explainability and verifiability of AI-driven decisions, fairness to mitigate discrimination, establishing open communication channels between all stakeholders, and providing internal governance practices to increase the accountability of AI-deploying organisations and to mitigate algorithmic discrimination [128]. The second version provides enhanced details on how organisations can decide on the degree of human involvement in AI decision-making, with examples provided, and provides additional best practices to mitigate data biases [128].

The EU passed the GDPR in 2016, which prohibits any automated decision that utilises sensitive personal data and that notably affects data subjects in the EU, and mandates a right to explanation that requires firms to provide data subjects with explanations for algorithmic decisions [129, 130]. In addition, the EU's new AI ethics guidelines outline seven requirements for the implementation of trustworthy AI, which include the requirement of ensuring diversity, non-discrimination and fairness [94]. They recommend steps to remove bias from the data and models, including implementation of 'oversight processes' to detect bias, and involving a wide range of stakeholders that are potentially affected throughout the system's life cycle, through long-term engagement mechanisms.

A few countries recognise the threat that AVs pose to employment, although they have yet to formulate detailed strategies to address them. The US government has begun conducting research on AVs' impacts on professional drivers; examples include the Federal Transit Administration's research on bus automation's impacts on labour and bus operations, along with collaborations between the DOT, the Department of Labour and other agencies [99]. Singapore's government has implemented programmes to retrain workers who will be potentially displaced by AI and new technologies through the national SkillsFuture initiative, to help them acquire new skills and adapt to inevitable disruption [131]. The Singapore government intends to transform the employment risks of AVs and AI into a beneficial opportunity for the nation's economy, as Singapore's manpower constraints can be met by AVs, for example, in the bus industry and in street-cleaning, where AVs are already under trial in Singapore [132].

4.3 Privacy and Cybersecurity

The benefits of AVs for smart cities can also be impeded if their communication networks are hacked for malicious uses or if AV passenger data are not secured against unauthorised access and misuse. This section first examines the sources of AV privacy and cybersecurity risks, before discussing the various measures taken by governments to address these risks.

4.3.1 Sources of Privacy and Cybersecurity Risks

While AVs' external connectivity is essential for their safe operation and to determine liability during collisions [133], unrestricted data sharing introduces privacy risks and unauthorised access to AV systems can compromise both cyber and physical security. Privacy can be undermined by the misuse of personal data collected in AVs that disadvantage AV passengers [134]. Examples include using AV passenger data to predict passengers' behaviour and harass them through tailored marketing [135, 136], and using the unique signatures of AV users in AV geographical data to re-identify individuals without authorisation [135, 137]. Insurance companies and credit rating agencies can also use personal information to calculate insurance premiums and credit scores associated with individuals. These can be inaccurate and discriminatory if existing datasets are biased and, consequently, could exacerbate power disparities and social inequalities [138, 139]. Scholars propose improving data privacy in AVs by disclosing the reasons behind information collection, boundaries to accessing and storing information, providing opt out options and requesting customer consent, although the latter is argued to be inadequate as customers accept terms and conditions without fully understanding them [34, 136, 140]. Furthermore, AVs could be used by public and private agencies to conduct widespread surveillance through the use of location tracking and audio/visual recording of passengers [34]. These risks of surveillance could potentially exacerbate power disparities and social unrest and curtail smart governance, whereby citizens have access to and actively utilise information to influence policy-making, which is an important aspect of smart cities [39, 141].

AVs' dependence on information communication technologies and external connectivity for operation also renders them more vulnerable to cyber threats than conventional vehicles, providing more avenues for third parties to hack and take control of safety-critical functions [8, 10, 142]. As cybersecurity encompasses the protection of both informational and non-information-based assets in cyberspace, such as personal and physical safety, societal values and national infrastructure [143, 144], AV cybersecurity risks can undermine safety and social stability, one of the key criteria for a sustainable city. Widely documented cybersecurity risks in AVs include hacking the AV's wireless Event Data Recorder system [135], jamming the AV's GPS signal for the purposes of theft [145, 146], modifying the AVs' sensors and maps to distort perceptions, conducting Denial-of-Service (DoS) attacks to prevent the reception of critical information [8, 145, 146] and selling the data for financial gain or to conduct crimes such as drug trafficking [142]. In addition, the huge costs that businesses incur from cyberattacks can undermine their productivity, their economic competitiveness and society's economic sustainability in the longer run [147].

4.3.2 Governance Responses to Privacy and Cybersecurity Risks

The European Parliament first highlighted the need to incorporate privacy into the design of Intelligent Transport Systems (ITS) in 2009 [148]. It launched the EU Cybersecurity strategy in 2013 and raised further awareness of cybersecurity risks in IoT systems in 2014 [148], before all member states committed to addressing data and cybersecurity issues through the Declaration of Amsterdam in 2016 [149]. The adoption of the GDPR in 2016 represents management of privacy risks by enhancing restrictions for all companies, regardless of their location, that process data from EU residents [150]. Some argue that stringent application of these rules may impede the commercial use and benefits of data sharing, which could undermine AV improvements and the competitiveness of European vehicle manufacturers [10]. The EU also aims to control cybersecurity risks by enacting the Directive on the security of network and information systems (NIS directive) [151], and released guidelines on the cybersecurity of connected vehicles in December 2016 to provide guidance on these issues [152].

China's new cybersecurity law, which came into effect in June 2017, details new requirements for personal data protection, such as the anonymisation of all personal information, the adherence of network operators to consumers' consent on data usage, and the deletion of personal information during unauthorised collection or use [153]. The law prioritises the protection of critical information infrastructure (CII) networks for the public interest, national economy and national security, the preservation of sensitive data within China, and restrictions on international data exchange [154–155], prompting concerns that the latter could discourage AV testing by foreign AV manufacturers [156].

In Singapore, amendments were made to the Computer Misuse and Cybersecurity (CMC) Act to criminalise the use of personal information that was 'obtained illegally from a computer' and of 'hacking tools' to commit crimes [157], while permitting exceptions for investigations or cases involving encrypted data [158]. In addition, the Public Sector (Governance) Bill was passed in 2017 to criminalise data misuse by government agencies [159, 160], which could address surveillance concerns. Adaptive approaches were also adopted, by raising cybersecurity awareness through local institutes of higher learning, forming partnerships between academia and the private sector to enhance Singapore's leadership as a cybersecurity service provider, and issuing public consultations in July 2017 to gather feedback on modifications to the Personal Data Protection Act (PDPA). The modifications to the PDPA include notification regimes during data breaches, enabling opt outs, 'risk and impact' assessments and new guidelines on exceptions to these rules when such data usage protects 'legitimate interests' [158].

In the United States, the NHTSA enhanced research on electronic vehicle systems safety and security in 2012 [161]. Subsequently, the SPY Car Act was enacted in 2017, which increases the transparency of and consumers' control over data usage in vehicles and prohibits manufacturers from using vehicle data for marketing and advertising purposes without vehicle users' consent. The Act stipulates vehicle cybersecurity requirements, such as penetration testing to evaluate vehicles' resilience to hacking, separating critical and non-critical software systems, and security measures during data collection, transmission and storage [162]. For AVs, the NHTSA released non-mandatory guidelines and recommends that AV entities follow the security measures outlined in international auto manufacturing standards, develop cyber incident response plans and publish a Voluntary Safety Assessment letter outlining their compliance, while emphasising that the latter is voluntary, to avoid delays in testing or deployment [96], which may reflect fears of stifling AV developments. To address AV privacy, the Federal Trade Commission will 'take action against a company' that makes 'deceptive claims' regarding the collection and use of consumer data retrieved from automated or connected vehicles or that has 'inadequate privacy or security practices' [99].

Eight AV privacy principles were released in August 2017 by the UK's DfT and the Centre for the Protection of National Infrastructure, which recommend that manufacturers follow international standards [163] and minimise 'shared data storage'. These principles are voluntary, which reflect the UK government's intention to avoid hindering AV developments, in order to enhance its technological leadership [164]. The government also seeks to strengthen the nation's adaptive capacity to address cybersecurity risks [165]. This is reflected in the National Cybersecurity Strategy 2016–2021, which aims to promote cybersecurity research with a focus on autonomous systems, stimulate growth in the UK's cybersecurity sector and enhance its citizens' responses to these threats, and in the DfT's recommendation to increase resilience through system design and AVs' responses to attacks [163].

5 Discussion

In this chapter, we explored AVs' potential as a solution for smart and sustainable development and analysed the technological risks associated with AVs: safety, socio-economic equity, privacy and cybersecurity. We then discussed the governance strategies adopted in various countries to address these risks.

Our research shows that safety risks can arise from ethical and technical issues associated with AVs' design and operation. Technical issues in the AV system range from the high costs and inaccuracies of sensors in the perception component in harsh driving conditions; decision-making algorithms' potential failure in correctly interpreting traffic and ethical rules, and understanding human–vehicle interactions; the various limitations of different control algorithms in modelling the AV's path and insufficient real-world testing to validate them; and the limitations of extensive test-driving and existing safety specification standards in validating AV safety. We also explored ethical issues, from using trolley problem experiments to formulate ethical rules in AVs and the safety risks arising from different methods, to programming ethics into AVs. Furthermore, we examined AV developers' incentives for product differentiation that could yield heterogeneous algorithmic preferences, reduced traffic coordination and new road safety risks, which necessitate more action to align AV stakeholders' incentives. In response, many governments have created committees to explore AVs' implications and adopted voluntary AV testing guidelines to avoid stifling AV development, while Singapore and China have implemented new regulations to ensure safety in AV testing. Most countries have released AI ethics guidelines to address ethical issues in AI systems that are not sector-specific, among which the EU's guidelines recommend practices for testing and verification to address the unpredictability of AI.

This chapter also examined how AVs can exacerbate socio-economic inequalities through algorithmic biases and unemployment. Data biases, the inclusion of sensitive characteristics in the data, biased models and the programming of utilitarian ethical rules can cause an algorithm to allocate safety risks unfairly and discriminate between different road users. More research should also be conducted to investigate systemic discrimination resulting from large-scale AV deployment, and how to overcome challenges in detecting biases, particularly given algorithms' opacity and unpredictability, and humans' excessive trust in them. New mechanisms are required to hold AV stakeholders accountable, given that they can also intentionally bias algorithms to maximise profits. To tackle bias and discrimination, governments have created committees to research AI-related risks and issued AI ethics guidelines. The EU's GDPR aims to address algorithmic discrimination, but the mandated explanations for individual aspects of algorithmic decisions may not effectively be understood by individuals [58] or effectively reveal discrimination between groups [166]. On the other hand, AI ethics guidelines and the AI governance framework issued by the EU and Singapore, respectively, outline steps to mitigate bias and increase the accountability of AI-deploying organisations. Scholars recommend that regulation should involve looking across the entire system's inputs, outputs and logic, to examine individual and group classifications and to formulate ethical standards to evaluate bias [117, 166]. Furthermore, AVs' displacement of jobs in incumbent industries can fuel income inequality, but most governments are still conducting research to explore these risks.

This chapter also highlighted the privacy and cybersecurity risks associated with AVs, which can impede smart city developments. Misuse of AV users' personal data can exacerbate social inequalities and dampen consumer trust in AVs, and unwanted surveillance of passengers through AVs could potentially undermine democratic processes and discourage citizens' active usage of connected platforms. Cyberattacks can also be conducted on AVs to commit crimes, undermine AV safety and disrupt critical infrastructure. Most of the surveyed responses on the management of privacy and cybersecurity risks apply to all computer systems and reflect command and control approaches with varying scope and objectives. For instance, the US SPY Car Act regulates vehicle data usage, the EU's GDPR establishes more restrictions around data privacy and transparency relative to other countries, the amended cybersecurity law in Singapore seeks to strengthen the public's response to data breaches, whereas China's new law seeks to preserve national sovereignty. Some have cautioned that the restrictions on international data exchange in China's cybersecurity law may discourage AV testing by foreign manufacturers, but the government also aims to drive innovation in transportation through its domestic technology companies [167]. The EU enacted the NIS directive to control cyber risks but also released voluntary cybersecurity guidelines. Cybersecurity guidelines specific to AVs were also released by the United States and United Kingdom. Lastly, the United Kingdom and Singapore also adopt adaptive approaches by consulting various stakeholders before updating their privacy laws, as well as investing in the cybersecurity industry and strengthening citizens' cybersecurity awareness to improve their nations' adaptability to cybersecurity risks.

The risks associated with AVs that can threaten public well-being, social equity, inclusivity and consumer trust in AVs illustrate the broader concerns in the smart cities literature that technological smartness alone may not lead to truly smart and sustainable cities [26]. Governments need to explicitly analyse how technological solutions such as AVs contribute directly to sustainability objectives, such as safety and equity, rather than follow smart mobility ideas for their technological capabilities alone, and carefully consider these potential risks in AV governance to achieve true urban smartness [12].

6 Conclusion

This chapter serves to inform policy-makers, scholars and various stakeholders in the automotive industry of the technological risks introduced by AVs – safety, socio-economic equity, privacy and cybersecurity– that have significant implications for smart cities, and highlight the emerging governance strategies to address them. The surveyed governments have largely rolled out AV testing guidelines and regulation to address the technical issues in AV safety, but the initiatives taken to address ethical issues and risks to socio-economic equity, privacy and cybersecurity are largely not specific to AVs, but general to AI and computer systems, with several exceptions. Other factors influencing AVs' promise for smart cities should also be examined in future, such as barriers to consumer acceptance and the role of other technologies in advancing smart and sustainable mobility [5, 61]. Future research could expand the analysis of algorithmic biases and discrimination during large-scale AV deployment; the trade-offs on individual versus collective safety involved in designing AVs with different types of ethical rules; AV stakeholder incentives and their implications; accountability mechanisms for AV stakeholders; the development of new standards for safety verification in ML systems; and conduct single and comparative case studies to analyse country-specific governance strategies for addressing AV risks.

Acknowledgement Araz Taeihagh is grateful for the support provided by the Lee Kuan Yew School of Public Policy, National University of Singapore through the Start-up Research Grant.

References

1. F.P. Appio, M. Lima, S. Paroutis, Understanding Smart Cities: Innovation ecosystems, technological advancements, and societal challenges. Technol. Forecast. Soc. Chang. **142**, 1–14 (2018)
2. J. Parkin, B. Clark, W. Clayton, M. Ricci, G. Parkhurst, Autonomous vehicle interactions in the urban street environment: A research agenda. Proc. Inst. Civ. Eng. Munic. Eng. **171**(1), 15–25 (2018)
3. S. Pettigrew, L. Fritschi, R. Norman, The potential implications of autonomous vehicles in and around the workplace. Int. J. Environ. Res. Public Health **15**, 1876 (2018)
4. C. Pugnetti, R. Schläpfer, Customer preferences and implicit tradeoffs in accident scenarios for self-driving vehicle algorithms. J. Risk Financ. Manage. **11**, 28 (2018)
5. M. Taiebat, M. Xu, Synergies of four emerging technologies for accelerated adoption of electric vehicles: Shared mobility, wireless charging, vehicle-to-grid, and vehicle automation. J. Clean. Prod. **230**, 794–797 (2019)
6. K. Pangbourne, D. Stead, M. Mladenović, D. Milakis, The case of mobility as a service: A critical reflection on challenges for urban transport and mobility governance, in *Governance of the Smart Mobility Transition*, (Emerald Publishing Limited, Bingley, 2018), pp. 33–48
7. R.I. Ogie, P. Perez, V. Dignum, Smart infrastructure: An emerging frontier for multidisciplinary research, in *Proceedings of the Proceedings of the Institution of Civil Engineers – Smart Infrastructure and Construction*, vol. 170, (ICE Publishing, 2017), pp. 8–16
8. J. Petit, S.E. Shladover, Potential cyberattacks on automated vehicles. IEEE Trans. Intell. Transp. Syst. **16**, 546–556 (2015)
9. W. Schwarting, J. Alonso-Mora, D. Rus, Planning and decision-making for autonomous vehicles. Annu. Rev. Control. Robot. Auton. Syst. **1**, 187–210 (2018)
10. D.M. West, *Moving Forward: Self-Driving Vehicles in China, Europe, Japan, Korea, and the United States*, Report (Brookings Institution, Washington, DC, 2016)
11. S. Li, P.C. Sui, J. Xiao, R. Chahine, Policy formulation for highly automated vehicles: Emerging importance, research frontiers and insights. Transp. Res. Part A Policy Pract. **124**, 573–586 (2018)
12. K. Noy, M. Givoni, Is "smart mobility" sustainable? Examining the views and beliefs of transport's technological entrepreneurs. Sustainability **10**, 422 (2018)
13. R. Leenes, F. Lucivero, Laws on robots, laws by robots, laws in robots: Regulating robot behaviour by design. Law Innov. Technol. **6**, 193–220 (2015)
14. S. Pendleton, X. Du, X. Shen, H. Andersen, M. Meghjani, Y. Eng, D. Rus, M. Ang, Perception, planning, control, and coordination for autonomous vehicles. Mach. Des. **5**, 6 (2017)
15. I. Coca-Vila, Self-driving cars in dilemmatic situations: An approach based on the theory of justification in criminal law. Crim. Law Philos. **12**, 59–82 (2018)
16. N.J. Goodall, Machine ethics and automated vehicles, in *Road Vehicle Automation. Lecture Notes in Mobility*, ed. by G. B. S. Meyer, (Springer, Cham/Berlin/Heidelberg, 2014), pp. 93–102
17. J. Himmelreich, Never mind the trolley: The ethics of autonomous vehicles in mundane situations. Ethical Theory Moral Pract **21**, 669–684 (2018)
18. H. Liu, Three types of structural discrimination introduced by autonomous vehicles. Univ. Calif. Davis Law Rev. Online **51**, 149–180 (2018)
19. S. Barocas, A.D. Selbst, Big data's disparate impact. Calif. Law Rev. **104**, 671–732 (2014)
20. D. Danks, A.J. London, Regulating autonomous systems: Beyond standards. IEEE Intell. Syst. **32**, 88–91 (2017)
21. N. Kalra, S.M. Paddock, Driving to safety: How many miles of driving would it take to demonstrate autonomous vehicle reliability? Transp. Res. Part A Policy Pract. **94**, 182–193 (2016)

22. N. Kalra, *Challenges and Approaches to Realizing Autonomous Vehicle Safety* (RAND, Santa Monica, 2017)
23. J.S. Brodsky, Autonomous vehicle regulation: How an uncertain legal landscape may hit the brakes on self-driving cars. Berkeley Tech. LJ **31**, 851 (2016)
24. Allianz, Autonomous cars in smart cities. *Telegr* (2017)
25. R.K.R. Kummitha, N. Crutzen, Smart cities and the citizen-driven internet of things: A qualitative inquiry into an emerging smart city. Tech. Forcasting Soc. Chang. **140**, 44–53 (2019)
26. H.S.M. Lim, A. Taeihagh, Algorithmic decision-making in AVs: Understanding ethical and technical concerns for smart cities. Sustainability **11**, 5791 (2019)
27. K. Anastasiadou, S. Vougias, "Smart" or "sustainably smart" urban road networks? The most important commercial street in Thessaloniki as a case study. Transp. Policy **82**, 18–25 (2019)
28. K. Borsekova, P. Nijkamp, Smart cities: A challenge to research and policy analysis. Cities **78**, 1–36 (2018)
29. D. Milakis, B. Van Arem, B. Van Wee, Policy and society related implications of automated driving: A review of literature and directions for future research. J. Intell. Transp. Syst. **21**, 324–348 (2017)
30. Z. Wadud, D. MacKenzie, P. Leiby, Help or hindrance? The travel, energy and carbon impacts of highly automated vehicles. Transp. Res. Part A Policy Pract. **86**, 1–18 (2016)
31. D.J. Fagnant, K.M. Kockelman, P. Bansal, Operations of shared autonomous vehicle fleet for Austin, Texas, market. Transp. Res. Rec. J. Transp. Res. Board **2563**, 98–106 (2015)
32. M. Maurer, J.C. Gerdes, B. Lenz, H. Winner, *Autonomous Driving: Technical, Legal and Social Aspects* (Springer, Berlin/Heidelberg, 2016) ISBN 9783662488478
33. G. Meyer, S. Beiker, *Road Vehicle Automation* (Springer, Cham, 2014) ISBN 3319059904
34. J. Schoonmaker, Proactive privacy for a driverless age. Inf. Commun. Technol. Law **25**, 96–128 (2016)
35. B. Paden, M. Čáp, S.Z. Yong, D. Yershov, E. Frazzoli, A survey of motion planning and control techniques for self-driving urban vehicles. IEEE Trans. Intell. Veh. **1**, 33–55 (2016)
36. D. González, J. Pérez, V. Milanés, F. Nashashibi, A review of motion planning techniques for automated vehicles. IEEE Trans. Intell. Transp. Syst. **17**, 1135–1145 (2016)
37. N.H. Amer, H. Zamzuri, K. Hudha, Z.A. Kadir, Modelling and control strategies in path tracking control for autonomous ground vehicles: A review of state of the art and challenges. J. Intell. Robot. Syst. **86**, 225–254 (2017)
38. International Standard J3016, Taxonomy and Definitions for Terms Related to On-Road Motor Vehicle Automated Driving Systems (2014)
39. M. De Jong, S. Joss, D. Schraven, C. Zhan, M. Weijnen, Sustainable–smart–resilient–low carbon–eco–knowledge cities; making sense of a multitude of concepts promoting sustainable urbanization. J. Clean. Prod. **109**, 25–38 (2015)
40. S.Y. Tan, A. Taeihagh, Smart city governance in developing countries: A systematic literature review. Sustainability **12**, 899 (2020)
41. M. Höjer, J. Wangel, Smart sustainable cities: Definition and challenges, in *ICT Innovations for Sustainability. Advances in Intelligent Systems and Computing*, (Springer, Cham, 2015), pp. 333–349
42. L.R. Suzuki, Smart cities IoT: Enablers and technology road map, in *Springer Optimization and Its Applications*, (Springer, Cham, 2017), pp. 167–190
43. C. Garau, F. Masala, F. Pinna, Cagliari and smart urban mobility: Analysis and comparison. Cities **56**, 35–46 (2016)
44. J.P. Skeete, Level 5 autonomy: The new face of disruption in road transport. Technol. Forecast. Soc. Chang. **134**, 22–34 (2018)
45. J. Sochor, A. Nikitas, Vulnerable users' perceptions of transport technologies. Proc. Inst. Civ. Eng. Urban Des. Plan. **169**, 154–162 (2016)
46. E.-K. Lee, M. Gerla, G. Pau, U. Lee, J.-H. Lim, Internet of vehicles: From intelligent grid to autonomous cars and vehicular fogs. Int. J. Distrib. Sens. Netw. **12**, 1550147716665500 (2016)

47. P. Cheney, How self-driving cars will ease traffic congestion. *Globe Mail* (2017)
48. J. Cooper, Driverless cars, hypersonic tunnels, no more traffic lights... Is this the future of travel? *Telegr* (2017)
49. L. Collingwood, Privacy implications and liability issues of autonomous vehicles. Inf. Commun. Technol. Law **26**, 32–45 (2017)
50. L. Hobert, A. Festag, I. Llatser, L. Altomare, F. Visintainer, A. Kovacs, Enhancements of V2X communication in support of cooperative autonomous driving. IEEE Commun. Mag. **53**, 64–70 (2015)
51. B. Chen, Z. Yang, S. Huang, X. Du, Z. Cui, J. Bhimani, X. Xie, N. Mi, Cyber-physical system enabled nearby traffic flow modelling for autonomous vehicles, in *Proceedings of the 2017 IEEE 36th International Performance Computing and Communications Conference (IPCCC)*, (IEEE, San Diego, 2017), pp. 1–6
52. M. Kane, J. Whitehead, How to ride transport disruption–a sustainable framework for future urban mobility. Aust. Plan. **54**, 1–9 (2018)
53. S. Hörl, F. Ciari, K.W. Axhausen, Recent perspectives on the impact of autonomous vehicles. Arbeitsberichte Verkehrs-und Raumplan. **1216**, 17609 (2016)
54. A. Nikitas, I. Kougias, E. Alyavina, E. Njoya Tchouamou, How can autonomous and connected vehicles, electromobility, BRT, hyperloop, shared use mobility and mobility-as-a-service shape transport futures for the context of smart cities? Urban Sci. **1**, 36 (2017)
55. T.J. Crayton, B.M. Meier, Autonomous vehicles: Developing a public health research agenda to frame the future of transportation policy. J. Transp. Health **6**, 245–252 (2017)
56. B.N. Silva, M. Khan, K. Han, Towards sustainable smart cities: A review of trends, architectures, components, and open challenges in smart cities. Sustain. Cities Soc. **38**, 697–713 (2018)
57. C. Olaverri-Monreal, Autonomous vehicles and smart mobility related technologies. Infocommun. J. **8**, 17–24 (2016)
58. L. Edwards, M. Veale, Slave to the algorithm? Why a "right to an explanation" is probably not the remedy you are looking for. Duke Law Technol. Rev. **16**, 1–65 (2017)
59. A. Chander, The racist algorithm. Mich. Law Rev. **115**, 1023–1045 (2017)
60. H. Liu, Three types of structural discrimination introduced by autonomous vehicles. Univ. Calif. Davis Law Rev. Online **51**, 149–180 (2018)
61. B. Herrenkind, A.B. Brendel, I. Nastjuk, M. Greve, L.M. Kolbe, Investigating end-user acceptance of autonomous electric buses to accelerate diffusion. Transp. Res. Part D Transp. Environ. **74**, 255–276 (2019)
62. O. Renn, C. Benighaus, Perception of technological risk: Insights from research and lessons for risk communication and management. J. Risk Res. **16**, 293–313 (2013)
63. Y. Li, A. Taeihagh, M. De Jong, The governance of risks in ridesharing: A revelatory case from Singapore. Energies **11**, 5 (2018)
64. A. Taeihagh, H.S.M. Lim, Governing autonomous vehicles: Emerging responses for safety, liability, privacy, cybersecurity, and industry risks. Transp. Rev. **39**, 103–128 (2019)
65. H. Cai, Z. Hu, G. Huang, D. Zhu, X. Su, Integration of GPS, monocular vision, and high definition (HD) map for accurate vehicle localization. Sensors (Switzerland) **18**, 3270 (2018)
66. Y. Gu, L.-T. Hsu, S. Kamijo, GNSS/onboard inertial sensor integration with the aid of 3-D building map for lane-level vehicle self-localization in urban canyon. IEEE Trans. Veh. Technol. **65**, 4274–4287 (2016)
67. T.S. Combs, L.S. Sandt, M.P. Clamann, N.C. McDonald, Automated vehicles and pedestrian safety: Exploring the promise and limits of pedestrian detection. Am. J. Prev. Med. **56**, 1–7 (2019)
68. U. Lee, J. Jung, S. Jung, D.H. Shim, Development of a self-driving car that can handle the adverse weather. Int. J. Automot. Technol. **19**, 191–197 (2018)
69. N. Carlini, D. Wagner, Adversarial examples are not easily detected: Bypassing ten detection methods, in *Proceedings of the Proceedings of the 10th ACM Workshop on Artificial Intelligence and Security*, (ACM, Texas, 2017), pp. 3–14

70. X. Huang, M. Kwiatkowska, S. Wang, M. Wu, Safety verification of deep neural networks, in *Proceedings of the Lecture Notes in Computer Science (Including Subseries Lecture Notes in Artificial Intelligence and Lecture Notes in Bioinformatics)*, (2017), pp. 3–29
71. N. Papernot, P. Mcdaniel, S. Jha, M. Fredrikson, Z.B. Celik, A. Swami, The limitations of deep learning in adversarial settings, in *Proceedings of the Proceedings – 2016 IEEE European Symposium on Security and Privacy, EURO S and P*, (IEEE, Saarbrücken, 2016), pp. 372–387
72. S. Shalev-Shwartz, S. Shammah, A. Shashua, Safe, multi-agent, reinforcement learning for autonomous driving. arXiv Prepr. arXiv1610.03295 (2016)
73. E. Ohn-Bar, M.M. Trivedi, Looking at humans in the age of self-driving and highly automated vehicles. IEEE Trans. Intell. Veh. **1**, 90–104 (2016)
74. R. Sparrow, M. Howard, When human beings are like drunk robots: Driverless vehicles, ethics, and the future of transport. Transp. Res. Part C Emerg. Technol. **80**, 206–215 (2017)
75. N.R. Kapania, J.C. Gerdes, Design of a feedback-feedforward steering controller for accurate path tracking and stability at the limits of handling. Veh. Syst. Dyn. **53**, 1687–1704 (2015)
76. C. Sun, X. Zhang, L. Xi, Y. Tian, Design of a path-tracking steering controller for autonomous vehicles. Energies **11**, 1451 (2018)
77. P. Koopman, M. Wagner, Autonomous vehicle safety: An interdisciplinary challenge. IEEE Intell. Transp. Syst. Mag. **9**, 90–96 (2017)
78. J. Stilgoe, Machine learning, social learning and the governance of self-driving cars. Soc. Stud. Sci. **48**, 25–56 (2018)
79. D. Hübner, L. White, Crash algorithms for autonomous cars: How the trolley problem can move us beyond harm minimisation. Ethical Theory Moral Pract **21**, 685–698 (2018)
80. J.C. Gerdes, S.M. Thornton, Implementable ethics for autonomous vehicles, in *Autonomous Driving: Technical, Legal and Social Aspects*, (Springer, Berlin/Heidelberg, 2016) ISBN 9783662488478
81. V. Bonnemains, C. Saurel, C. Tessier, Embedded ethics: Some technical and ethical challenges. Ethics Inf. Technol. **20**, 41–58 (2018)
82. A. Wolkenstein, What has the Trolley Dilemma ever done for us (and what will it do in the future)? On some recent debates about the ethics of self-driving cars. Ethics Inf. Technol. **20**, 163–173 (2018)
83. J. Fleetwood, Public health, ethics, and autonomous vehicles. Am. J. Public Health **107**, 532–537 (2017)
84. A. Etzioni, O. Etzioni, Incorporating ethics into artificial intelligence. J. Ethics **21**, 403–418 (2017)
85. W. Wallach, C. Allen, I. Smit, Machine morality: Bottom-up and top-down approaches for modelling human moral faculties. AI & Soc. **22**, 565–582 (2008)
86. S.M. Thornton, S. Pan, S.M. Erlien, J.C. Gerdes, Incorporating ethical considerations into automated vehicle control. IEEE Trans. Intell. Transp. Syst. **18**, 1429–1439 (2017)
87. S. Tzafestas, Roboethics: Fundamental concepts and future prospects. Information **9**, 148 (2018)
88. A. Johnsen, C. Kraetsch, K. Možina, A. Rey, *D2. 1 Literature Review on the Acceptance and Road Safety, Ethical, Legal, Social and Economic Implications of Automated Vehicles* (2017)
89. J. Danaher, M.J. Hogan, C. Noone, R. Kennedy, A. Behan, A. De Paor, H. Felzmann, M. Haklay, S.-M. Khoo, J. Morison, et al., Algorithmic governance: Developing a research agenda through the power of collective intelligence. Big Data Soc. **4**, 1–21 (2017)
90. C. Allen, I. Smit, W. Wallach, Artificial morality: Top-down, bottom-up, and hybrid approaches. Ethics Inf. Technol. **7**, 149–155 (2005)
91. J. D. Stemwedel, *Forbes*
92. C. Neels. *The Importance of Incentives in the Development of Autonomous Vehicles*. Available online: https://medium.com/@cneels/the-importance-of-incentives-in-the-development-of-autonomous-vehicles-967409458597. Accessed 10 Jan 2019
93. A. Dafoe, *AI Governance: A Research Agenda* (Centre for the Governance of AI Future of Humanity Institute University of Oxford, Oxford, 2018)

94. L. Hio, New advisory council set up to look into ethics of artificial intelligence. *The Straits Times* (2018)
95. J. Ding, *How China Seeks to Govern AI*. Available online: https://medium.com/@ChallengesFnd/how-china-seeks-to-govern-ai-baf1c0cd1a54. Accessed 5 Jan 2019
96. High level Expert Group on Artificial Intelligence, *Ethics Guidelines for Trustworthy AI* (Brussels, Belgium, 2019)
97. A. Halsey, 'We're listening,' Department of Transportation says on the future of driverless cars. *Washington Post* (2018)
98. U.S. Department of Transportation, *Automated Driving Systems 2.0 a Vision for Safety* (Washington, DC, 2017)
99. National Science and Technology Council; U.S. Department of Transportation, *Ensuring American Leadership in Automated Vehicle Technologies Automated Vehicles 4.0* (United States of America, 2020)
100. B. Canis. *Issues in Autonomous Vehicle Deployment* (2018)
101. Department for Transport, *The Pathway to Driverless Cars: Summary and Action Plan* (2015)
102. Centre for Connected and Autonomous Vehicles, *Code of Practice: Automated Vehicle Trialling* (London, 2019)
103. Crown Government sets out next steps in establishing the UK as global leader in connected and autonomous vehicles
104. Crown Industrial Strategy: The Grand Challenges. Available online: https://www.gov.uk/government/publications/industrial-strategy-the-grand-challenges/industrial-strategy-the-grand-challenges
105. F. Fan. *Guidelines to Ensure Safe Self-Driving Vehicle Tests*. Available online: http://english.www.gov.cn/state_council/ministries/2018/04/13/content_281476110368454.htm. Accessed 20 Feb 2020
106. X. Zhao, Ministry to speed development of self-driving vehicles. *China Dly* (2018)
107. S. Dai, 'China's Google' says government support and population size matters in global AI race. *South China Morning Post* (2018)
108. C. Cadell, A. Jourdan, *Reuters* (Beijing, July 2017)
109. S. Nicola, E. Behrmann, M. Mawad, *Bloomberg* (March 2018)
110. European Commission Guidelines on the Exemption Procedure for the EU Approval of Automated Vehicles. Available online: https://ec.europa.eu/growth/content/guidelines-exemption-procedure-eu-approval-automated-vehicles_en. Accessed on 2 Jan 2020
111. Government of Singapore, *Road Traffic (Amendment) Bill* (Legislation Division of the Singapore Attorney-General's Chambers, Singapore, 2017)
112. Channel News Asia Regulations in Place to Ramp Up Driverless Vehicle Trials in Singapore
113. J.A. Kroll, S. Barocas, E.W. Felten, J.R. Reidenberg, D.G. Robinson, H. Yu, Accountable algorithms. Univ. Pa. Law Rev. **165**, 633 (2016)
114. J. Garcia-Gathright, A. Springer, H. Cramer, Assessing and addressing algorithmic bias-but before we get there. arXiv Prepr. arXiv1809.03332 (2018)
115. B. Lepri, N. Oliver, E. Letouzé, A. Pentland, P. Vinck, Fair, transparent, and accountable algorithmic decision-making processes. Philos. Technol. **31**, 611–627 (2018)
116. T. Zarsky, The trouble with algorithmic decisions: An analytic road map to examine efficiency and fairness in automated and opaque decision making. Sci. Technol. Hum. Values **41**, 118–132 (2016)
117. D. Danks, A.J. London, Algorithmic bias in autonomous systems, in *Proceedings of the IJCAI International Joint Conference on Artificial Intelligence*, (AAAI Press, Melbourne, 2017), pp. 4691–4697
118. P. Lin, Why ethics matters for autonomous cars, in *Autonomous Driving: Technical, Legal and Social Aspects*, (Spinger, Berlin/Heidelberg, 2016), pp. 69–85. ISBN 9783662488478
119. M. Janssen, G. Kuk, The challenges and limits of big data algorithms in technocratic governance. Gov. Inf. Q. **3**, 371–377 (2016)
120. N.J. Goodall, Machine ethics and automated vehicles, in *Road Vehicle Automation*, ed. by G. Meyer, S. A. Beiker, (Springer, Berlin, 2014)

121. S.A. Cohen, D. Hopkins, Autonomous vehicles and the future of urban tourism. Ann. Tour. Res. **74**, 33–42 (2019)
122. Association, I.S, *Ethically Aligned Design a Vision for Prioritizing Wellbeing with Artificial Intelligence and Autonomous Systems* (IEEE, 2018)
123. R. McAllister, Y. Gal, A. Kendall, M. Van Der Wilk, A. Shah, R. Cipolla, A. Weller, Concrete problems for autonomous vehicle safety: Advantages of Bayesian deep learning, in *Proceedings of the 26th International Joint Conference on Artificial Intelligence*, (AAAI Press, Melbourne, 2017), pp. 4745–4753
124. H.Y. Liu, Irresponsibilities, inequalities and injustice for autonomous vehicles. Ethics Inf. Technol. **199**, 193–207 (2017)
125. M. Alonso Raposo, M. Grosso, J. Després, E. Fernández Macías, C. Galassi, A. Krasenbrink, et al., *An Analysis of Possible Socio-Economic Effects of a Cooperative, Connected and Automated Mobility (CCAM) in Europe* (Publications Office, Brussels, 2018)
126. E.L. Groshen, S. Helper, J.P. MacDuffie, C. Carson, *Preparing US Workers and Employers for an Autonomous Vehicle Future* (W.E. Upjohn Institute for Employment, Kalamazoo, 2019)
127. M. Gittleman, K. Monaco, Truck-driving jobs: Are they headed for rapid elimination? ILR Rev. **73**, 3–24 (2020)
128. Personal Data Protection Commission of Singapore; Infocommunications Media Development Authority; SG:D Empowering Possibilities Model Artificial Intelligence Governance Framework Second Edition (2020)
129. B. Goodman, S. Flaxman, European Union regulations on algorithmic decision-making and a "right to explanation". AI Mag. **38**, 50–57 (2017)
130. H.S.M. Lim, A. Taeihagh, Autonomous vehicles for smart and sustainable cities: An in-depth exploration of privacy and cybersecurity implications. Energies **11**, 1062 (2018)
131. A. Teng, More than 90 per cent of Skills Future credit claims are related to jobs: Ong Ye Kung. *The Straits Times* (2020)
132. Z. Abdullah, Driverless road-cleaning vehicles may make debut here. *The Straits Times* (2019)
133. V. Dhar, Equity, safety, and privacy in the autonomous vehicle era. Computer (Long. Beach. Calif). **49**, 80–83 (2016)
134. K. Rannenberg, Opportunities and risks associated with collecting and making usable additional data, in *Autonomous Driving*, (Springer, Cham, 2016), pp. 497–517
135. W.J. Kohler, A. Colbert-Taylor, Current law and potential legal issues pertaining to automated, autonomous and connected vehicles. St. Cl. Comput. High Tech. LJ **31**, 99 (2014)
136. D.J. Glancy, Privacy in autonomous vehicles. St. Cl. L. Rev. **52**, 1171 (2012)
137. C.Y.T. Ma, D.K.Y. Yau, N.K. Yip, N.S. Rao, V privacy vulnerability of published anonymous mobility traces. IEEE/ACM Trans. Netw. **21**, 720–733 (2013)
138. J. Berkeley, What machines can tell from your face. Economist **413**(9054) (2017)
139. E. Horvitz, AI, people, and society. Science **357**(6346), 7 (2017)
140. P. Pype, G. Daalderop, E. Schulz-Kamm, E. Walters, M. von Grafenstein, Privacy and security in autonomous vehicles, in *Automated Driving*, (Springer, Cham, 2017), pp. 17–27
141. A. Caragliu, C. Del Bo, P. Nijkamp, Smart cities in Europe. J. Urban Technol. **18**, 65–82 (2011)
142. C. Lee, Grabbing the wheel early: Moving forward on cybersecurity and privacy protections for driverless cars. Fed. Comm. LJ **69**, 25 (2017)
143. R. Von Solms, J. Van Niekerk, From information security to cyber security. Comput. Secur. **38**, 97–102 (2013)
144. M.E. López-Lambas, The socioeconomic impact of the intelligent vehicles: Implementation strategies, in *Proceedings of the Intelligent Vehicles*, (Elsevier, New York, 2018), pp. 437–453
145. D. Dominic, S. Chhawri, R.M. Eustice, D. Ma, A. Weimerskirch, Risk assessment for cooperative automated driving, in *Proceedings of the 2nd ACM Workshop on Cyber-Physical Systems Security and Privacy*, (ACM, Vienna, 2016), pp. 47–58
146. S. Parkinson, P. Ward, K. Wilson, J. Miller, Cyber threats facing autonomous and connected vehicles: Future challenges. IEEE Trans. Intell. Transp. Syst. **18**, 2898–2915 (2017)

147. S.J. Shackelford, T.L. Fort, D. Charoen, Sustainable cybersecurity: Applying lessons from the green movement to managing cyber attacks. U. Ill. L. Rev. **2016** (1995)
148. S. Pillath, E. Parliament, Service, E.P.R, *Automated Vehicles in the EU* (European Parliament, 2016)
149. What developments are driving the deployment of Connected and Autonomous Vehicles? (2017)
150. European Parliament Council of the European Union, *Regulation (EU) 2016/679 of the European Parliament and of the Council of 27 April 2016 on the Protection of Natural Persons with Regard to the Processing of Personal Data and on the Free Movement of Such Data, and Repealing Directive 95/46/EC (General Da*; 2016)
151. The Directive on security of network and information systems (NIS Directive). Available online: https://ec.europa.eu/digital-single-market/en/network-and-information-security-nis-directive
152. ENISA, *Cyber Security and Resilience of Smart Cars* (Heraklion, Greece, 2017)
153. Xia, S. China Cybersecurity and Data Protection Laws: Change Is Coming 2017
154. S. Yan, China's new cybersecurity law takes effect today, and many are confused. *CNBC* (2017)
155. H.F. He, Cybersecurity law causing 'mass concerns' among foreign firms in China. *South China Morning Post* (2018)
156. *KPMG Overview of China's Cybersecurity Law* (KPMG Advisory (China) Limited, 2017)
157. Ministry of Home Affairs, G. of S, *Amendments to the Computer Misuse and Cybersecurity Act to Take Effect from 1 June 2017* (2017)
158. Response to Feedback on the Public Consultation on approaches to managing personal data in the Digital Economy (2018)
159. J. Seow, Parliament: New laws on data sharing between public sector agencies. *The Straits Times* (2018)
160. *Public Sector (Governance) Act 2018* (Singapore, 2017)
161. Administration, N.H.T.S, *NHTSA and Vehicle Cybersecurity* (National Highway Traffic Safety Administration, 2018)
162. *SPY Car Act of 2017, 115th Congress*. https://www.congress.gov/bill/115th-congress/senate-bill/680/text (United States of America, 2017)
163. *The Key Principles of Vehicle Cyber Security for Connected and Automated Vehicles* (Department for Transport: United Kingdom, 2017)
164. CCAV; Vehicles, C. for C.& A. *UK Connected and Autonomous Vehicle Research and Development Projects 2017* (Centre for Connected & Autonomous Vehicles: United Kingdom, 2017)
165. Cabinet Office HM Treasury, and The Rt Hon Philip Hammond MP, N. security and intelligence, *National Cyber Security Strategy 2016–2021* (2016)
166. T. Henderson, *Does the GDPR Help or Hinder Fair Algorithmic Decision-Making?* (2018)
167. A. Minter, China could steer self-driving cars. *Bloomberg* (2018)

Part IV
CAV Developments and Experiments

Novel Hybrid-Testing Paradigms for Automated Vehicle and ADAS Function Development

Selim Solmaz, Franz Holzinger, Marlies Mischinger, Martin Rudigier, and Jakob Reckenzaun

1 Introduction

The increasing challenge in the calibration, testing, and validation of ADAS functions, active safety systems, and technologies for autonomous driving in the different development cycles requires novel and alternative solutions. Especially the verification and testing of the ADAS/AD functions with respect to the expected vehicle behavior poses a major difficulty during the development of such systems. In order to ensure accurate vehicle behavior, several tests must be conducted under safe driving conditions. The conventional approach for doing this is to test the developed vehicle within closed proving grounds accompanied by limited performance evaluations and tests on public roads. However, the coverage and extent of these evaluations are often very limited, as they are in general too costly and often infeasible to perform for verifying every ADAS function before commissioning. The current approach in the automotive industry is to employ simulation tools to verify such systems. Due to the lack of one-to-one correlation between simulation models and real vehicle measurements, simulation-only approaches are not sufficient for verification and testing. Given these facts, there is an increasing interest for novel testing concepts combining real-world testing and simulation, particularly for the development of automated vehicles and related technologies. Motivated by these facts, we introduce two novel testing concepts proposed for the development and testing of ADAS/AD functions in this chapter.

S. Solmaz (✉) · F. Holzinger · M. Mischinger · M. Rudigier · J. Reckenzaun
Virtual Vehicle Research GmbH, Graz, Austria
e-mail: selim.solmaz@v2c2.at; franz.holzinger@v2c2.at; marlies.mischinger@v2c2.at; martin.rudigier@v2c2.at; jakob.reckenzaun@v2c2.at

U. Z. A. Hamid, F. Al-Turjman (eds.), *Towards Connected and Autonomous Vehicle Highways*, EAI/Springer Innovations in Communication and Computing,
https://doi.org/10.1007/978-3-030-66042-0_8

The first testing concept involves a novel steerable rolling test bench (DÜRR X-Road-Curve System; see [1]) with a corresponding methodology for development, testing, and validation of ADAS/AD systems utilizing a real vehicle in the loop. This novel test bench allows independent steering of the front wheels by rotating the rollers about respective vertical axis. The regulation of speed and angle of these rollers keeps the vehicle position relatively fixed on the rolling test bench. This consequently enables lateral and longitudinal inputs by the driver or the automated vehicle control system. In this chapter, a practical implementation of this test bench for the functional test approach for camera-based ADAS/AD systems is demonstrated. Particularly, it is shown how Lane Keeping Assistance (LKA) and Adaptive Cruise Control (ACC) functions are tested on this vehicle-in-the-loop setup [2]. In the suggested solution, Model.CONNECT™ co-simulation platform is used as the central medium of integration [3]. As part of the solution, the measured driving status from the test bench (in the form of wheel speeds and steering angle) is fed into a vehicle dynamics simulation, which in return calculates the dynamic response of the vehicle. The vehicle states and dynamic parameters are fed into an environment simulation software, which generates a video output in a virtual world and models a driving scenario. The video output is then shown on a monitor in front of the intelligent traffic camera installed in the car. This represents an over-the-air (OTA) stimulation of the camera input [4]. The camera provides object lists as input for LKA and ACC algorithms running on the embedded ECU on the car [2, 5]. Based on the simulated virtual world, the car can keep to the lane or the distance between ego and target vehicle autonomously, solely relying on the video stream shown on the monitor. The concept was introduced recently in a seminal paper by the authors in [6].

The second testing concept on the other hand is an alternative concept for ADAS/AD system testing solution named the "hybrid testing," where a real vehicle is combined with virtual ones in a co-simulation framework. Hybrid testing is a novel testing methodology, which was developed in the scope of the EU-H2020 project INFRAMIX to test several mixed traffic scenarios involving autonomous and manual driven vehicles and is based on the Model.CONNECT™ co-simulation platform. In some respects, hybrid testing is similar to the vehicle-in-the-loop (i.e., SIL, VEHIL, or VIL) systems, where real hardware and virtual components are combined in the same simulation framework, which in this case is based on the co-simulation platform Model.CONNECT™ at its core. This is also a generalization of hardware-in-the-loop (HIL) testing, specifically involving a test vehicle as the "hardware" in a feedback loop with the simulated modules. Hybrid testing methodology involves the real vehicle driving in an enclosed proving ground using the real ADAS functions running on its real-time ECU, whereas the driving scenarios and the surrounding traffic as well as the sensor data are simulated on the co-simulation platform using dedicated software components.

Figure 1 shows how these new test concepts can be integrated into the verification and validation process of ADAS/AD development, from simulation-only tests, via

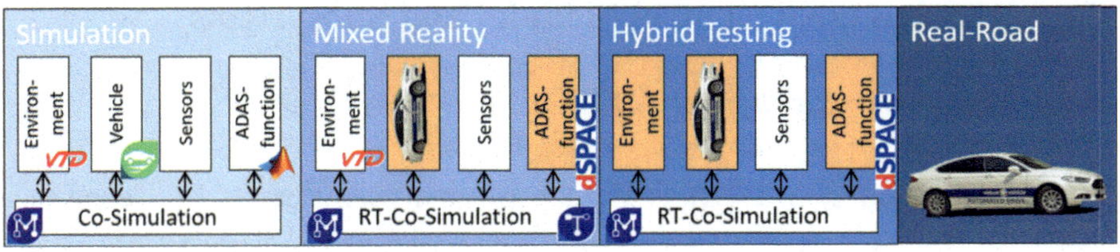

Fig. 1 Consistency of the development process with staggered evaluation steps

mixed-reality tests on a test bench and hybrid tests with the vehicle and simulated sensors on the road. Simulation-only concepts are fast and, compared to test bench and road test, a cheap method to verify ADAS/AD functions. Multiple parallel simulations allow to test a large amount of test scenarios in a short time. Thereby, all functions can be integrated in one simulation environment like CarMaker [7]. An alternative approach is the modular integration of various simulation components with the use of co-simulation.

Selected and critical scenarios can be tested with mixed-reality methods on a test bench, where the real vehicle is integrated in a virtual environment via co-simulation framework. This allows to analyze the behavior on real vehicle dynamics in a limited extent. The limitation comes from the fact that the longitudinal dynamics are fully represented, whereas the lateral dynamics are represented only in the tire forces. In contrast to hybrid tests or testing on a proving ground, the mixed-reality tests have the advantage to reproduce test scenarios in a very flexible and continuous fashion. In a next step, the vehicle with stimulated sensors can be verified on a real road with hybrid testing. This is an extension of the test bench tests and allows additional analyses, especially regarding vehicle dynamics, before finally the vehicle with real components is tested on the road. This continuous testing enables consistent validation and verification during the development of ADAS/AD systems.

This chapter is organized as follows: First we introduce the ADAS functions utilized in the test vehicle in Sect. 2. We then give an overview of the conventional simulation-based ADAS development in Sect. 3. Next, in Sect. 4, we introduce our novel mixed-reality test bench implementation utilizing a co-simulation platform and a rolling chassis-dynamometer. Subsequently, we introduce the hybrid testing methodology in Sect. 5. Based on the introduced two novel testing paradigms, we describe how the proposed techniques can be utilized during the development and verification of the ADAS/AD functions before they are deployed on the end product, in both Sects. 4 and 5, respectively. It is discussed finally how these systems are likely to reduce the amount and complexity of the on-road testing of ADAS/AD systems thereby increasing the overall safety, security, and quality while also reducing the time to market of these systems.

2 ADAS Function Control Modules and Subsystems

2.1 Motorway Chauffeur (MWC)

Motorway Chauffeur (MWC) is an experimental SAE Level 3/3+ equivalent autonomous driver assistance technology developed in-house at the Virtual Vehicle Research GmbH. The MWC combines a number of separate control subsystems or advanced driver assistance (ADAS) technologies or Automated Driving (AD) functions, which include a longitudinal and a lateral guidance system. Speed and target distance regulation in the form of Adaptive Cruise Control (ACC) and lane and target trajectory following in the form of Lane Keeping Assistance (LKA) was developed in-house. An essential component of the MWC is the trajectory planner (TP) performing the strategic planning of the lane change maneuvers while utilizing the lateral and longitudinal tracking controllers. The general structure and architecture of the MWC is shown in Fig. 2.

The implementation of the MWC controller is based on MATLAB/Simulink, where the ADAS components were developed in a simulation environment involv-

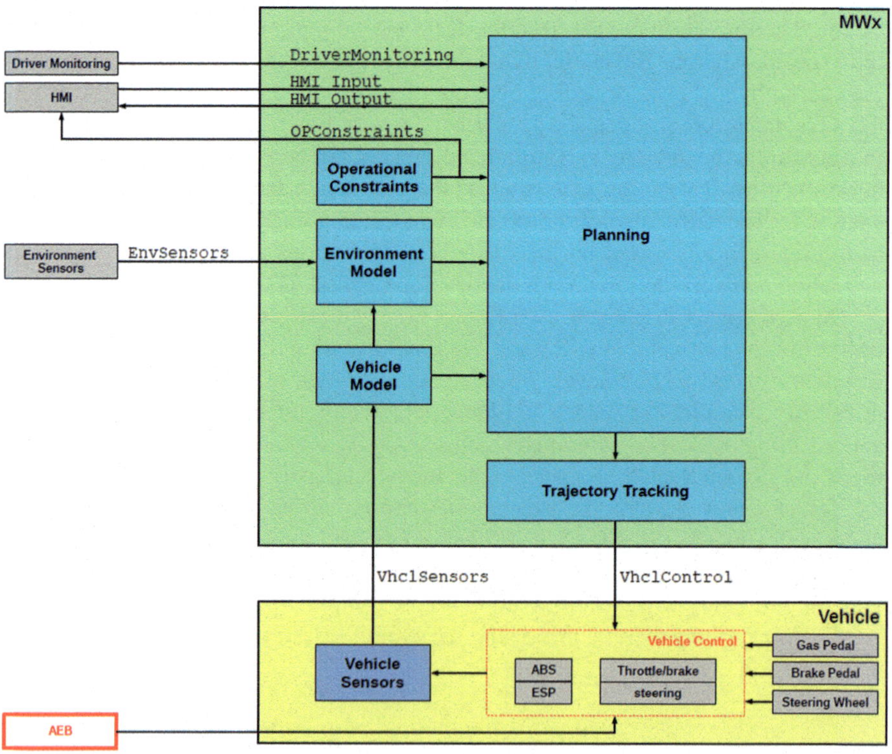

Fig. 2 Virtual Vehicle Research Gmbh Motorway Chauffeur (MWC) Architecture

ing MATLAB/Simulink and CarMaker [7], which was quite useful for virtual development and testing before implementation on the test vehicle. Once the respective ADAS components comply with certain performance criteria in simulation, they are exported to embedded C/C++ code to run on DSpace MicroAutobox-II real-time target ECU hardware found on the test vehicle. See Sect. 5.2 for a detailed description of the test vehicle setup. As was experienced during the development process, there are however many limitations and differences imposed by a simulation-only development, particularly in modeling sensor signals with regard to uncertainties, biases, as well as limited and varying data rates. This, on the other hand, results in different behavior and/or reduced performance between simulation-developed and real-life implemented algorithms.

2.2 Adaptive Cruise Controller (ACC) Subsystem

The Adaptive Cruise Control is an ADAS function to assist the driver in longitudinal control of the vehicle. The ACC adapts its driving behavior to the current traffic situation by keeping a safe distance to a possible leading vehicle or driving with a desired speed. As the ACC is a well-developed system, almost every major automotive manufacturer offers nowadays such an ADAS function. An overview over the existing types and technologies can be found in recent compilation books such as [8, 9].

The current implementation of the ACC was developed and integrated as part of the MWC according to the international standard ISO 22179 (intelligent transport systems, full speed range adaptive cruise control (FSRA) systems, performance requirements and test procedures). The core of the ACC controller consists of a state machine that is responsible for switching between different modes. These modes include car-following, cruising, standby, and inactive. For each of these specific modes, standard pole placement based on P- and PI-controllers was developed, respectively. Additionally, models for sensor fusion and target selection complement this ADAS function.

2.3 Lane Keeping Assistance (LKA) Controller Subsystem

A flatness-based bumpless controller design is used for the LKA. Therefore, a linear single track vehicle model is used for the lateral vehicle dynamics [10] in combination with a linearized lane keeping model [11]; more details and analysis can be found in [12, 13].

The LKA controller developed in MATLAB/Simulink was compiled on a MicroAutobox-II hardware and utilized the Dataspeed ADAS Kit [14] for controlling the vehicle steering. The measured variable of the controller was the relative position of the ego vehicle from the MobilEye ADAS camera [4].

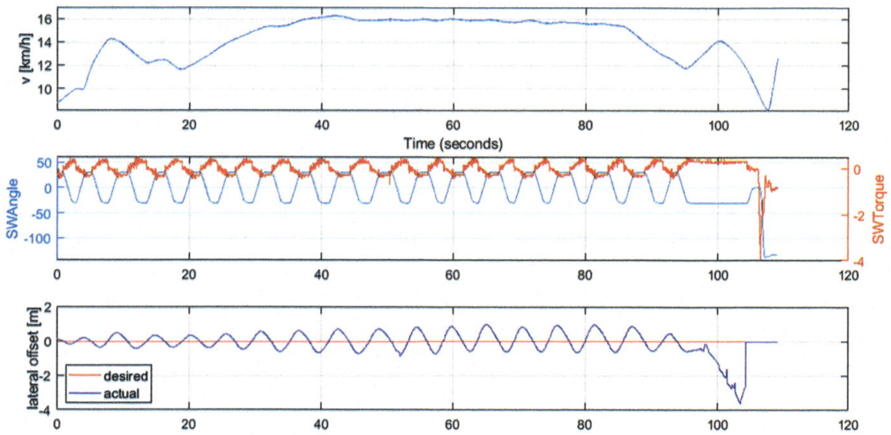

Fig. 3 The log of the variables relevant to LKA function during an early successful test run

The first tests on the test bench with the real vehicle indicated an insufficient dynamic behavior. The steering wheel angle and the steering wheel torque in the second plot of Fig. 3 show oscillations with a vehicle speed greater than 30 km/h. The third plot in Fig. 3 illustrates an unstable behavior of the lateral vehicle position. The oscillations are caused by structural delays of the closed loop in the range of 300 ms and could not be reduced by adjusting the parameters of the controller.

2.3.1 Delay Compensation Utilizing Kalman Predictor

To improve the LKA controller behavior, i.e., to reduce the lateral oscillation due to the additional delay in the closed loop, a Kalman filter prediction based on the lane tracking model [11] is used. Based on the estimated states and the last measured values, the 300 ms delay can be compensated by an n-step predictor. During the prediction of the n-steps, it is assumed that the input data does not change. According to the measurements, this assumption is generally not valid. Nevertheless, the implementation shows significant improvements of the closed loop control behavior. A detailed analysis of the Kalman filter implementation was discussed in [15].

2.4 Trajectory Planner

The trajectory planner uses inputs from several sources to compute a speed and position trajectory. There are three sources for this: the ego vehicle dynamics, the environment objects, and the traffic objects. From the ego vehicle, it uses vehicle dynamics variables like longitudinal speed v_x, lateral speed v_y, accelerations, and

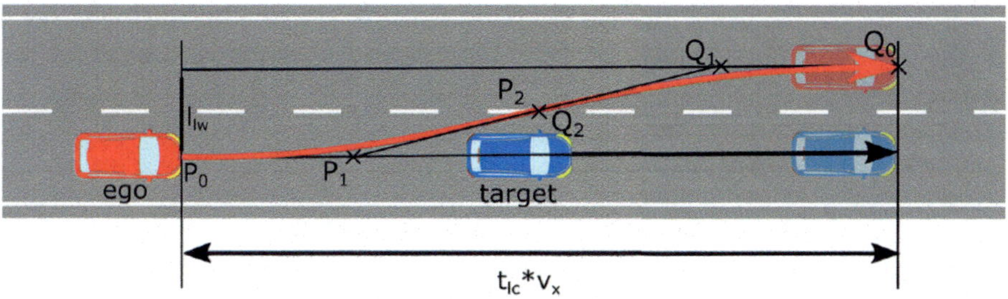

Fig. 4 Lane change: Bezier curve based on lane change trajectory

yaw rate $\dot{\varphi}_z$. The vehicle sensor provides information about road marks like position, line type, and visibility as well as information about the nearby traffic objects (surrounding other vehicles) like position, velocity, and type. The trajectory planner described here was designed for simulation of motorway driving with well-defined and visible road markings. As a simplification, the trajectory planner drives the AV in the middle of the lane. Only when a lane change is performed, the AV leaves the middle of the lane and is transferred to the new lane. This transfer follows trajectory based on a Bezier curve.

The LKA controller of the MWC is used to follow the actual lane and also to perform the lane changes. Besides the vehicle dynamic variables, the reference trajectory that implies the description of the line to follow and an offset y_L are the inputs for the LKA. With these two inputs, the lane change is generated. While $t < t_{LC}/2$, the reference trajectory follows lane 1 (old lane), and with $t_{LC}/2 < t < t_{LC}$, it follows lane 2 (new lane), and the demanded offset y_L follows a Bezier curve. Figure 4 shows such lane change maneuver. At time t_0 the ego vehicle starts the lane change maneuver at position P_0, and completes the manuever at time t_1 and position Q_0. The lane change duration t_{lc} is a parameter of the trajectory planner.

Using the symmetry with respect to $P_2 = Q_2$ and splitting the curve in two parts leads to two quadratic Bezier curves. The trajectory of the centerline of the lane and the spline are superposed to form the resulting trajectory. This works fine for straights and curves with low curvature. This does not work however for narrow curves that are not part of a highway and also for evasive maneuvers. Yet these extreme situations are outside the scope of this implementation. The decision if a lane change is initiated is taken by a network of tactical decisions. The basis of these decisions is a matrix, which describes where a target vehicle from the surrounding traffic is located. Each matrix element holds the ID of the nearest vehicle in this area. First, each of the surrounding vehicles is assigned to a lane with the help of the road markings. After that, this information is used to allocate the vehicles in the obstacle matrix. With the known surrounding, lane change requests are generated, and predefined rules are checked to ensure that a lane change will not cause an accident.

The trajectory planner uses the rightmost lane if it is free from other vehicles. If a slower vehicle prohibits the AV from reaching its free flow speed, a lane change request to the left lane is generated. When not driving on the rightmost lane, the trajectory planner forces the AV to change to the right lane, should this lane be free in front of the AV. Another reason for a lane change is merging into the main traffic on an on-ramp. There is also the possibility of an externally triggered lane change, e.g., from a traffic management center via I2V communication.

In case of a lane change request, a set of rules has to be checked. This set of rules consists of discrete decisions like the availability of a lane or free space on the desired lane. These rules can also be parameterized, e.g., the necessary time gap to a leading vehicle can be chosen by setting the appropriate parameter. When these checks for a lane change are passed, the lane change will be started. The trajectory planner passes a desired lateral lane position to the trajectory execution controller of the MWC. The core of the speed controller is the ACC function of the MWC. While using tactical decisions and the obstacle matrix of the lateral controller, an appropriate target is selected. Usually, the target vehicle is the vehicle on the same lane in front of the ego vehicle. However, in case of a lane change, the desired lane is taken into account. If there is a vehicle on the target lane, this is chosen as the target vehicle, or if there are no vehicles on the desired lane, the desired velocity for speed following (i.e., cruise control function) is chosen. In special situations, separate controllers override the ACC function, e.g., an emergency brake controller.

3 Simulation-Based Development of ADAS Functions

Simulation has become an indispensable part of vehicle development in general. Specifically, in the case of ADAS and AD function development, it is a necessary technique. The number of tests to verify and validate the functionality and safety-relevant features of ADAS/AD functions in every conceivable driving scenarios is not feasible without simulation. Therefore, virtual vehicle prototypes in traffic and environment platform are used. A list of simulation tools, which provide 3D-vehicle dynamics systems, is given in [7]. These provide an internal simulation environment and traffic simulation as well as sensor simulation and also support the possibility to integrate these models from third-party tools.

Modular integration of simulation components utilizing co-simulation, e.g., with [3], is an alternative concept. This modular approach allows a consistent usage of components from simulation-only test, via mixed-reality and hybrid tests, as illustrated in Fig. 1. The adjustments of the simulation components from one test phase to the next can thus be kept to a minimum.

It is clear that the amount of tests required to verify ADAS/AD systems cannot be handled without simulation. Specifically, in the early phases of the ADAS/AD development process, the benefits of the simulation is apparent. The interactions of sensors, actuators, control strategies, and vehicle dynamics can be analyzed by utilizing corresponding mathematical models, and without needing real components. Also, instead of having to test the components on a real vehicle on the road, the interaction can be pre-tested on any scenario in the simulation. Nevertheless, the more progress is made in the development process, the more difficult it becomes to ensure validation with simulation-only concepts. The use of simulation-only concepts is therefore no longer sufficient, especially at the limits of vehicle dynamics, which makes the verification and validation with real components essential. Since simulation-based development is not the main focus of this chapter, and also the fact that there is a vast literature on this topic, we conclude its discussion here after its brief introduction.

4 A Framework for ADAS Function Development Utilizing Chassis Dynamometers

Engineering and development of ADAS functions as well as AD technologies need extensive testing and validation cycles, much more so than traditional automotive safety systems, before their approval and deployment to the end users. This process is usually quite time- and cost-intensive utilizing the traditional automotive testing and validation approaches. Therefore, it is in the interest of ADAS/AD system suppliers and OEMs alike to speed up the time-to-market of such systems, without jeopardizing the compliance with expected safety and performance levels [16]. According to the traditional methods, the required testing and validation effort for the ADAS/AD systems is often performed in enclosed proving grounds, as well as utilizing limited trials on public roads. Such conventional approaches however can only provide limited performance evaluations since the associated costs of such testing methods are quite high, while their corresponding coverage is low. As a remedy, simulation techniques are quite often employed by system developers to extend the scenario testing of the ADAS/AD functions, which help to enrich performance evaluations as well as to provide testing possibilities for safety critical edge scenarios. For a sufficient validation, only a subset of scenarios is mandatory, due to the assumption that only a small number of test scenarios lead to potentially faulty behavior of the ADAS/AD functions [17]. Regardless, simulative analysis alone cannot sufficiently answer the testing and validation requirements of ADAS/AD systems for every possible driving scenario before their commissioning, since there are various nonlinearities and uncertainties that cannot be modelled in simulation.

To cater for the testing and validation task of ADAS/AD functions, novel testing approaches have been proposed in the recent years, where simulation is combined with hardware components in various applications. For this purpose, hardware-in-the-loop (HIL) techniques are commonly utilized for the development and testing of automotive internal combustion engines [2], electronic-control-units (ECUs) [5, 18], as well as components for electric vehicles and electrified drive-trains [19]. The integration of an entire automotive vehicle into a simulation environment is called "vehicle-in-the-loop" (or VEHIL) demonstrated in a seminal work by TNO [20]. This concept was used in a follow-up study to develop and test automotive longitudinal and lateral control systems; for more details, see [21]. Therefore, an entire vehicle on a rolling test bench is combined with a simulated environment. Based on this concept, a research, development, and testing platform for emergency braking and pre-crash systems was derived [22–24]. Similar VEHIL concepts have been developed in recent years by different suppliers, e.g., the DRIVINGCUBE from AVL [25] with a focus on the development and validation of ADAS/AD functions.

Besides the concept of testing the entire vehicle on the test bench, VEHIL tests are performed on a test track (proving ground), where the sensor data are provided by the simulation of the environment; details can be found in [22–24].

Consequently, the aim of this section is to introduce a new steerable rolling-dynamometer test system along with the software components as well as the demonstration of a suitable testing methodology, which can cater for the testing and validation requirements for ADAS/AD systems and automotive active safety systems. Conventional rolling test benches do not support the steering of the vehicle. The additional integrated rotatable front rollers of the novel test bench allows a steering of the front wheels and enables the longitudinal and lateral road force emulation on the vehicle. This testing system allows, for instance, the tuning of Lane Keeping Assistance (LKA) and Adaptive Cruise Control (ACC) functions, which was demonstrated with the *VIRTUAL VEHICLE* Automated Drive Demonstrator (ADD); see [26]. Therefore, the driving scenario was simulated, and the environment was visualized on a screen to stimulate the camera system of the ADD vehicle with the driving functions on the test bench. This framework allows to simulate different driving conditions and to reproduce critical driving scenarios for tuning and validation of ADAS functions.

4.1 X-Road-Curve Steering Chassis Dynamometer

In order to guarantee the quality and functionality of vehicle features of series vehicles, testing at the end of the production line is essential. In addition to quality control tests, the alignment of the chassis and wheels, the adjustment of the headlamps, and driving safety tests like the check of the ABS system are performed. Therefore, a rolling test bench (i.e., chassis dynamometer) is used at the end of the assembly line (i.e., end-of-line, EOL) to test the series vehicle. Typically, these tests of the EOL control step take about 1.5 to 5 min per vehicle. If problems are detected on the vehicle during the EOL inspection, they are examined at the rework station and repaired if necessary. The development of ADAS functions as well as automated driving functions is tremendously challenging as well as a complex process, and therefore new approaches are urgently needed. These new methods have to be incorporated in different stages of the development starting from design, calibration, test procedure, up to the production and the EOL processes [27].

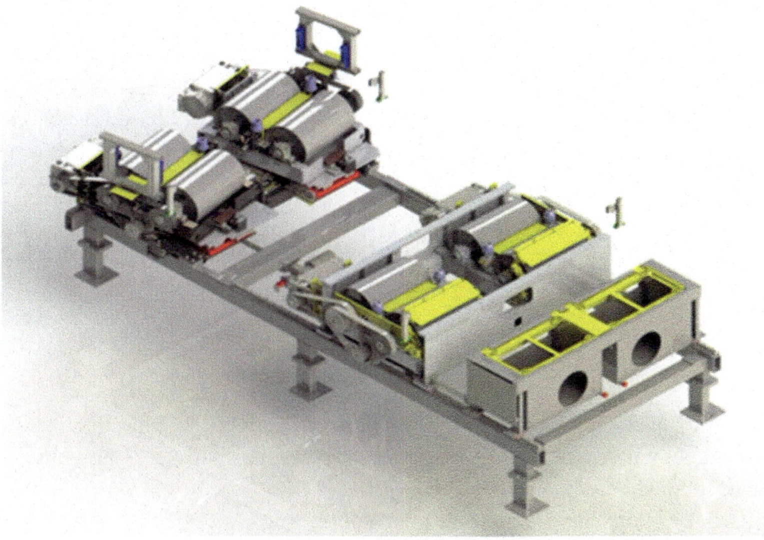

Fig. 5 DÜRR X-Road-Curve Test Bench platform 3D-CAD drawing

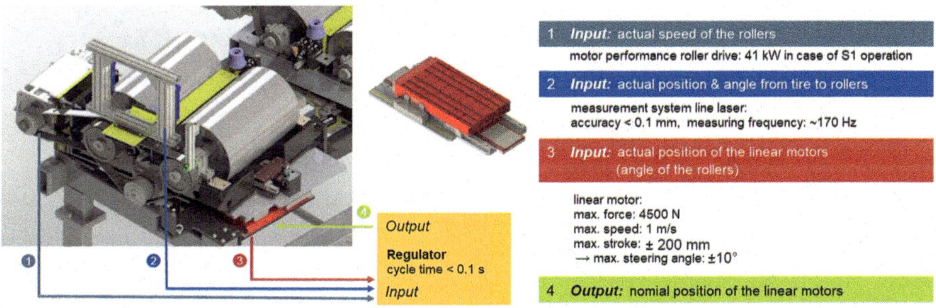

Fig. 6 Detailed view of the steering front roller set together with the system input-output limitations

In order to handle these new challenges, DÜRR Industrial Products Company has developed a novel chassis dynamometer system called X-Road-Curve [1]. Therefore, the traditional EOL brake/ABS chassis dynamometer has been extended to test more vehicle functions. A 3D CAD drawing of this new dynamometer is shown in Fig. 5. The new test bench has independent rotating rollers on the front axle. These automatically keep the vehicle in the center of the test bench without the necessity for a driver to be in the vehicle. In addition, it enables tests to be carried out with steering movements. These novel test bench technique allows testing and developing of driving functions for autonomous vehicles.

Figure 6 shows the steering front rollers. The cycle time of the controller is less than 0.1 s. In Fig. 7, the top-down view of the chassis dynamometer with the rotating plates is illustrated. Linear actuators are used to move the plates. Distance sensors

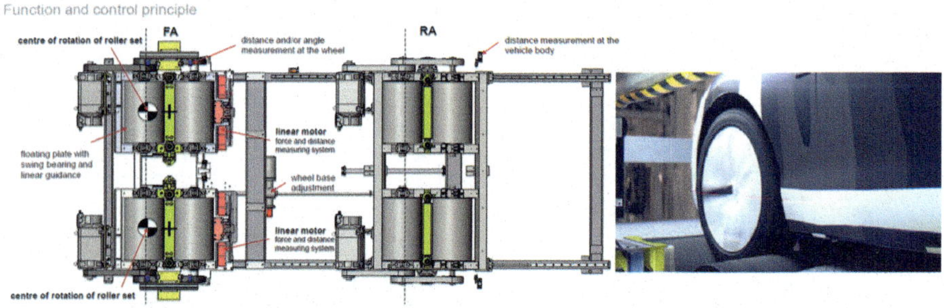

Fig. 7 Functional and control principle of the steering dynamometer

at the side of each tire measure the position of the wheels. This distance is used as feedback signal for the controller to rotate the plates and keep the vehicle in the middle of the test bench.

The front rollers have a maximum rotation angle of ±10°, each. Depending on the steering ratio of the vehicle, this is about 200° at the steering wheel. The maximum velocity of the vehicle on the test bench can be up to 170 km/h for driving straight and is limited to 130 km/h with steering movements. Each roller is powered by a 40 kW motor with a peak force of 6000 N.

4.2 ADAS/AD Test Bench System Description

At the beginning of the section, a brief description of the used components in the DÜRR X-Road-Curve test bench is given. Then, various development methods (Simulation, HIL, VIL) are described, and their characteristics are discussed. In the end, the actual testbed for ADAS function development, testing, and calibration is described as combination of the different components and application of the described methods.

The first key element of the DÜRR X-Road-Curve is the *Virtual Vehicle* Automated Drive Demonstrator (ADD). The underlying platform of the ADD is a standard 2016 Ford Mondeo where additional hard- and software components were mounted to enable automated control of the vehicle (shown on Fig. 8). Automated drive capabilities are primarily enabled using the Data Speed ADAS Kit [14] and various other HW components depicted. The driving functions are running on a D-Space MicroAutobox-II [28] control unit (not shown on Fig. 8) that is used as the onboard processing unit. Out of the whole set of sensors, in this specific application, only the Mobileye traffic camera [4] was used.

The second key component of the test bench is the DÜRR chassis dynamometer that was described in the previous section in more detail.

The third key element is the Model.CONNECT™ [3] co-simulation framework to link the physical components with software and simulation modules.

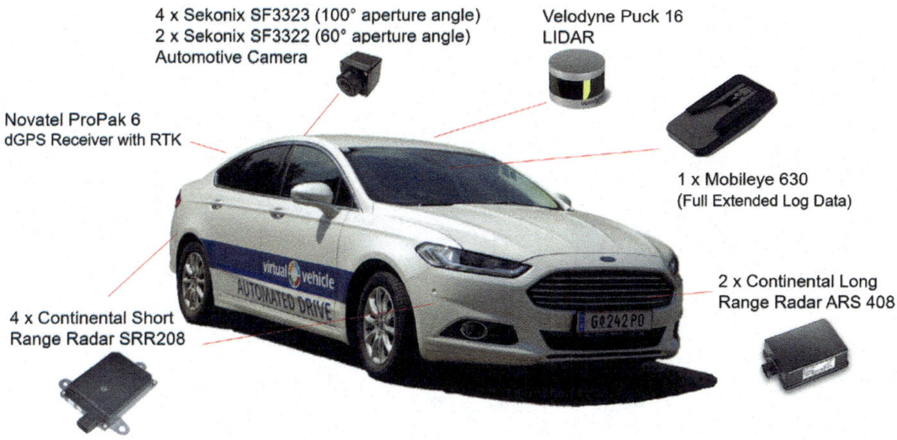

Fig. 8 Virtual vehicles Automated Drive Demonstrator (ADD) with the hardware breakdown layout

When automotive engineers need to choose the appropriate testing and development method, the development stage and the development scope define the requirements regarding testing cycle time and accurateness. In early development stages for both – system- and competent-level development – usually pure simulation tools are used. Simulation is limited as for every model assumptions are made. All models are abstractions from reality due to an intentional or non-intentional neglect of physical processes and effects happening in reality.

For prototype testing or if certain effects cannot be modeled in the desired accuracy in simulation, HIL methodology could be applied. In a HIL setup, a simulation model is replaced by a physical component. In this setup, the input that goes into the component is emulated in simulation, while the output of the real component is again fed into and continuously processed by the coupled simulation [2, 18]. HIL testing enables to take a detailed and accurate look on a certain component, with the opportunity of a high number of repetitions in a short time. As HIL testing is usually done in a lab setting, HIL comes with a relatively low cost. Apart from component behavior, it also gives evidence on the overall system behavior, within the rage of the capability of the surrounding simulation.

A generic mixed-reality testing HIL setting is shown in Fig. 9, where real hardware components are embedded into a co-simulation framework enabling a bi-directional communication between the virtual environment and physical components. Over the last decades, HIL setups have often been used in the design phase of novel products and components, especially in the design and development of various active safety systems. In these lab studies, vehicle dynamics signals output from simulation are transmitted to the controller to test the response of the controller and/or the embedded software stacks running on the controller [29].

More than one component can replace simulation models. When the whole vehicle is tested, as an ultimate extension of HIL, the term vehicle-in-the-loop

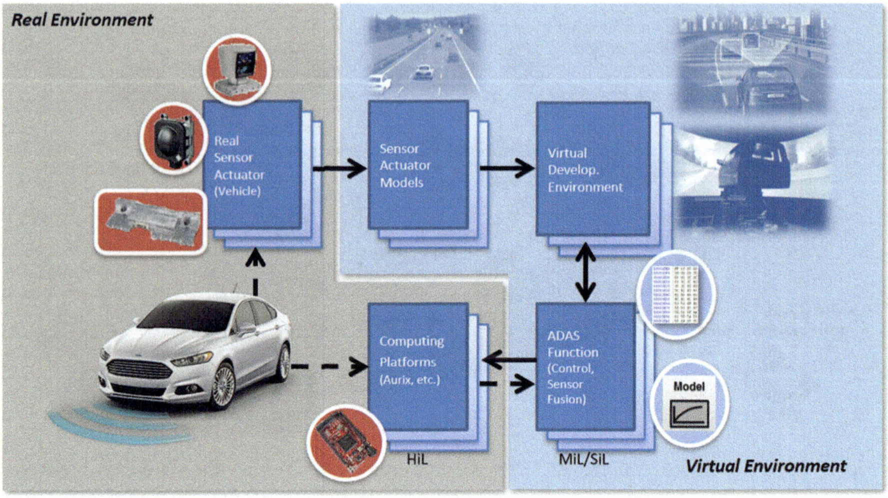

Fig. 9 Mixed-reality testing concept of real and virtual components

(VIL) is commonly used. In this specific testing methodology, a real vehicle on a test bench and a simulation environment are coupled and exchange data during every simulation time step. The idea behind this is based on the direct link between vehicle movement in simulation when operating the vehicle on the test bench. One requirement for a test bench setup is the ability to mimic road properties as indicated and given by the virtual environment. If an ADAS function shall be tested in such a framework, conventional test benches are often not sufficient due to the lack of emulating steering behavior. For simpler active safety systems, the necessary signals for the system to operate can simply be acquired from the virtual simulation environment. In contrast, automated driving functions need to calculate a model of their surrounding environment to plan and execute suitable reaction.

The implementation of the perception testing was a major challenge in the development of the test bench. AD functions require relevant input signals about its surroundings through sensors, so HIL testing of sensors, sensor components, and perception systems became an important work package. To create the needed input signals, different techniques have been proposed in the literature [25]. All examples from literature first simulate the static (road, lanes) and dynamic (objects, vehicles, VRUs) ground truth data in a 3D virtual environment.

From the simulated ground truth, there are two approaches to generate the sensor input for the driving function in a lab or on a test bench. The first approach is to utilize the simulated sensor output. Thereby the sensor model derives the input for the ADAS function in simulation and provides this information via an interface. In the second approach, a physical sensor is stimulated using the physical stimulation of the real vehicle sensor and passes the signal to the platform running the driving function. This approach is used in this demonstration.

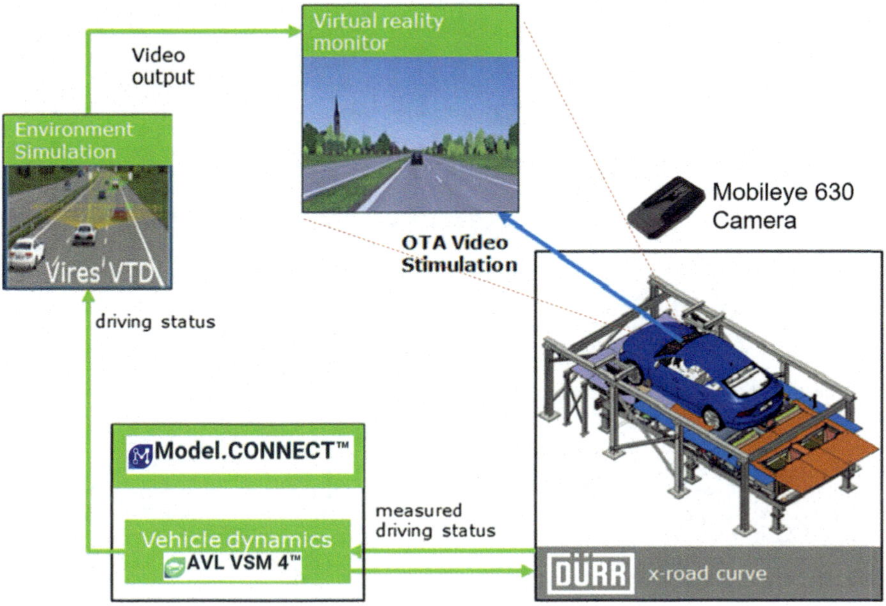

Fig. 10 DÜrr X-Road-Curve VIL ADAS function testing and calibration setup

In the DÜRR X-Road-Curve, a Mobileye camera provides the input for the lateral and longitudinal control of the vehicle. In this testbed, a big screen showing a representation of the scenery was used. The distance to the screen needs to be calibrated, so that the screen fits the camera's field of view and that the distance to the objects shown on the screen fits the estimates from the camera sensor. The environment perception of a driving function is an important element of its architecture. For all automated vehicles that follow the sense-plan-act cycle, the decision-making step usually follows the perception of the surrounding environment. After calculating a model of the environment, the path planning can be initiated on strategic (route), tactical (e.g., lane selection), and reactive (e.g., immediate obstacle avoidance) level.

The full architecture of the mixed-reality VIL simulation system implementation is shown in Fig. 10. The core of the testbed is the DÜRR X-Road-Curve steerable chassis dynamometer. The physical testbed is coupled with a Model.CONNET™ co-simulation framework and a VR monitor to create stimuli for the Mobileye camera. This integrated mixed-reality concept of three different components enables testing of ADAS functions.

Signals from the real car and the physical part of the test bench drive the vehicle dynamic simulation. In this specific application use case, the vehicle dynamics is simulated in AVL-VSM software as part of the Model.CONNECT™ environment. The vehicle dynamics block controls the environment simulation too, to emulate the driving scenario and a real-time 3D visualization of the virtual environment. This 3D virtual twin is shown on a television screen, which is placed in front of the vehicle. It is also placed in the field of view of the Mobileye camera that is mounted

Fig. 11 Mixed-reality testing system implementation visuals

on the windshield. When all components are properly calibrated to an integrated, mixed-reality simulation test bed, the vehicle assumes that it drives on a real road.

By connecting all these individual parts, the speed and directional vehicle motion on the testbed is realized and duplicated within the digital twin. In this specific test setup, two ADAS functions were running on the vehicle under test.

- *Adaptive Cruise Control (ACC):* vehicle holds a pre-set velocity when no car drives in front and adapts its velocity to a lead car
- *Lane-Keeping Assistant (LKA):* holds the vehicle within a specific lane

In this demonstration project, we used the Dataspeed ADAS Kit [14] to control the ADDs steering actuator enabling a lateral and longitudinal movement of the vehicle. A Mobileye ADAS camera [4] was used for receiving the relative position information of the ego-vehicle with regard to the lanes markings as well as distance and speed of target vehicles in front. The ACC and LKA functions were written in MATLAB/Simulink code and exported to real-time C/C++ for the specific DSpace MicroAutobox-II hardware that they run on.

We demonstrated that the proposed setup is well suited for consistent testing of ADAS functions during development. The same setup could be utilized in the end-of-line production process, where the final checks of the system are performed before delivering the car to its customer. A further possible application could be yearly inspections performed by car repair shops. For a deeper understanding of the test setup implementation, visuals showing the DÜRR X-Road-Curve test bench in its demonstrated version are given in Fig. 11.

4.3 Cross-Domain Integration

Compared to real road ADAS/AD testing, the vehicle on the test bench has a reduced dynamic behavior. Due to the reduced vehicle body movement, physical system signals like longitudinal and lateral acceleration or yaw rate are not measurable. These physical signals have to be provided by simulation, where the remaining signals are modelled in a dynamic subsystem or via stimulation, where pre-recorded measurements are used to stimulate the missing signals. For the integration of the simulation models and the real vehicle on the test bench, a co-simulation framework (Model.CONNECT™, [3]) is used. It allows the integration of different simulation tools with different communication media and interfaces to the vehicle and the testbed. In addition to the simulation of the missing physical sensor signals, the co-simulation framework also provides the environment of the vehicle via a screen and the additional road specific loads for the test bench (e.g., road gradient). The overall integration consists of the mechanical integration of the vehicle and the test bench between the wheels and the rollers and the steering capability of the X-road-Curve for the steering maneuvers. In addition to that, there is a visual integration between the simulated environment (road, traffic, traffic sign, etc.) and the camera system of the ADAS/AD function. The several hardware components (ADAS-KIT, MWC, camera) and the software components (vehicle dynamic simulation, environment) communicate via an electric-bus integration with different protocols (CAN, Ethernet). Figure 12 shows the schematic integration of the VIL. The missing physical signals of the vehicle are simulated in a vehicle dynamic model in VSM. Therefore, VSM is integrated into the co-simulation platform Model.CONNECT™, which ensures the data exchange between the components at certain coupling times. The VSM model acts like an observer and provides the missing signal via CAN

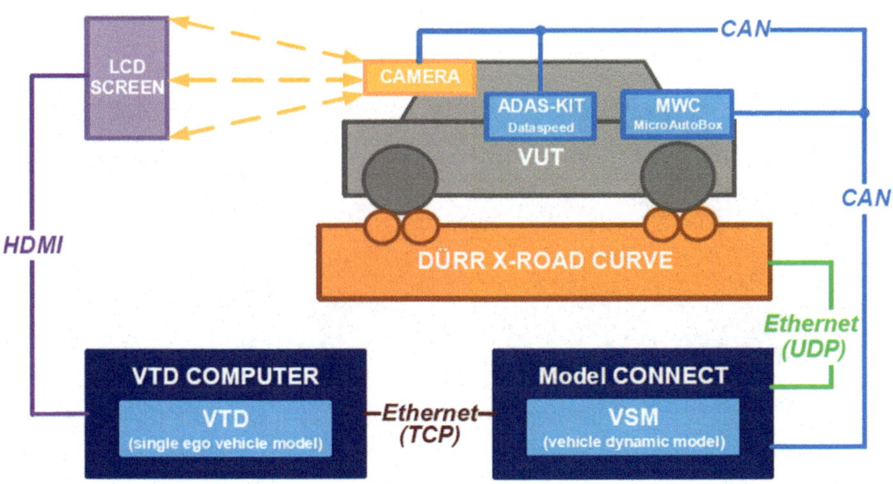

Fig. 12 Topology of the VIL test bench setup

to the control units for the ADAS function. The current speeds of the test bench rollers are sent to the dynamic vehicle via an Ethernet/UDP communication. The environment is simulated in VTD on a separate workstation and communicates via Ethernet/TCP with dynamic vehicle model. VTD also provides the road and traffic information to the real vehicle on the test bench via an LCD screen, where environment from the vehicle point of view is illustrated. The camera in the vehicle films the screen, and the ADAS function acts regarding the information from the camera and dynamic behavior of the vehicle.

The integration of software with real-time components leads to a number of challenges regarding the coupling (e.g., round-trip-time, coupling error, etc.). These challenges were already discussed by the authors; see [6].

4.4 Implementation and Test Results

For testing the self-developed ADAS/AD functions described in Sect. 2, a test scenario was defined, which demonstrates the ACC functionality on the one hand and shows the LKA function on the other. For this purpose, a driving cycle involving a complete virtual circuit was created. The driving cycle modelled in *VDT* [30] is illustrated in Fig. 13. The straight lane in the top section of the scenario is for the ACC tests and the bottom track with a curvy road is for demonstrating the LKA functionality.

4.4.1 Adaptive Cruise Control (ACC) Testing Scenario

The ACC function is evaluated on a straight lane in the top section of the test scenario in Fig. 13. The test consists of two vehicles. The first vehicle is the

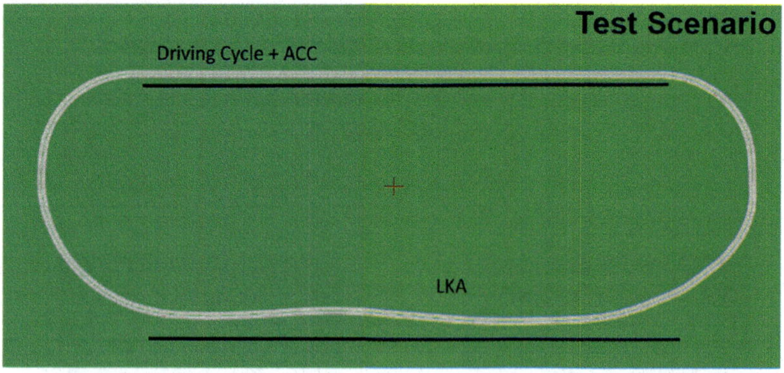

Fig. 13 Driving cycle to test ACC and LKA functions on the mixed-reality test bench

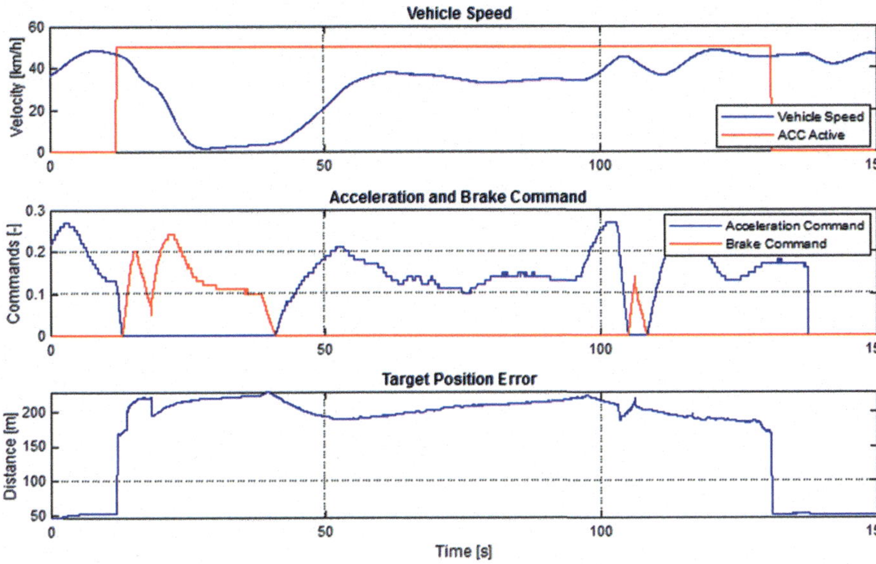

Fig. 14 Vehicle ACC function test results on the DÜRR X-Road-Curve mixed-reality test bench

Vehicle under Test (VuT), with activated ACC function and a demand velocity of $v_d = 50\,\text{km/h}$. In front of the VuT stands a second vehicle (target vehicle) on the same lane. When the VuT approaches the target vehicle, it decelerates close to $v = 0\,\text{km/h}$ until the target vehicle accelerates first to $v = 30\,\text{km/h}$ and after $t = 50\,\text{s}$ to $v = 100\,\text{km/h}$. The VuT follows the target vehicle with its velocity until the target vehicle accelerates and leaves the visual range of the VuT. Due to the lower-demand velocity, the VuT accelerates to $v_d = 50\,\text{km/h}$. The results of the ACC test scenario are shown in Fig. 14. The first plot shows the velocity of the VuT and the ACC activation signal. The acceleration and braking commands of the ACC function are shown in the second plot, and the last plot shows the distance to the target vehicle (position error). After activation of the ACC function and identification of the target vehicle, the VuT reduces the velocity and stands almost still behind the target vehicle. The VuT accelerates and follows the target vehicle with $v = 30\,\text{km/h}$ at about $t = 50\,\text{s}$. The target vehicle speeds up to $v = 100\,\text{km/h}$ at about $t = 100\,\text{s}$, and the VuT accelerates to the demand velocity $v_d = 50\,\text{km/h}$ and remains at this speed.

4.4.2 Lane Keeping Assistance (LKA) Testing Scenario

On the track at the bottom of the driving cycle in Fig. 13, the LKA scenario is tested. The part of the track consists of slightly curved sections. Figure 15 shows the results of the LKA test scenario on the DÜRR X-Road-Curve test bench. The plot on the top

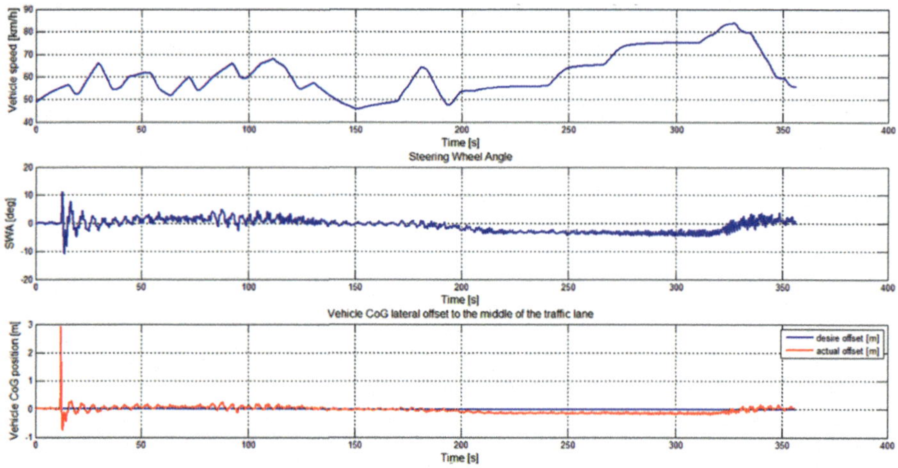

Fig. 15 Vehicle LKA test on DÜRR X-Road-Curve mixed-reality test bench utilizing vehicle steering angle

shows the vehicle velocity of the VuT. The steering wheel angle (SWA) is illustrated in the second plot. The last plot shows the position of the vehicle (center of gravity) relatively to the lane-center line. The LKA is activated at about $t = 15$ s after a long curve. The scenario starts with a right turn, changes into a left turn, and ends again in a right turn. The VuT accelerates at the second half of the track. The VuT follows the course of the road, but a high-frequency oscillation is observed at the steering angle, which rises with increasing speed. It has been shown that this oscillation does not occur during real road tests even at higher speeds; see [6]. This leads to the conclusion that the oscillations are caused by imperfections of the coupling between the vehicle with the test bench and the simulation (e.g., communication delay, measurement noise, etc.). Additional investigations are necessary for a detailed analysis of these imperfections.

5 A Vehicle-in-the-Loop Testing Method for the Development of Automated Driving Functions

The development of automated driving functions and technologies poses various challenges even at the development stages. Testing the automated or ADAS function and verifying the vehicle behavior in multiple traffic scenarios poses a major difficulty during the development cycle of such systems. In order to ensure accurate vehicle behavior, several tests must be conducted under safe driving conditions. However, these are in general too costly and are often infeasible to perform for verifying every ADAS function before deploying such systems on production vehicles. The current approach in automotive industry is to employ simulation tools to verify such systems against certain performance criteria. However, simulation-only verification and testing also have their limitations.

Advanced testing methodologies that bring together the advantages of virtual and real tests have been developed in the past. Real-time HIL simulation is commonly used in automotive engineering, e.g., in engine development [2] and electronic control unit (ECU) development [5, 18] as well as for the development and testing of electric vehicle components [19]. An extension of the HIL-simulation technique on the vehicle level was proposed as vehicle-in-the-loop (VEHIL) simulation in the seminal work by TNO [20] and conceptualized as testing option for longitudinal and lateral control systems in [21].

Several examples of combined virtual and real testing have been reported in the literature [16, 31–33]. Some end products are also available offered for this purpose by commercial vehicle dynamics software producers [34]. Also there are some recent use case demonstrations for the utilization of the combined virtual and real testing concept by some automotive manufacturers in a series of publications [35–39].

As an alternative implementation of the combined or mixed reality for automotive testing, we developed in the scope of the EU-Horizon 2020 funded project [40] an alternative testing concept for ADAS and automated driving systems testing named the "hybrid testing" methodology, where a real vehicle can be combined with virtual ones in a co-simulation framework for analyzing ADAS system performance in virtually created traffic scenarios. Hybrid testing is a novel testing methodology based on the Model.CONNECT™ co-simulation platform [41–44], which was implemented and extended in the scope of the INFRAMIX project to test several mixed traffic scenarios involving autonomous and manually driven vehicles.

In some respects, hybrid testing is similar to the VEHIL or VIL system solutions, where real components and simulated components are combined, however with a distinct difference in utilization of the co-simulation platform Model.CONNECT™ at its core. This is also a generalization of hardware-in-the-loop (HIL) testing, specifically involving a test vehicle as the "hardware" in a feedback loop with the simulated modules. Other publications suggest calling it scenario-in-the-loop testing. In the hybrid testing methodology, a real car drives in an enclosed proving ground using the real ADAS functions running on its real-time ECU (dSPACE-MicroAutobox-II), whereas the driving scenarios and the surrounding traffic as well as the sensor data are simulated on the co-simulation platform using dedicated software components.

In the developed hybrid testing solution, a real-life automated vehicle (AV), which is a generic test vehicle, is driven on a proving ground, where the required perception sensors inputs are provided by a real-time co-simulation of static and dynamic environment using SUMO [45] traffic simulation software. The utilized automated driving function is an in-house developed SAE Level 3 ADAS function with lateral and longitudinal tracking as well as lane change decision capabilities (i.e., Motorway Chauffeur or MWC), enabling the AV to have Adaptive Cruise

Control (ACC), Lane Keeping Assistance (LKA), and trajectory planning (TP) functionalities.

In this chapter we describe the implementation details of the hybrid testing methodology and its corresponding components together with the use case implementation for testing and evaluation of a trajectory planning algorithm.

5.1 Submicroscopic Co-simulation Framework and Hybrid Testing Methodology

In the scope of the hybrid testing methodology, one of the core components is the test vehicle driven in an enclosed proving ground and is interacting, via a coupled submicroscopic co-simulation framework, inside a virtual traffic environment. The sensor signals of the vehicle are simulated. That means, virtual static objects such as lane markings, traffic signs, and dynamic virtual traffic are provided by the submicroscopic co-simulation. The current vehicle position is measured by a GPS sensor to localize the ego vehicle on a corresponding digital HD map, which is linked by the co-simulation framework. The position and other vehicle states of the VuT are transmitted to the traffic simulation component, and based on this, the virtual traffic reacts accordingly. The concept of the hybrid testing methodology is shown in Fig. 16, where virtual and real components are illustrated as two interconnected parts of the environment. The coupling of physical and virtual components enables a testing environment for evaluating ADAS functions under realistic dynamic traffic conditions.

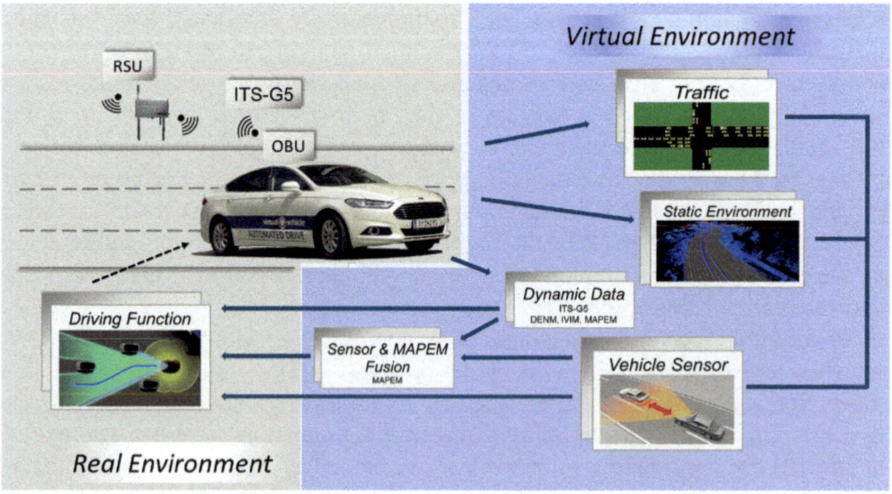

Fig. 16 Hybrid testing concept

The test vehicle is equipped with an ADAS-Kit that enables controlling the car from an algorithm running on a development ECU. The details of the test vehicle and its hardware setup are described in Sect. 5.2. In addition to the test vehicle, there are two further components that belong to the real environment: the *Road Side Unit* (RSU) [46] and the *On-Board Unit* (OBU) [46]. Both are ITS-G5 communication devices from Siemens Mobility and enable connectivity as well as information support to the VuT. The RSU is placed on the proving ground and sends out C-ITS messages over the ITS-G5 communication channel. The OBU that is integrated inside the car receives and interprets the messages and directs them to the driving function.

Since the most important physical components are described in separate sections, the following subsections explain specifically the submicroscopic co-simulation framework and its various components. Each of these software components was developed for a specific task and has particular responsibilities as described in detail below.

5.1.1 Traffic Simulation

As microscopic simulation module, the SUMO traffic simulator is used. SUMO provides realistic mobility pattern for a multitude of vehicles and generates the overall traffic flow in our investigations. Each vehicle in the simulation is modeled as an individual agent using simplified car-following models. Additionally, SUMO provides an interface to control individual vehicles. Therefore, the current vehicle position, the vehicle speed, and the orientation have to be transmitted to SUMO. The most important prerequisite for the exchange of the vehicle position to SUMO is the match of the SUMO coordinate system, the map which is used in the submicroscopic co-simulation as well as the GPS coordinates of the real car.

5.1.2 Static Environment

The main challenge of the *Static Environment* is to provide relevant road infrastructure information to the ego vehicle. This means a highly accurate course of all lanes on the current road section and positions of the traffic signs in front of the ego vehicle. A widely used format for the description of static infrastructure elements, like lane courses and their attributes as well as traffic signs, is the OpenDRIVE [2] file format, which is also the input for the *Static Environment*. The *Static Environment* can be parameterized by the length l of a searching box which is depicted in Fig. 17. According to the input of the *Static Environment* and the parameterization of the dashed box, the model will filter all information about the lanes inside the dashed box such as the x/y coordinates, the road mark type, the lane id, and the road id. This detailed description of the lanes inside the dashed box as well as the signals will be sent to the *Sensorsystem*.

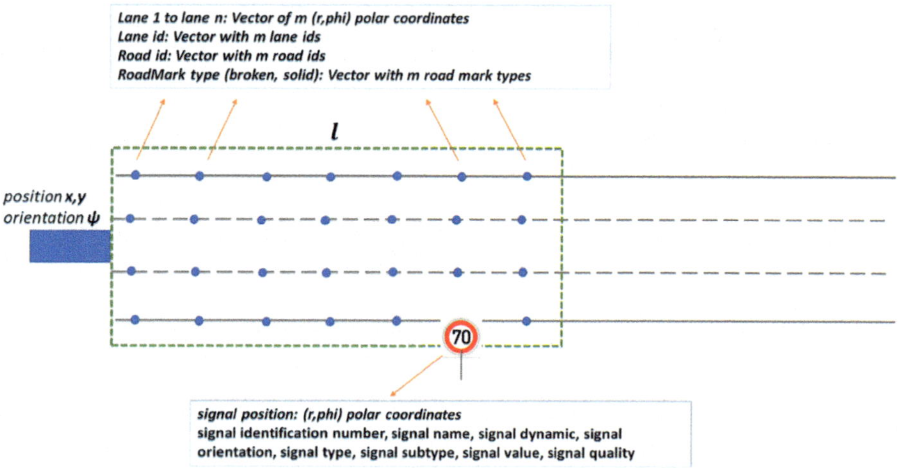

Fig. 17 Output of the *Static Environment*

Figure 17 shows the output of the *Static Environment* which is transmitted to the *Sensorsystem*. Every lane is described by a set of polar coordinates and specific attributes such as the lane id, the road id, and the road mark type. There are nearly the same parameters for the description of traffic signals such as polar coordinates for the signal position, signal ids, and so on.

5.1.3 Sensorsystem

The *Sensorsystem* is used for ongoing monitoring of the vehicle's environment. The *Sensorsystem* provides continuously updated information regarding the traffic situation to the automated driving function so that the trajectory planning algorithm can be adapted to the current traffic situation. The *Sensorsystem* is a 2D sensor with ideal characteristics. That means neither non-linearity, signal noise, nor environment influences are modeled. The *Sensorsystem* in the submicroscopic co-simulation includes two sub-sensors. The first one is the *Static Environment Sensor*, and the second one is the *Dynamic Environment Sensor*. The *Static Environment Sensor* can be parameterized by its range r and azimuth angle α, as can be seen in Fig. 18. The horizontal field of view of the *Static Environment Sensor* corresponds to the real field of view of a camera inside the car which detects lanes and signs in front of the car. This sensor is parameterized with a range of 150 m and an azimuth angle of 40°.

The input from the *Static Environment* can be described as a part of a static map which includes the lanes and signs in front of the vehicle. The main task of the *Static Environment Sensor* is to decide which of the information from the map is located inside the sensors field of view and can be seen by the vehicle. The output of the *Static Environment Sensor* are lanes and signs which are inside its field of view; see Fig. 18.

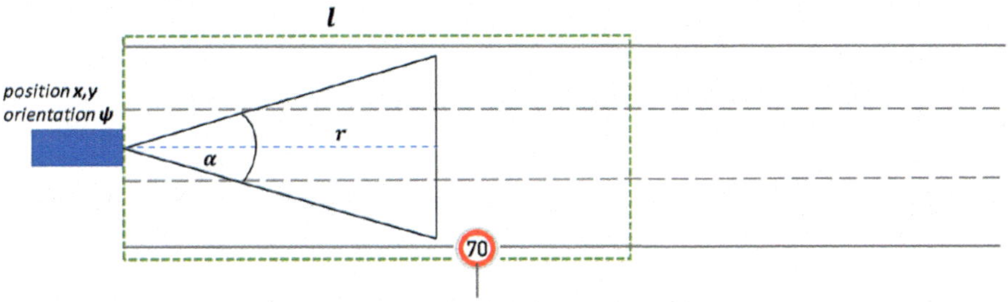

Fig. 18 Output of the *Static Environment Sensor* is the information inside the rectangular shape

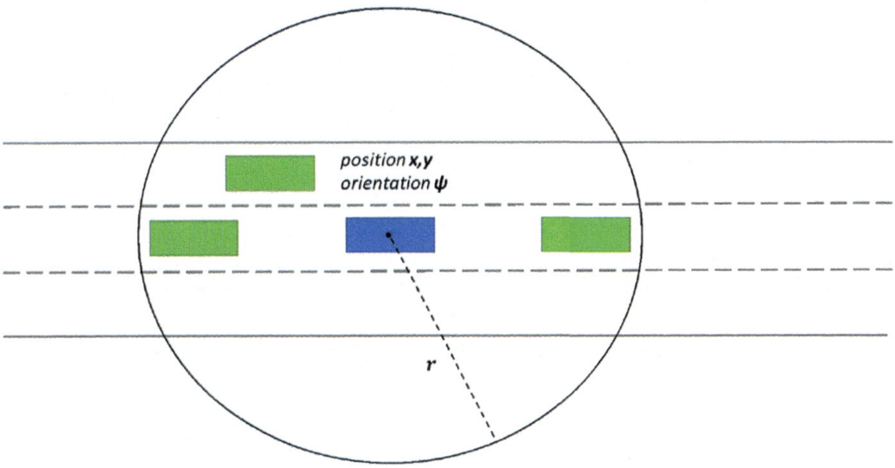

Fig. 19 *Dynamic Environment Sensor* – field of view

The *Dynamic Environment Sensor* receives dynamic input via the coupling interface from the SUMO simulation. Since the driving function needs the information about vehicles in front of the vehicle as well as in the rear section of the vehicle, this sensor is parameterized in a way that vehicles can be detected in a range of 200 m with a field of view of 360°; see Fig. 19. Additionally, to each vehicle position (x, y), the velocity v_x in longitudinal direction and the velocity v_y in lateral direction and the vehicle width are transmitted via the interface to the microscopic simulation SUMO. To decide which information is inside the sensors' visual range is the main task of the vehicle sensor.

5.1.4 Dynamic Data

The *Dynamic Data* model is responsible for the handling of dynamic data information which is received via OBU. The tasks of the *Dynamic Data* model range from revision of received messages to transmission of ADAS-relevant information from that messages to the real environment. This model has a mechanism which continuously revises an interface for new incoming messages. In case of a received message, the next step is to parse the message and extract particular information of it. After verification of the message content and its validity inspection for the current vehicle state, the information and a validity flag are forwarded to the driving function of the real environment, and the VuT will act accordingly to that message.

5.2 *Test Vehicle and Experimental Setup*

The test vehicle used for the testing of in-house developed ADAS functions is based on a Hybrid Ford Mondeo MY2016 platform. This vehicle has an ADAS kit build by Dataspeed Inc. [14] and is a comprehensive add-on hardware and software solution that allows the full control of the throttle, brake, steering, and shifting of the test vehicle. The picture of the vehicle along with the installed sensor hardware is shown in Fig. 8. Among the indicated sensors in the figure, only the dual-antenna RTK-GPS is utilized in the scope of hybrid testing. The test vehicle also has many computational hardware for implementation, where most of the development ECUs are installed at the trunk of the vehicle. Of primary importance in this list is the Dataspeed CAN interface, which enables the access to the onboard vehicular sensors and controls. Using this interface, the throttle, brake, and steering as well as other parameters can be controlled using a secondary ECU. The data rates for the control-specific parameters provided by the Dataspeed CAN interface can be seen in Table 1.

Table 1 Dataspeed CAN data and the corresponding data rates

Feature	Specification
Platform	Ford CD4
Initial release date	Oct 2015
Throttle control frequency	50 Hz
Brake control frequency	50 Hz
Steering control frequency	50 Hz
Steering angle control	Yes
Steering torque control	No
Gear shift control (PRNDL)	Yes
Turn signal control	Yes
ULC (speed control)	Yes
Individual wheel speeds	100 Hz
3-Axis accelerometer	100 Hz
Roll and yaw rate gyro	100 Hz
Parking SONAR Sensors	5 Hz
Tire pressures	2 Hz
GPS	100 Hz

5.2.1 Test Vehicle Setup

The test vehicle also houses a dSPACE MicroAutobox-II (ds1401) real-time ECU [28] that runs the ADAS functions in real time based on the provided sensor data from onboard sensors as well as the simulated one from the co-simulation platform. The ADAS functions for ACC, LKA, and TP are implemented in MATLAB/Simulink and later exported to C++ code that is automatically generated by the MATLAB/Simulink embedded coders to run on the dSPACE MicroAutobox real-time hardware. Therefore, the ADAS functions running on the MicroAutobox ECU provide the driving commands to the vehicle's actuators through the Dataspeed CAN interface.

In hybrid testing, the co-simulation framework runs on the Nuvo 7006 PC that is located inside the vehicle, and the RSU (Road Side Unit) is located outside the vehicle as part of the test-site infrastructure. The PC in the VuT is responsible for running the traffic and environment simulation, sensor models, and object list generation algorithms in a co-simulation environment while also being connected in real time to the CAN bus of the VuT to collect variables relevant for representation of the VuT in the traffic simulation on the PC. During hybrid testing, the ADAS control functions are based on the vehicle (specifically running on the dSPACE MicroautoBox-II platform) as opposed to having an ADAS function block along with vehicle dynamics simulator in the submicroscopic co-simulation framework. Additionally, the car is equipped with an OBU for bidirectional communication with the RSU. In this context, the communication between the RSU and OBU is unlimited, and there can be persistent bidirectional link between the RSU and the OBU during the demonstrations of the specific hybrid testing use case demonstrations planned. That is, when a pre-specified traffic control message is sent by the RSU to the VuT, it will act according to this message as soon as the message is received by the OBU in VuT. The complete hardware and software architecture and the internal communication structure implemented on the Ford Mondeo test vehicle for realizing the hybrid testing methodology are seen in Fig. 20.

In the scope of the hybrid testing methodology, the sensor signals of the test vehicle are simulated, and they sense the virtual objects created in the simulated static environment as well as the dynamic traffic components. The simulated sensor signals exclude the GPS sensor data, which provides survey-grade (i.e., cm-level accurate) position and heading information from the real vehicle to localize the ego vehicle on the digital HD map (i.e., the static environment) used in the co-simulation platform. The position and other vehicle states of the VuT are sent to the traffic simulation, where the simulated vehicles in the traffic simulation react to the VuT. The environmental information such as the lane markings, road curvatures, etc. as well as other traffic participants are only virtually present. The virtual dynamic objects consist of the vehicles of the surrounding traffic, while the static environment consists of features such as the road markings and traffic signs. The concept of the hybrid testing methodology is shown in Fig. 16, where virtual and real components are illustrated as lumped at the right- and left-hand sides of the figure, respectively.

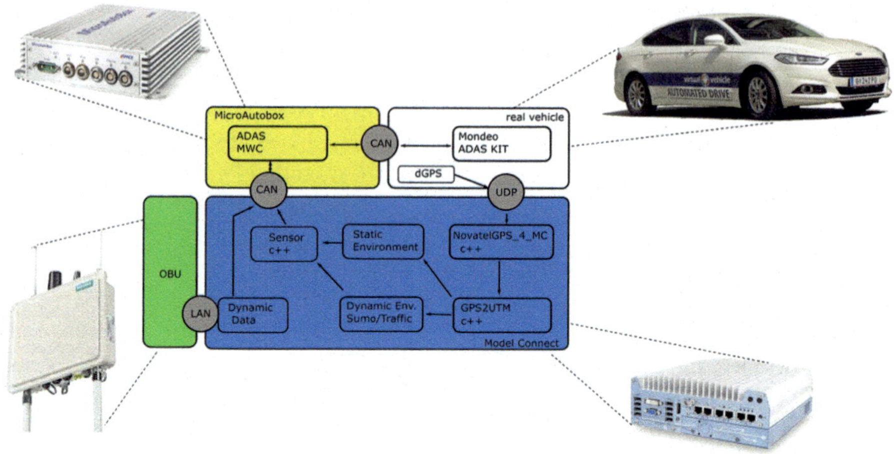

Fig. 20 Hardware and software architecture for hybrid testing

The use of virtual sensors naturally imply that the vehicle is not able to sense real objects and real road markings in the testing area (i.e., proving ground). Therefore, hybrid testing needs to be conducted on an open area, large enough to cover the virtual test track with additional buffer zones around the road layout, for safety purposes. Also, real infrastructure components are present in the form of RSU that sends out C-ITS messages over the ITS-G5 communication channel to an OBU that is integrated to the vehicle. The real vehicle reacts to the virtual elements accordingly while also taking into account the real-world communication via RSU. Moreover, using the OBU, the status of the real vehicle can be sent back to the RSU for further processing as may be required from the use case scenario. When approaching the area of interest (e.g., the road works zone), the automated vehicle receives a C-ITS message from the RSU, which will then react according to this message, considering the surrounding virtual traffic.

The results of hybrid testing are logs of positions and velocities of the VuT and the simulated traffic, as well as the logs of the C-ITS message transmissions. Out of these logs, behavior of the VuT can be investigated. The logged data either come from the simulation tool used for the virtual traffic (SUMO) or the vehicle sensors itself (inertial measurements such as acceleration, yaw rate, steer angle, etc.). During the test runs, the vehicle states (position, velocity, acceleration, heading, yaw rate, etc.), the states of the surrounded vehicles, as well as the C-ITS messages, are recorded and can be used for adapting and validating simulation parameters. Signal processing and evaluation tools are implemented in MATLAB and Microsoft Excel. The resulting investigations can be used to verify the vehicle behavior (both VuT and the simulated vehicles) in the simulation (parameters of the car-following and lane change model in SUMO) or lead to adaptions of the vehicle behavior in the real vehicle. With the findings of hybrid testing, both the simulation and the automated driving function can be validated and improved.

The time frame for hybrid testing scenarios is between 30 and 60 s depending on the desired velocity and as restricted by the test site physical limitations. The actual useable stretch of the road section (ÖAMTC Lang/Lebring test track) used during the hybrid testing demonstrations is about 250 m, and assuming an average speed of 50 km/h during a hypothetical test, the entire stretch is covered in less than 30 s.

5.2.2 HD Map Generation as Preparation for Hybrid Testing

Before starting with hybrid testing, the preparation of an HD map was required. The OpenDRIVE standard, which is an xml-compliant scheme, provides a detailed representation of static road infrastructure elements such as lane markings as well as traffic signs and enables safe vehicle guidance along pre-calculated trajectories. Due to the high level of detail in this format and also SUMO's import mechanisms for that file format, the OpenDRIVE standard is used as basis in the submicroscopic co-simulation. After local reference measurements at the proving ground, the OpenDRIVE file was created semi-automatically based on these measurements, as well as OpenStreetMap™and Google Maps™. As microscopic simulations do not require that high-level detailed maps in order to model traffic, SUMO provides a conversion tool for OpenDRIVE files to generate SUMO-specific net files. Figure 21 shows the test area as satellite picture and the corresponding SUMO net file. One of the fundamental challenges is the mapping of the current vehicle position, measured by a GPS sensor, as the OpenDRIVE file uses UTM coordinates and the traffic simulation uses a SUMO-specific coordinate system.

The preparation of the test vehicle and the HD maps of the corresponding proving ground enabled the conduction of different scenarios. A description of these scenarios together with an analysis of the results is given in the following section.

5.3 Scenario Description, Implementation Results and Analysis

Various number of use case studies were conducted during the hybrid testing trials, and two specific use case examples are discussed in this section. The following experiments are focusing specifically on lane change and merging maneuvers in combination with C-ITS messages. The initiation of lane changes is triggered by the trajectory planner as a result of the road layout as well as the traffic situation at hand. While the vehicle drives on the real test track, it is fully aware of the presence of virtual vehicles close to it and takes their behavior consequently into account. The use case studies were extended with communication elements to investigate the reaction of the vehicle on specific C-ITS messages for speed advices. Besides the speed information, this messages include also GPS positions of a *relevance zone* as seen in Fig. 21. Here, what is meant by the *relevance zone* is the zone for which the C-ITS message information is valid. In what follows, we describe two example

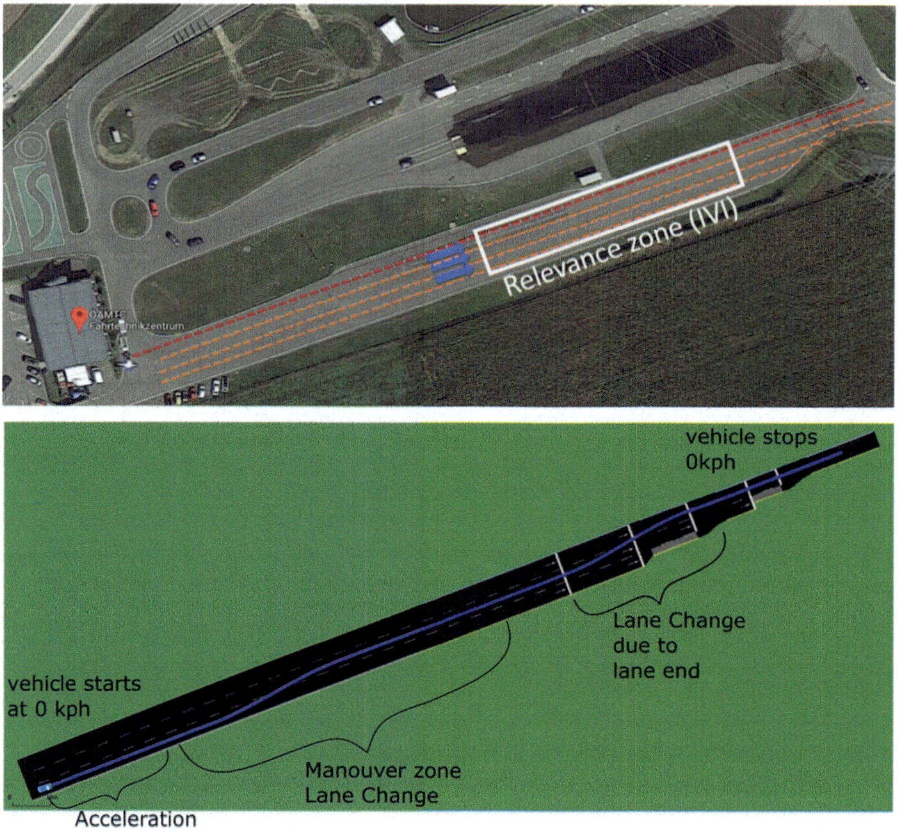

Fig. 21 Maneuver zone satellite picture and SUMO net file

scenarios and the corresponding result as part of this study. In these example test runs, the VuT starts on the second lane (from the right), which corresponds to the first lane of the main road, and receives speed advice and acts accordingly.

5.3.1 Scenario I

This scenario shows that the VuT is capable to react accordingly to a speed advice when the VuT is driving on the main road. In this experiment, virtual traffic is not present. The VuT drives initially with a speed of 30 km/h on the main road when it receives a message via the OBU, which was sent from the RSU. In Fig. 22a, we see the desired and the actual velocity history of the ego vehicle during this experiment. We can see that the VuT adapts accordingly to the speed advice of 50 km/h when it reaches to the relevance zone. In Fig. 22b, we show the commended steering angle output from the vehicle as a result of the trajectory planner control function. Here the lane change action is clearly visible.

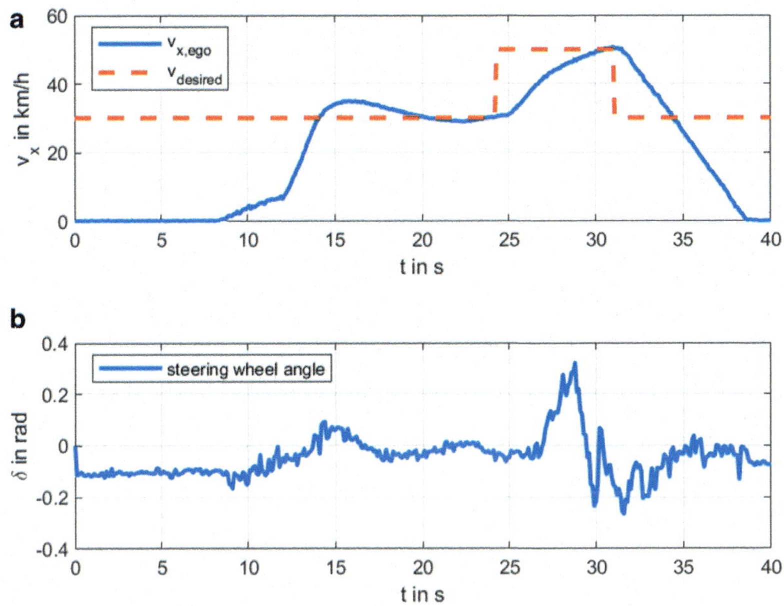

Fig. 22 Vehicle dynamic parameters of scenario I of hybrid testing. (**a**) Velocity (**b**) Steering angle

5.3.2 Scenario II

The next scenario shows that the VuT can react to a C-ITS message also when traffic is present in the near vicinity. As in the previous scenario, the VuT is driving with a speed of 30 km/h initially on the main road when it receives a message with a speed advice of 50 km/h. But now, a leading vehicle hinders the VuT from reaching the speed advice since it is driving slower. As a result, the VuT was not able to reach the speed advice of 50 km/h, and it adapted its speed to the leading vehicle. Moreover, in Fig. 23b, c, we observe that the VuT conducted an overtaking maneuver and passed the leading target vehicle at the end of the stretch as was expected from the trajectory planning algorithm. The resulting steering angle variation during this experiment is shown in Fig. 23b. Finally, in Fig. 23c, we illustrate the path and a snapshot of the positions of the ego vehicle (VuT) and the leading target vehicle during the test scenario, where the solid rectangle depicts the VuT and the empty rectangle designates the target vehicle.

In the second experiment, the trajectory planner had to take into account the distance to the leading target vehicle and had to adapt the VuT's driving behavior to keep a safe distance. Therefore, further parameters of the trajectory planner were measured during the test and evaluated afterward utilizing the co-simulation framework. Specifically, for the trajectory planner, we evaluated the *minimum distance gap* in meters, the *minimum time gap* in seconds, as well as the *minimum time to collision* in seconds. The results of these parameters are gathered in Table 2.

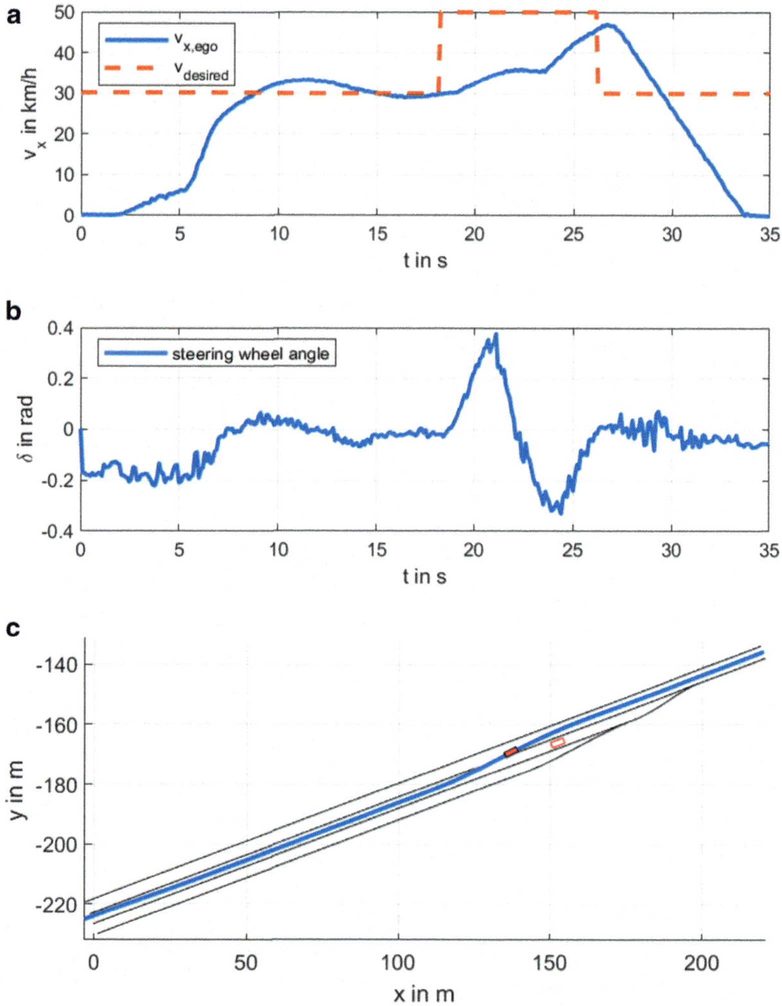

Fig. 23 Vehicle dynamic parameters of scenario II of hybrid testing. (**a**) Commanded and actual velocity variations (**b**) Measured steering wheel angle (**c**) Vehicle path along the HD-Map

Table 2 Distance measurements – scenario II

Minimum distance gap [m]	Minimum time gap [s]	Minimum TTC [s]
11.7	1.17	4.85

Depending on the KPIs for the ADAS function under test, many test metrics and parameters can be calculated and evaluated during hybrid testing to ensure conformity to the performance objectives.

Following the process from simulation to real-word testing presented in Sect. 1, this section described the real-world testing aspects. The following section concludes the outcome from this and all previous sections and provides an outlook to future work.

6 Conclusions and Outlook

In this chapter, we presented the implementation details of two novel testing and verification methods for the development and testing of ADAS/AD functions, both utilizing and expanding the idea of combining real-world testing with simulation elements.

The first testing/verification concept involves a steerable rolling dynamometer (DÜRR X-Road-Curve chassis test bench) system together with a vehicle-in-the-loop solution with OTA camera stimulation. It has been demonstrated that the suggested VEHIL solution can potentially be utilized to replace on-road tests and enable repeatable batch scenario testing opportunities. It is also possible to use the proposed solution to check ADAS/AD functions for robustness, integrity, and functional safety, which is normally not easily achievable with traditional road testing method. This test bench solution and the corresponding methodology can also be used for endurance testing of ADAS/AD functions. Therefore, the suggested mixed-reality test bench system can provide an intermediate step in ADAS/AD system development, testing, and validation.

In the second testing/verification method, we presented the generalization of the co-simulation framework together with a vehicle-in-the-loop testing concept and named this "hybrid testing," where a real vehicle can be combined with virtual ones on a real test track (or proving ground) to evaluate and validate ADAS functions in realistic traffic scenarios. Specifically, we have described the co-simulation framework with all its sub-components, which enables real-time integration between the traffic scenario simulation and the real-life test vehicle along with the ADAS functions running on it.

There are open problems related to the test bench implementation that clearly need further investigation. First of all, there is an observed oscillatory behavior, which is possibly caused by the multiple feedback loops that are an inherent part of the system. These feedback loops exist within the X-Road-Curve chassis dynamometer, the Model.CONNECT™ co-simulation platform that interface the environment simulation and vehicle dynamic components, as well as in the intelligent ADAS camera (MobilEye) utilized as part of the driving function. The inherent complex interactions among these multiple dynamic and interlocking loops are an open scientific problem that needs closer investigation. Another open problem is related to the utilized camera stimulation. Many ADAS/AD systems utilize not only cameras but also combinations of other sensors such as radars, lidars, and/or ultrasonic sensors. Stimulation and/or simulation of other sensors in the scope of a consistent testing methodology is another open problem. These are therefore some of the possible directions of research on this specific implementation.

The hybrid testing implementation on the other hand demonstrated that successful and repeatable virtual driving scenarios are achievable on a proving ground for the functional testing of ADAS/AD functions. The suggested testing method can be automated and performed as batch tests, if the effect of various input parameters needs to be evaluated and/or a statistical evaluation is required. Moreover, scenarios involving the interaction of VuT with new road control elements (e.g., ITS-G5, 5G, and/or VMS) and traffic management centers can also be set up and evaluated. Hybrid testing can also be extended, likewise, in various directions. This proof-of-concept implementation utilized only a traffic simulator (SUMO) for the scenario modeling, which as a first step can be improved by the integration of a photo-realistic environment simulator (such as CARLA, Unity 3D, LG Simulator, etc.) to the co-simulation framework. Additionally, it is possible to include physical sensor models to the co-simulation framework for use of the methodology for perception algorithm development. As a further extension, it is possible to include real sensor equipment to the co-simulation framework to perform enhanced reality scenario testing of ADAS/AD functions for standard compliance evaluations and validation. Finally, it is also interesting to compare the same test scenario in simulation, test bench, and hybrid testing evaluations to analyze the quantitative correlation between these testing options. These possible extensions shall be investigated for future work.

Acknowledgments This work is partially funded by the INFRAMIX project and was accomplished at the VIRTUAL VEHICLE Research GmbH in Graz, Austria. INFRAMIX has received funding from the European Union's Horizon 2020 research and innovation program under grant agreement no 723016. The authors would like also to acknowledge the partial financial support of the COMET K2-Competence Centers for Excellent Technologies program of the Austrian Federal Ministry for Transport, Innovation and Technology (BMVIT), the Austrian Federal Ministry of Science, Research and Economy (BMWFW), the Austrian Research Promotion Agency (FFG), the Province of Styria, and the Styrian Business Promotion Agency (SFG). Content of the paper reflects only the authors' view. The Innovation and Networks Executive Agency (INEA) is not responsible for any use that may be made of the information it contains. The authors would like to thank all partners within INFRAMIX consortium for their cooperation and valuable contributions. The authors would furthermore like to express their thanks to their supporting industrial project partners, namely, AVL List GmbH and DÜRR Assembly Products GmbH.

References

1. X-Road-Curve dynamically tests autonomous cars, (2018). https://www.durr-group.com/en/duerrmore/autonomous-driving/x-road-curve/, Accessed on 11 Apr 2018
2. T.E. Passenbrunner, M. Sassano, H. Trogmann, L. del Re, M. Paulweber, M. Schmidt, H. Kokal, Inverse torque control of hydrodynamic dynamometers for combustion engine test benches, in *Proceedings of the 2011 American Control Conference*, 2011, pp. 4598–4603
3. AVL-Model.CONNECT^{TM}, the neutral model integration and co-simulation platform connecting virtual and real components. 2018. https://www.avl.com/-/model-connect. Accessed on 11 Apr 2018

4. MOBILEYE Inc., sensing the future. 2018. https://www.mobileye.com/. Accessed on 11 Apr 2018
5. D. Fodor, K. Enisz, Vehicle dynamics based ABS ECU verification on real-time hardware-in-the-loop simulator, in *2014 16th International Power Electronics and Motion Control Conference and Exposition*, Sept 2014, pp. 1247–1251
6. S. Solmaz, F. Holzinger, A novel test bench for development, calibration and functional testing of ADAS/AD functions, in *2019 IEEE International Conference on Connected Vehicles and Expo (ICCVE)*, Nov 2019, pp. 1–8
7. IPG Automotive, *Reference Manual*, 4th edn. (2012)
8. D. Watzenig, M. Horn, *Automated Driving. Safer and More Efficient Future Driving* (Springer, 2017)
9. H. Winner, S. Hakuli, F. Lotz, C. Singer, *Handbook of Driver Assistance Systems* (Springer International Publishing Switzerland, 2016)
10. R. Rajamani, *Vehicle Dynamics and Control* (Springer US, 2012)
11. J. Kosecka, R. Blasi, C.J. Taylor, J. Malik, A comparative study of vision-based lateral control strategies for autonomous highway driving, in *Proceedings. 1998 IEEE International Conference on Robotics and Automation (Cat. No.98CH36146)*, vol. 3, May 1998, pp. 1903–1908
12. G. Nestlinger, M. Stolz, Bumpless transfer for convenient lateral car control handover, *IFAC-PapersOnLine*, vol. 49, no. 15, pp. 132–138, 2016. 9th IFAC Symposium on Intelligent Autonomous Vehicles IAV 2016
13. G. Nestlinger, Modellbildung und Simulation eines Spurhalte-Assistenzsystems, Master's thesis, Techniche Universität Graz, 2013
14. Dataspeed Inc. We make robots move and cars drive by themselves. http://dataspeedinc.com/. Accessed 04 Nov 2018
15. S. Solmaz, G. Nestlinger, G. Stettinger, Improvement of lane keeping assistance ADAS function utilizing a Kalman filter prediction of delayed position states, in *2019 IEEE International Conference on Connected Vehicles and Expo (ICCVE)*, Nov 2019, pp. 1–8
16. M. Paulweber, K. Lebert, *Mess- und Prüfstandstechnik: Antriebsstrangentwicklung · Hybridisierung · Elektrifizierung*. Der Fahrzeugantrieb (Springer Fachmedien Wiesbaden, 2014)
17. H. Beglerovic, S. Metzner, M. Horn, Challenges for the validation and testing of automated driving functions, in *Proceedings of the AMAA 2017 Conference*, 2017
18. T. Hwang, J. Roh, K. Park, J. Hwang, K.H. Lee, K. Lee, S. Lee, Y. Kim, Development of HILS systems for active brake control systems, in *2006 SICE-ICASE International Joint Conference*, Oct 2006, pp. 4404–4408
19. J.D. Moore, G.M. Molen, SIL, HIL, and vehicle fuel economy analysis of a pre-transmission parallel PHEV, in *2013 World Electric Vehicle Symposium and Exhibition (EVS27)*, pp. 1–7, Nov 2013
20. L. Verhoeff, D.J. Verburg, H.A. Lupker, L.J.J. Kusters, Vehil: a full-scale test methodology for intelligent transport systems, vehicles and subsystems, in *Proceedings of the IEEE Intelligent Vehicles Symposium 2000 (Cat. No.00TH8511)*, pp. 369–375, Oct 2000
21. D.J. Verburg, A.C.M. van der Knaap, J. Ploeg, Vehil: developing and testing intelligent vehicles, in *Intelligent Vehicle Symposium, 2002. IEEE*, vol. 2, pp. 537–544, June 2002
22. O.J. Gietelink, D.J. Verburg, K. Labibes, A.F. Oostendorp, Pre-crash system validation with PRESCAN and VEHIL, in *IEEE Intelligent Vehicles Symposium, 2004*, pp. 913–918, June 2004
23. J. Ploeg, F.M. Hendriks, N.J. Schouten, Towards nondestructive testing of pre-crash systems in a HIL setup, in *2008 IEEE Intelligent Vehicles Symposium*, pp. 91–96, June 2008
24. F. Hendriks, M. Tideman, R. Pelders, R. Bours, X. Liu, Development tools for active safety systems: PRESCAN and VEHIL, in *Proceedings of 2010 IEEE International Conference on Vehicular Electronics and Safety*, pp. 54–58, July 2010
25. C. Schyr, A. Brissardt, Drivingcube – a novel concept for validation of powertrain and steering systems with automated driving, in *Proceedings of the* 13th *International Symposium on Advanced Vehicle Control AVEC2016*, Sept 2016

26. VIRTUAL VEHICLE automated drive demonstrator (ADD), an open research platform for automated driving, 2018. https://www.v2c2.at/automated-driving/. Accessed on 11 Apr 2018
27. T. Weck, J. Neumann, How to calibrate and test advanced driver assistance and autonomous driving systems EoL, in *Proceedings of the 19. VDI-Kongress SIMVEC – Simulation und Erprobung in der Fahrzeugentwicklung*, 2018
28. DSPACE Inc., Microautobox 2 Kompaktes, robustes Prototyping-System für den Fahrzeugeinsatz. 2018. https://www.dspace.com/. Accessed on 11 Apr 2018
29. O. Gietelink, J. Ploeg, B.D. Schutte, M. Verhaegen, Development of advanced driver assistance systems with vehicle hardware-in-the-loop simulations. Veh. Syst. Dyn. **44**, 569–590 (2006)
30. VTD – Vires Virtual Test Drive, 2018. https://vires.com/. Accessed on 11 Apr 2018
31. T. Bock, M. Maurer, F. van Meel, T. Müller, Vehicle in the Loop, Ein innovativer Ansatz zur Kopplung virtueller mit realer Erprobung, *ATZ – Automobiltechnische Zeitschrift*, vol. 110, pp. 10–16, Jan 2008
32. S.A. Fayazi, A. Vahidi, Vehicle-in-the-loop (vil) verification of a smart city intersection control scheme for autonomous vehicles, in *2017 IEEE Conference on Control Technology and Applications (CCTA)*, pp. 1575–1580, Aug 2017
33. H. Németh, A. Háry, Z. Szalay, V. Tihanyi, B. Tóth, Proving ground test scenarios in mixed virtual and real environment for highly automated driving, in *Mobilität in Zeiten der Veränderung*, ed. by H. Proff (Springer Gabler, Wiesbaden, 2019), pp. 200–2009
34. VIL-Systeme: Einfach reale mit virtuellen Fahrversuchen verbinden, 2019. https://ipg-automotive.com/de/produkte-services/test-systems/vil-systeme/. Accessed on 23 Aug 2019
35. T. Kim (Hyundai MOBIS), Vehicle-in-the-loop (Vil):Validating driver assistance systems with synchronous virtual and real test drives, *dSpace Magazin*, vol. 2, 2019
36. U. Eberle, S. Berger, S. Hallerbach, Erprobung eines Fahrzeugs mit simulierten Verkehrsteilnehmern. Patent, Oct 2019. Patentnumber: DE102018003130A1
37. S. Hallerbach, Y. Xia, U. Eberle, F. Koester, Simulation-based identification of critical scenarios for cooperative and automated vehicles. SAE Int. J. CAV **1**(2), 93–106 (2018)
38. D. Augustin, J. Schucker, J. Tschirner, M. Hofmann, U. Konigorski, A simulation-based reinforcement learning approach for long-term maneuver planning in highway traffic scenarios, in *VDI Berichte Nr. 2349*, pp. 47–60, 2019
39. S. Hallerbach, U. Eberle, Simulation-based testing of cooperative and automated vehicles. Conference-Presentation at KO-HAF final event, 09 2018. Accessed 21 Jan 2020
40. Horizon 2020-INFRAMIX, Preparing road infrastructure for mixed vehicle traffic flows. https://www.inframix.eu/. Accessed 10 Jan 2020
41. G. Stettinger, M. Benedikt, M. Horn, J. Zehetner, C. Giebenhain, Control of a magnetic levitation system with communication imperfections: A model-based coupling approach. Control. Eng. Pract. **58**, 161–170 (2017)
42. G. Stettinger, M. Horn, M. Benedikt, J. Zehetner, A model-based approach for prediction-based interconnection of dynamic systems, in *53rd IEEE Conference on Decision and Control*, pp. 3286–3291, Dec 2014
43. M. Tranninger, G. Stettinger, M. Benedikt, M. Horn, Diagnosis of interconnected systems via well tuned model-based coupling algorithms, *IFAC-PapersOnLine*, vol. 51, no. 24, pp. 1271–1277, 2018, in *10th IFAC Symposium on Fault Detection, Supervision and Safety for Technical Processes SAFEPROCESS 2018*
44. G. Stettinger, M. Horn, M. Benedikt, J. Zehetner, Model-based coupling approach for non-iterative real-time co-simulation, *2014 European Control Conference (ECC)*, pp. 2084–2089, 2014
45. P.A. Lopez, M. Behrisch, L. Bieker-Walz, J. Erdmann, Y. Flötteröd, R. Hilbrich, L. Lücken, J. Rummel, P. Wagner, E. Wiessner, Microscopic traffic simulation using sumo, in *2018 21st International Conference on Intelligent Transportation Systems (ITSC)*, pp. 2575–2582, Nov 2018
46. Siemens Mobility GmbH, Sitraffic Vehicle2x. https://new.siemens.com/global/de/produkte/mobilitaet/strassenverkehr/connected-mobility-solutions/sitraffic-vehicle2x.html. Accessed 23 Jan 2020

Autonomous Driving in the Real-World: The Weather Challenge in the Sohjoa Baltic Project

Mauro Bellone, Azat Ismailogullari, Jaanus Müür, Oscar Nissin, Raivo Sell, and Ralf-Martin Soe

1 Introduction

According to [1] in the United States, there are 6,301,000 vehicle crashes every year, 24% of which are related to road weather conditions. The major cause, for weather-related crashes, is the wet road surface, followed by ice and snow, revealing an important safety aspect to be considered in the future development of autonomous driving. In Europe, 29% of fatalities happen in nondry conditions (including rain, fog, snow, etc.), according to [2] showing data from 2016. The same data correlating weather conditions and autonomous driving are yet not available as most of the driverless vehicles are driving for test purposes. Furthermore, the number of crashes is not statistically significant to show, and, in fact, there are millions of vehicles driving billions of kilometers every day without crashing, while only a few of them are autonomous (e.g., the Google's fleet is currently composed of 55 vehicles driving one million of kilometers per year). This means that many more testing

M. Bellone (✉)
Chalmers University of Technology, Gothenburg, Sweden
e-mail: mauro.bellone@chalmers.se

A. Ismailogullari · O. Nissin
Metropolia University of Applied Sciences, Vantaa, Finland

J. Müür
Ragnar Nurkse Department of Innovation and Governance, Tallinn University of Technology, Tallinn, Estonia

R. Sell
Department of Mechanical and Industrial Engineering, Tallinn University of Technology, Tallinn, Estonia

R.-M. Soe
Smart City Center of Excellence, Tallinn University of Technology, Tallinn, Estonia

U. Z. A. Hamid, F. Al-Turjman (eds.), *Towards Connected and Autonomous Vehicle Highways*, EAI/Springer Innovations in Communication and Computing,
https://doi.org/10.1007/978-3-030-66042-0_9

vehicles would be required to effectively demonstrate safety in autonomous driving [3]. Supporting the correlation between weather and crashes, in [4] the authors investigate the impact of meteorological conditions on the frequency of road-crashes in urban environments, concluding that daily precipitation and mean temperature below 10 Celsius present a positive correlation with the number of daily crashes. There are two possible reasons: human error and machine error. Manufacturers and developers are working constantly to reduce the machine error to the minimum, and to build driving assistance systems to help to limit the human error. The effects of weather on autonomous driving and the effect of weather conditions on sensors such as GPSs, cameras, lidar, and radar are also discussed in other research studies [5, 6]. From a different perspective, standardization organizations are also contributing to create advanced standards for road vehicles functional safety (ISO26262) and sensors for autonomous driving in adverse weather (ISO/WD 24650 – under development).

Developing an autonomous vehicle (AV) involves the design of a control system able to behave according to predefined rules. This can be done in two ways, classical control theory or data-driven controllers. The classical method to address the control problem is to build a precise analytical model of the vehicle (including driving environment) and to design a controller in such a way that the process (vehicle) follows the desired behavior. Even though successful autonomous driving tests were conducted in restricted areas and test tracks in the last decades, the classical control theory approach was found as not effective to solve the autonomous driving problem in the real world. The main barrier resides in finding an analytical model that describes the real world in its entire complexity, including moving obstacles, urban traffic, and weather conditions. While those three items are all major issues deserving thorough individual studies, this chapter will mainly focus on the weather conditions.

In contrast to classical control theory, data-driven approaches seem to be effective in many real-world scenarios. Data-driven approaches are based on model identification methods that strongly depend on the big quantities of data acquisition and selection of scenarios for training and testing. However, it is often forgotten that the majority of safety-critical situations are very rare and thus hard to demonstrate in reality by including them into the testing scenarios. The main strength of data-driven methods is that they work very well with artificial intelligence (AI) and neural network-based methods.

Driving under adverse weather conditions is challenging even for experienced drivers for two main reasons (among others), namely, friction loss and sensing loss. By friction loss, one can refer to the change in adherence between the tyre and the asphalt, due to special conditions that make the road surface more slippery (for instance, snow, rain, dust, etc.). In cases of low friction, the velocity of the vehicle must be reduced to ensure safety and comfort for passengers. Sensing loss is more related to the loss of visibility (for example, in case of fog, heavy rain, and snow) or better to lack in incoming information about the external environment, resulting in a challenging issue for autonomous driving. This is a major problem as algorithms strongly rely on sensors to grab information about the environment and to derive the best action (driving decision) accordingly. The main sensors used for this purpose are image cameras and range sensors, and the process of grabbing and interpreting sensory information is typically referred to as *perception*. The reliability of sensory information is, indeed, fundamental for each control system to work properly as it affects the reliability of the controller. In the classical approach, the measure of the robustness is carried out by the designer by simulating several noise levels in different parts of the system (for instance input, control output or feedback loop), and then measuring the sensibility and sensitivity index of output signals to the noise. This method has the advantage of providing precise information about the operational range in which the system will work reliably. In AI-based autonomous driving, the system is already trained with noisy data to obtain more general-purpose controllers able to perform well also in such noisy conditions. However, these methods have a strong drawback of lacking information about the real operational range, the system has a nonzero probability to fail in any condition even without any mechanical or electronic failure.

A new trend of research is looking toward interpretable models for autonomous driving [7], meaning that driving models generated using AI should provide easy-interpretable rationales for their behavior in such a way that passengers, insurance companies, developers, etc. can understand the input–output relation for specific behaviors. This is in response to a specific change in the legislation, which is included in the GDPR directive, namely, the *right of explanation* [8]. This can be considered analogous to the *Algorithmic Accountability Act* [9] recently introduced in the legislation of the United States. The rationality of this concept is that many algorithms present abstract decision-making capabilities, and, according to the recent regulation, a user can ask for an explanation about an algorithmic decision. This is in direct conflict with pure AI-based algorithms that are able to take decisions without any rational explanation.

This chapter will further describe sensors and related issues in detail, starting from a description of our use-study case project (Sohjoa Baltic) in Sect. 2, then the most recent technologies for sensors and intelligent driving are reviewed in Sect. 3, with specific reference to weather-related issues, including a description of typical testing environments that are used in the research phase to measure the performance of sensors and algorithms. As most of the technologies available today are based on data coming from sensors, the problem of intelligent perception and how to interpret such a big quantity of data is described in Sect. 4, leading the reader to the most valuable asset in today's technology: data sets. Many research and industrial players are acquiring and providing publicly available data sets to be used by anyone who is willing to contribute to the field. A list of data sets available today is given in Sect. 5, with a specific distinction between real road data and simulated data. Finally, Sect. 6 provides a qualitative description of the pilot studies in Sohjoa Baltic project and how the weather affected the performance during the pilot studies. Relevant conclusions and ideas for further studies are given in Sect. 7.

2 Piloting Autonomous Electric Minibuses: Sohjoa Baltic

Sohjoa Baltic (EU's Interreg BSR funding) researches, promotes, and pilots automated, driverless electric minibuses as part of the public transport chain, especially for the first/last-mile connectivity. The project studies the environmentally friendly and smart automated public transport solutions, also providing guidelines on legal, regulatory, and organizational setup needed for running such a service in an efficient way in the Baltic Sea Region. Driving in the Baltic Sea Region is challenging for different reasons, the legislative aspect was investigated in several project studies resulting in nonharmonized legislation and standardization among different states in the European Union [10]. Finland and Norway have been reliable testbeds as it was possible to acquire special permits from local transportation authorities for autonomous driving on public roads.

In order to have sufficient diversity, during the project (2017–2020), six partner cities planned to take on the autonomous public transportation trials, open to the public and running in mixed traffic on a regular timetable with different schedules, routes, and their own bus stops. The cities of Kongsberg (NO), Helsinki (FI), Gdansk (PL), and Tallinn (EE) launched their pilots in 2018–2019. Tallinn pilot was paused for the winter and will continue in spring 2020. Unexpected Danish regulatory issues lead to the cancellation of the pilot in Vejle (DK).

The schedule of the pilots was chosen also in consideration of the climate conditions in the specific locations and the technical requirements of the buses. The underlying principle for the scheduling of the pilots in different locations is related to the investigation of the impact of weather conditions in different seasons, providing diversity of information for the study. According to the Finnish Meteorological Institute (see Fig. 1), wintertime provides less precipitation probability while having some problems with possible snow on the road, whereas summertime is characterized by higher precipitation and higher temperature. A conclusion from this analysis is that hardware itself, including sensors and electronic components, must be robust enough to work in a wide temperature range (from -20 to $+50$).

Rain and snow are critical problems for autonomous vehicles, precipitations occur all year long while the temperature may vary in a wide range. Furthermore, in conditions with extreme air temperature (below -20 °C and over $+50$ °C), many technical issues may arise with AVs. Automated vehicles are often fully electric, which may be subject to overheating in warm environments, especially if the operated route has large elevation gradients. In addition to the stress caused to the electric drivetrain, warmer temperatures dictate the need to increase the use of air conditioning, which negatively affects the operational range of the batteries.

In cold environments, unheated vehicle battery packs may show reduced performance, and the need for heating inside also negatively affects the range that can be operated between charging the vehicle. Cold weather, and changes in ambient temperature, may cause ice formation on surfaces and equipment, and removing liquid water (plumbing and seal design) is a very important design parameter in vehicles that will be operated in cold environments. Formation of ice or packed snow

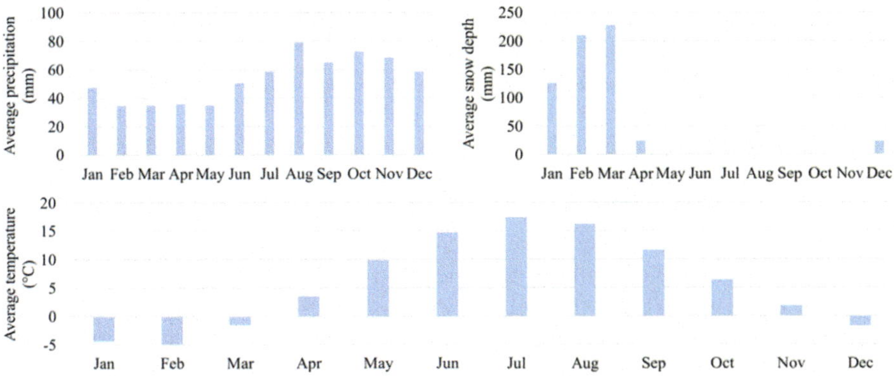

Fig. 1 Average precipitation, snow depth, and monthly temperature in Helsinki between 1961 and 2019. Data source: Ilmatieteenlaitos.fi

to the spatial awareness sensors needs to be considered in the vehicle (or operation process) design, as it might prevent the operation of the vehicle completely. Even door operation, floor heating, etc. need to be carefully studied in locations where packed snow or ice can be an issue.

Below-freezing temperatures on the road surface may cause the formation of an ice layer on top of the tarmac, yielding to lowered tyre-road friction. This is a normal phenomenon in areas, where temperatures regularly drop to freezing conditions. With regular cars and human drivers, the effects of this phenomenon are mitigated in various ways: driver training and experience, decreasing operational velocity (both by the driver voluntarily and via lowered top speed regulations), using specialized winter tyres (studded and/or high-friction rubber mixtures). These methods can be partly applied to robotic vehicles as they can use winter tyres, and they can be connected to a centralized meteorological station that can impose automatic speed limitations according to local weather conditions. The operation during low-friction conditions should take into account the vehicle speed, lowering the operational velocities. Low friction yields to increased stopping distances, as shown in (1), which generalizes the calculation for stopping distance as a factor of vehicle initial speed and friction coefficient between the tyre and the road.

$$d = \frac{v_o^2}{2\mu g} \tag{1}$$

where d (m) is the stopping distance, v_0 (m/s) is the initial speed, μ is the friction coefficient, and finally g (m/s^2) the gravitational acceleration constant. In consideration of this relevant issue, in [11] the authors study a machine learning model for road surface and friction estimation using cameras. Their results show that a neural network-based model leads to 94–99% classification accuracy for dry, wet/water, slush, and snow/ice conditions. During the design of automated vehicles that will operate under nonoptimal weather conditions, designers need to take the road friction issue into account, by lowering the operational speeds and/or increasing the safety margins of reaction when an obstacle is detected, increase safety distance between vehicles, and decrease maximum deceleration (to avoid hard braking).

In areas, where snowfall is significant, the road characteristics (i.e., lane measurements and roadside profile) will change due to packed snow. If robotic vehicles operate on a road from which the accumulated snow is removed by plowing, the snowbanks on the roadside could be interpreted as a modified environment or obstacle by the vehicles, causing the operation to slow down or stop, or causing the vehicle to lose localization accuracy due to changed environmental landmarks. Furthermore, the measurements of the road lane might change because of the snow, and this may cause all traffic to travel closer to the centerline of the road, limiting the lateral distance for the robot vehicle to operate.

3 Weather-Related Effects on Sensors

The main prerequirement to achieve weather-independent autonomous driving is a proper sensor input and dependable algorithm designed to target the operational domain. All system elements, hardware and software must be reliable, properly interfaced, and designed in a way that safety is always considered as the highest priority. However, both hardware and software have limitations particularly in case of adverse weather conditions and system design requirements put additional pressure to optimize certain criteria, quite often, for example, the cost. Finding an optimal solution in these conditions without compromising safety is a challenging task. Sensor weather-limitations can be roughly divided as follows:

Functional and Parametric Limitations All sensors have functional technical parameters that are limited in their nature, alike human sensing. For example, range sensors have a minimum and a maximum distance they are able to measure or maximum sampling frequency. All these parameters may be affected by weather conditions in a different way.

Operational Limitations Every piece of hardware is always affected by the environment they are operating. For example, sensors have minimum and maximum operational temperature, humidity, etc., which cannot be violated.

Reliability and Robustness Limitations Even the highest quality hardware is not 100% guaranteed in harsh weather conditions. Sensors' signals can interrupt during the operation, or in the worst case provide a high level of noise in the data signal. Furthermore, robustness is often affected by the design requirements, such as cost optimization, thus cost-effective sensors may be preferred resulting in reduction robustness and reliability.

Interface and Communication Limitations Interfaces between sensors and computers have limitation in terms of communication bandwidth and reliability that can also be triggered by weather conditions (e.g., too hot or humid climate can influence the performance of onboard computers; heavy wind can influence the mobile network connection between the vehicle and the operation center). If the communication channel is overloaded, sensor reading may not reach the control algorithm in a proper time resulting in a full system failure.

Software Limitations Software limitations are more weather-independent, that is, affected by the system logic and may be hidden. Eliminating or at least taking into account these limitations is harder than it looks as they can appear only in worst-case scenarios. Proper simulation models and big data training sets may help find out edge cases and algorithm limitations.

Due to hardware and software limitations, it is very hard to design and implement fault-tolerant systems and ensure that autonomous driving works in every weather condition and situation. At the current state, the most advanced autonomous driving archived is SAE level 4 [12], which means full autonomy in a limited operational domain. The operational domain defines the terms and conditions when the vehicle is considered reasonably safe and can operate in autonomous mode. Beside the operational domain for the vehicles also the infrastructure may require some change to allow autonomous driving on public road, the levels are better defined as infrastructure support of automated driving, aka ISAD [13]. Operational domains can be defined according to weather conditions, traffic conditions, geographical area, etc. As an example, a level 4 AV shuttle—operational domain definition is presented in Table 1. In principle, this applies to all popular shuttles available in the market, Navya, and Easymile, just to name a few, but also to more recent ones like ISEAUTO and Gacha. Easymile EZ10 and Navya shuttles are the most common autonomous shuttles, they are both manufactured in France, and they are used in a wide range of applications (including the pilots in Sohjoa Baltic). They are both equipped with standard sensors for perception and localization, and utilize ROS (Robot Operating System) as middleware and bus interface. ISEAUTO is a last-mile AV shuttle designed and manufactured in Tallinn, Estonia, in cooperation by TalTech and the university and AuVe Tech [14, 15]. ISEAUTO is equipped with a standard set of sensors for a low-speed AV—cameras, lidars, and radar supported by GNSS, IMU, and ultrasonic sensors. The autonomy is archived by open-source software stack Autoware.AI utilizing ROS. Figure 2 shows the sensors' placement and the visual shape of the vehicle.

Sensor and algorithm limitations must be properly taken into account to implement safe autonomous vehicles. It is important to define the operational domain and if operational domain limitations are violated, for example, in extreme weather conditions, autonomous functionality must be limited, in case of SAE level 4+ [12]. The vehicle may re-adapt the level of automation to SAE level 3 or 2, and, in extreme cases, it should stop the operation and switch to manual driving or taken over by remote control.

Autonomous vehicles are constantly localized in real time through a combination of satellites, for example, Global Navigation Satellite System (GNSS) [16, 17], and odometer sensors that provide data on wheels' velocity, also referred to as

Table 1 Example of an operational domain definition for a Level 4 AV shuttle

Operational domain	Limitations	Risks
Geographic location	Predefined route with pre-recorded and clean lidar maps are required	Due the weather and season changes the maps lose accuracy
Roadway type	Paved roads only without unknown objects and potholes	Unexpected changes of the road structure due harsh weather events or reconstruction works
Speed	Cruising speed up to 30 km/h	Due the higher speed limits, vehicle may lower traffic flow
Day/night	Direct sunlight towards cameras is not allowed Driving in nighttime only allowed in urban environment with public illumination	Direct sunlight toward camera can generate loss in object detection accuracy
Weather conditions	Up to moderate rain, fog, and snow	Rapid changes of weather conditions
Traffic conditions	Light traffic only	Interference with other vehicles and road users
Network conditions	Constant online connection is required	Overload of mobile network, malfunction of network service

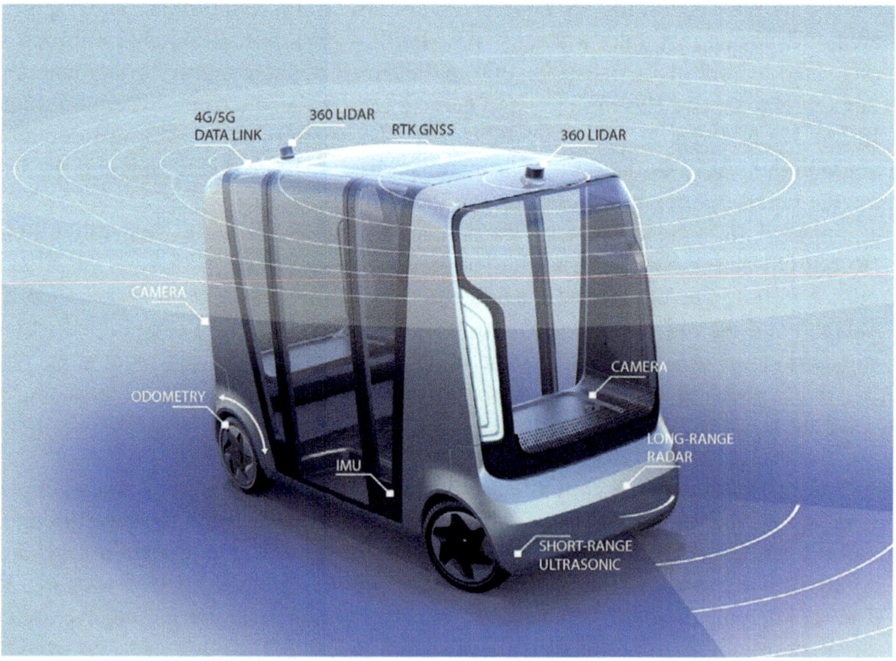

Fig. 2 Example of a last-mile shuttle with sensor setup, ISEAUTO

encoders; that is, mechanical motion sensors that generate digital signals in response to the motion. As satellite-localization requires a direct connection between a spot-on Earth and satellites, in the case of disrupted connection (e.g., being indoors or between tall urban buildings or extreme weather conditions), odometers help to estimate the position of the vehicle via a process commonly known as dead-reckoning [18]. Automated vehicles also need to report constantly on their exact pose and for this, GNSSs are integrated with Inertial Measurement Units (IMUs) that help to measure the vehicle's orientation using specific sensors (accelerometer, gyroscope, and magnetometer). The IMUs are typically located inside the vehicle and mounted on the chassis. They are typically not affected[1] by weather conditions. This combination helps to localize the vehicle also in tunnels and areas where satellite vision is not enough. In the case of low-speed self-driving vehicles, the localization is mostly calculated based on prerecorded point cloud map real-time distance measurements. These methods can also be combined to ensure more precise results and redundancy in case of loss of one sensor input.

The main sensors used to acquire information from the environment in autonomous driving are cameras, lidars, and radars. All these sensors have different working principles and work in different ranges in terms of illumination condition, temperature, and visibility, and they are, by far, the most affected by weather-related issues. Front and back cameras could be used to improve vehicles' global positioning, via map localization based on visual data, also referred to as visual simultaneous localization and mapping or visual-SLAM in short [19]. However, such techniques are under development in a very active research community. The real-time analysis of the surrounding environment is performed via cameras and 3D sensors with the main goal of detecting objects on the road rather than localization. The cameras are sensitive to weather conditions, for example, rain, snow, fog, and so on, and this effect has been already studied in several works. In [20], a method to benchmark image sensors' behavior in adverse weather, focusing explicitly on fog, is shown. The study provides measures for cameras tested into a fog chamber, highlighting performance at different visibility distance. Following the same line of evidence, in [21] two methods to detect fog in image cameras in night scenarios are presented. However, these methods are simply aimed at giving, to the vehicle or the driver, the information about the presence of fog, which can be used to regulate velocity or (in principle) improve perception. Working on the perception line of research, in [22] the authors propose a neural network-based algorithm to enhance visibility from cameras in poor weather with the final aim to detect and track vehicles on road, and they use simulation to show the performance under snowstorm conditions. However, monocular cameras are still tied to 2D images thus providing no information about distance.

With the final objective of having more reliable measures of the distance between the vehicle and any surrounding object in the environment, light detection and

[1]Very rare events such as magnetic storms, affecting the magnetometers inside IMUs, are not considered here.

ranging sensors, commonly known as LIDARs, are commonly used for autonomous vehicles. The working principle of this type of sensors is the measure of the time of flight of a light wave emitted and received back, providing an indirect measure of distance. LIDARs improve issues related to weather and illumination change, though subtle distortion may occur due to fog and rain, and particularly snow. Indeed, as other spatial awareness sensors, lidars are susceptible to weather effects producing opaque barriers for light to travel, resulting in a loss of intensity.

These barriers include, but are not limited to, liquid and solid water particles, in the form of ice crystals, fog and rain droplets, solid particles, such as dust, and other debris, e.g. fallen tree leaves. In these cases, the main spatial awareness sensors, the LIDARs will receive reflections from these barriers and most often vehicle systems will interpret these reflections as obstacles and initiate obstacle avoidance protocol, namely velocity reduction. Most of the current robotic bus manufacturers define operating conditions so that, during limited visibility, the autonomous operation is not permitted. Only little study has been done in research in this field, for example in [23] the authors show the effect of fog and rain on the energy attenuation of LIDARs emitted and received light, but only in simulation using synthetic data. Similar results are shown in [24], where the authors used synthetic data to predict the performance of LIDARs in the rain, underling its effect on driving assistance systems, showing a systematic reduction of object detection distance as a function of the rain rate. The same effect is not shown under snowy conditions, in this case, a snowflake hit by the laser provides an echo to the sensor resulting in false object detections [25]. This effect has been widely experienced during the pilot studies in the Sohjoa Baltic project.

A robust sensor widely used in many vehicles, for tasks such as adaptive cruise control, is the RADAR. It uses the measure of the time of flight of radio-frequency waves to build an indirect measure of distance [26]. Though based on the same principle, LIDARs and RADARs provide different accuracy as the higher frequency of light also has the advantage of having a smaller beamwidth of the wave (i.e., the opening angle from which most of the wave is emitted). The higher beamwidth of RADAR constitutes a strength in case of bad weather, making the wave able to pass through small objects [27].

By now, it should be evident that each sensor and technology has strengths and drawbacks, hence using a variety of different sensors, operating on different wavelengths in the electromagnetic spectrum, and using data combining algorithms, technique called sensor fusion would be a reasonable choice to increase the overall robustness of the vehicles [28]. Supporting this hypothesis, in [29] the authors show the performance of an algorithm that integrates information coming from LIDAR, RADAR, and camera to achieve all-weather object tracking and classification with over 80% accuracy on their benchmark. Furthermore, in [30] the authors deal safety-critical situations in automated vehicles resulting in a robust RADAR–LIDAR sensor fusion method.

3.1 Typical Testing Environments

Reliability in autonomous driving is considered a serious challenge to overcome, especially for public transportation that requires to work in any condition, including adverse weather. Weather-related issues are tested in the literature using the following methods: virtual simulation, indoor simulation, test tracks, and real environment.

In virtual simulation, it is possible to test the performance of algorithms on synthetic data. In this category, it is possible to count simulation on virtual data sets or simulated environment conditions on real data sets. For example, in [31] the authors use the data set cityscapes [32] to test the robustness of recognition algorithms by simulating fog on the real image data set. On the same line of research, in [5] the authors simulate fog on the images coming from the Kitti data set [33]. Rain is also commonly used in simulated environments to predict its effects on sensors, for instance, LIDARs [24] or cameras [34]. However, it must be noted that these techniques are not fully reliable, for example in [34], rain is simulated by simply adding vertical lines on camera images, while it is clear that the effects of rain on cameras are much more complicated than that, causing refraction and blurring on camera lenses that cannot be simply described as vertical lines.

An alternative way is to simulate weather conditions and analyze their influence on sensors by using climate chambers [35]. This means that specific laboratories must be built featuring special equipment to simulate temperature variations, rain, snow, fog, or dust. This is a reasonable compromise to have cost-effective realistic data, though not able to grab the motion effects of the vehicle. Supporting the same topic, [36] introduces a benchmark data set recorded in well-defined weather conditions in a climate chamber. In [27], the authors simulate dust (for example, during strong wind in the desert of Australia) to evaluate the performance of a radar. Whereas in [37], the authors simulate rain in a climate chamber to analyze the effect on a LIDAR and a camera positioned at the front of a vehicle. With a specific focus on LIDARs' performance or more in general on point cloud sensors, also including in the set the depth-sensing cameras, in [38] a testing methodology involving a climate chamber to validate fog and turbulent snow performance was shown, where the results clearly indicate a consistent drop of performance.

Only a little work has been done in real environments such as test tracks and public roads due to the intrinsic difficulties and costs of data acquisition. Furthermore, vehicles are typically manually driven to acquire data and test algorithms offline. One example is shown in [39], where the authors show the performance of a deep learning-based algorithm on a data set composed of 10,000 km driving in the northern countries in any weather condition. The data set used in [39] is not yet publicly available at the time of writing this chapter.

4 Intelligent Control

Human eyes have evolved in structure through the eons to improve the overall sensing capability, but this is not yet enough to justify our extraordinary ability to distinguish objects, interpret daily scenarios, and take reasonable decisions in any circumstance. Along with our eyes, neural connections and brain plasticity have also developed, resulting in a highly complex system capable of performing high-level perception. On the same line, autonomous vehicles are following two paths of evolution: the former is the hardware, sensors are becoming increasingly more accurate, whereas the latter is the software, performing an incredible amount of calculations, nearly real-time, enabling perception capabilities.

Back in the 90s, industry and academia realized that driving a vehicle was not only a matter of accelerating, braking, and steering (these operations can easily be done with high accuracy using classical control theory and feedback loops with simple controllers). The most challenging part concerns the decisional process. Given the desired vehicle speed and steering angle, a well-designed closed-loop control system is capable of following the desired behavior, but the real question is: which is the desired behavior?

The answer to this question involves the study of perception, cognition, and decisional processes. An example is given in Fig. 3, here it is possible to see different elements: what is indicated as "system" is, in fact, the vehicle that responds to the input u (the driver pushing the pedals or the steering wheel) with different velocity v and steering angle θ. Assuming that the vehicle is autonomous, there is a controller computing the control input to the vehicle according to the desired velocity v_{des} and steering angle θ_{des}. The controller simply reacts to the measure of the error between the desired velocity and the actual velocity (given from the measurements of the current output). Up to this point, the classical approach of control theory could provide a reasonable control law to drive the vehicle at a desired speed and steering angle.

Everything changes when one considers that the desired control input cannot be derived regardless of the driving scenario; the picture in the bottom part of Fig. 3, for example, shows a child following a ball. What will this child do in a few seconds? Will he stop? How should the vehicle react in such a situation? What is the quantity that should be measured? What is the relation between the speed of the vehicle and the approaching child?

Note that the figure is intentionally built to remind an external control loop acting on a subsystem. The external loop is dedicated to the high-level interpretation of images (perception), providing such information to a decisional process that calculates the desired control input for the subsystem. These two blocks, perception, and decisional process constitute the fundamental keys to build intelligent vehicles. Remaining on the analogy to the classical control systems, perception can be seen as the measurement process, whereas the decisional process can be seen as the controller.

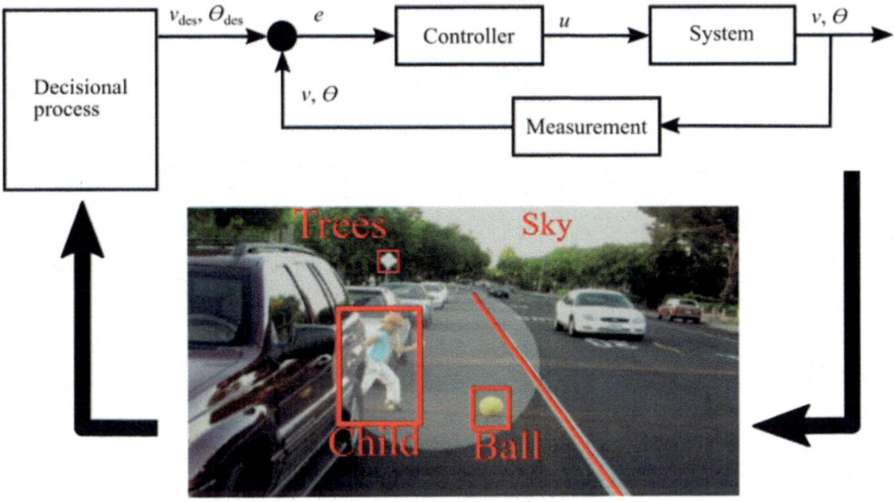

Fig. 3 Example of perception system in a feedback loop. Velocity and steering angle can be measured in an inner loop, while perception and environmental information must be processed at a higher level of the feedback loop where the decisional process takes place

Table 2 indicates the capabilities of different sensors in comparison with human eyes in specific tasks such as object detection, classification, and so on. Clearly, each sensor has strengths and drawbacks, including human eyes. The mainstream in research supports the idea that sensor fusion could be the best solution to combine information coming from different sources and reach high performances in any condition, despite weather, daylight, and other limitations [40, 41]. It is also assumed that, with proper selection of sensors and fusion algorithms, the potential perception capabilities may surpass any given single sensor performance. This aspect is very important to ensure that perception in self-driving vehicles can perform better than humans and, in turn, be much safer in any weather conditions, eventually leading to vision zero in terms of fatalities in traffic.

Assuming that the perception process provides reliable results, the decisional process could calculate the desired velocity and steering angle values for the vehicle. However, in the case of disturbances and uncertainties, like weather conditions or sensor failures, etc., the information content for the decisional process decreases leading to a high probability of global failures.

In the research community, there are two mainstream views to solve this problem: the modular approach and the end-to-end approach [42]. The former attempts to solve small problems in many separate modules that can be based on different intelligent control techniques, whereas the latter, a.k.a. *end2end*, considers the vehicle in its entirety as a black box providing a full driving model. Supporters of end2end learning propose to build a full backpropagation-based model having all sensor data as input and velocity/steering wheel angle as output. Whereas modular approaches aim at building a pipeline of individual blocks connected in a predefined

Table 2 Summary of performance for sensors in specific tasks

Performance aspect	Human	Automated Vehicle			
	Eyes	Radar	Lidar	Camera	Sensor fusion
Object detection	Good	Good	Good	Fair	Good
Object classification	Good	Poor	Fair	Good	Good
Distance estimation	Fair	Good	Good	Fair	Good
Edge detection	Good	Poor	Good	Good	Good
Lane tracking	Good	Poor	Poor	Good	Good
Visibility range	Good	Good	Fair	Fair	Good
Weather	Fair	Good	Fair	Poor	Good
Low illumination	Poor	Good	Good	Fair	Good

manner. It is clear that end2end learning can work in many circumstances providing a very high capacity of generalization and abstraction. The biggest concern with end2end is that they tend to oversimplify the problem and an error in the control system is unpredictable and very hard to detect in a testing stage, leading to untestable AI [43]. This is the main argument to support modularity in systems that can be better interpreted and debugged.

Please note that it is not in the objective of this chapter to provide theoretical knowledge about AI and data-driven tools for control and perception, for which the reader can refer to [44], containing a thorough explanation of many data-driven techniques for control, and [45] for a theoretical background about deep learning.

5 Data Hunger

Along with the chapter, the shift from the classical approach for vehicle control to the machine learning-based approach was mentioned; the main reason is that ML is expected to solve the problem of perception in autonomous driving using data as the main driving force. As past applications of ML shown, the more the data the better the solution that any AI-based system can provide. The result is that research and industry require massive quantities of data to work with, generating a race toward big data acquisition that has seen a continuous increment of both open and privately owned data sets.

Nowadays, data are considered a valuable asset, generating massive investments though not yet enough to feed the data hunger. Indeed, the real question is: how much data should autonomous vehicles collect to generate a reasonable driving model? Currently, Google (with its subsidiary Waymo) has a fleet composed of roughly 55 vehicles tested for over one million kilometers per year, corresponding roughly to 30,000 h of driving, which is more or less what one taxi driver does in his/her entire work life. Such data cover most of the common scenarios, different illumination conditions, and weather, but still not enough to be considered safe [3]. The reason why autonomous driving is not considered safe yet is to be found by analyzing the driving statistics, which, for most of the situations, involves previously seen and predictable scenarios. However, unpredictable events, though part of the real driving scenarios, hence probable, cannot be considered as outliers, as they can generate catastrophic events. This concept is known in economics as "the black swan,", but often neglected in AI systems, though fundamental to reach a high level of safety. The black swan is an example of an event that can occur with low probability, thus part of the distribution, and with major effects on the system. Swans are white, should a black one still be considered as a swan? For an intelligent system to recognize unpredictable events effectively, it is necessary to acquire as much information as possible regarding the occurrence of such events that for autonomous driving correspond to safety loss. To answer the initial question, we do need more data describing unpredictable events and variability, but many hours of driving are required to find a black swan.

Table 3 provides a review of the most common data sets publicly available today with relative literature references. The items in the list have been categorized by year of acquisition (or publication), sensors available, illumination, and weather scenarios. In [46], a table listing data set for autonomous driving from 2001 to 2007 is available as a sign of the activities of the last 20 years in the field; those data sets have been omitted in Table 3.

The first important evidence from Table 3 is that the sensor configuration is not consistent, only some of the data set provides visual information, depth information, and geo-localization. This means that this data can hardly be integrated into each other to abstract more knowledge for the learning procedures of the algorithms. Not all of them contain scenes in daylight and nightlight, resulting in a lack of generalization of illumination during the driving data. As expected, most of the data sets contain sunny scenes (typically including also cloudy scenarios), but not many include rain, fog, or snow, and only a few of them contain all those scenarios. Exploring the data sets in detail, it is also possible to note inconsistencies in scenes labeling, which results in poor algorithmic performance.

However, it is important to emphasize that all these data sets have been acquired with a big effort from providers. This must be acknowledged as a relevant result without which the most recent results in perception would not have been achieved. Besides the technical issues to overcome, also legal restrictions are preventing big data sets to be recorded and used. For this reason, a recent trend of research is investigating the use of virtual data sets (see Table 4), which are cost-effective and prevent any legal issue related to driving autonomous vehicles in the urban environment. This trend started in the last 5–10 years with the objective to improve consistency and generality in the data, and the trend keeps growing. In virtual data sets, one can find all scenarios, including rain, fog, and snow, generated by simulating cars driving in realistic game-like environments. It has been proven that ML algorithms are able to generalize fairly well on this data, but they are less performant real scenarios.

Table 3 List of open data sets available

Data set			Sensors			Weather			
	Year	Day/night	Camera	Lidar	Geo location	Sunny	Rain	Fog	Snow
VidCam [46]	2008	D	y	n	n	y	n	n	n
Ford Campus Vision and Lidar [47]	2009	D	y	y	y	y	n	n	n
Kitti [33]	2012	D	y	y	y	y	n	n	n
Malaga urban data set [48]	2013	D	y	y	y	y	n	n	n
EISATS [49]	2014	D/N	y	n	y	y	y	y	y
KAIST multispectral [50]	2015	D/N	y	n	y	y	n	n	n
Cityscape [32]	2015	D	y	n	y	y	n	n	n
Udacity	2016	D	y	y	y	y	n	n	n
Vistas [51]	2017	D/N	y	n	y	y	y	n	n
Oxford robotcar [52]	2017	D/N	y	y	y	y	y	n	y
BDD100k [53]	2018	D/N	y	n	y	y	y	n	n
Apolloscape [54]	2018	D	y	y	y	y	y	y	n
NuScenes [55]	2018	D/N	y	y	y	y	y	n	n
A*-3D [56]	2019	D/N	y	y	y	y	y	n	n
D2city [57]	2019	D	y	n	y	y	y	n	n
A2D2 [58]	2019	D	y	y	y	y	n	n	n
KAIST Urban [59]	2019	D	y	y	y	y	n	n	n
Waymo	2019	D/N	y	y	y	y	y	n	n
Unsupervised Llamas [60]	2019	D	y	y	y	y	n	n	n

6 Experience from the Pilot Studies

The pilots in Sohjoa Baltic have provided first-hand experiences for thousands of passengers, most of them being introduced to automated vehicle transportation for the first time. To ensure safety onboard, an operator was always on board to be able to take control of the vehicle in case of an emergency. The passengers have taken part in an anonymous feedback survey, and these results indicate that traveling on a small, self-driving electric shuttle is a positive experience. It is interesting to note that roughly 80% of passengers answering the question "would you use the service with no operator on board?" were willing to consider the idea. Precisely, 35% answered "yes, definitely" and 44% answered "yes, but not now," demonstrating that all these pilot projects help to improve technology and to build the right environment for communities to accept innovation. However, during these

Table 4 List of open virtually generated data sets available

			Sensors				Weather		
Data set	Year	Day/night	Camera	Lidar	Geo location	Sunny	Rain	Fog	Snow
Synthia [61]	2016	D/N	y	y	y	y	y	y	y
P.F.B [62]	2017	D/N	y	y	y	y	y	y	y
Virtual Kitti [63]	2017	D/N	y	y	y	y	y	y	n
Virtual Kitti 2 [64]	2020	D/N	y	y	y	y	y	y	n
Marulan [65]	2010	D/N	y	y	y	y	y	y	y
PixelAccurate DepthBenchmark [66]	2019	D/N	y	y	y	y	y	y	n
CARLA [67]	2017	D/N	y	y	y	y	y	y	y

pilots, a responsible and qualified operator has been always onboard explaining the technology answering passengers' questions regarding the functions of the vehicle. This may have some impact on the passengers' feedback.

Three big pilots have been implemented along with the Sohjoa Baltic project and here the experience will be shared with specific focus to the city of Helsinki in Finland, Tallinn in Estonia, and Kongsberg in Norway, the route for each pilot is shown in Fig. 4. Although these pilots are not enough to fully demonstrate safety in autonomous driving in the urban environment for public transportation purposes, the pilots surely contribute to the development of the technology, build trust in future users of the public transportation and helps to identify possible causes of failures that can be debugged by manufacturers.

Three small pilots are also planned into the project, but only one of them already took place. The small pilot in Gdansk (Poland) was active for a month, from September 6th to October 4th, 2019, on the bus line 322 going to the city zoo. The automated bus was active 7 days a week and 5 h per day free of charge for passengers and transported over 3300 travelers during the operation. As indicated in Table 5, the length of the path was 1.8 km for a round trip, going back and forth between the two different points shown in the map in Fig. 4d. In this case, differently from the others, the bus was going back and forth on the same route, while in Helsinki and Tallinn the route was designed as a closed loop. The maximum speed for all the pilots was limited to 15 km/h, the average speed during the pilot in Gdánsk was 8.22 km/h, and the bus had three stops during the path.

6.1 Observation from Helsinki Trials

The Sohjoa Baltic pilot in Finland took place in Helsinki, more specifically at the Aurinkolahti residential area. The pilot route went from Vuosaari metro station

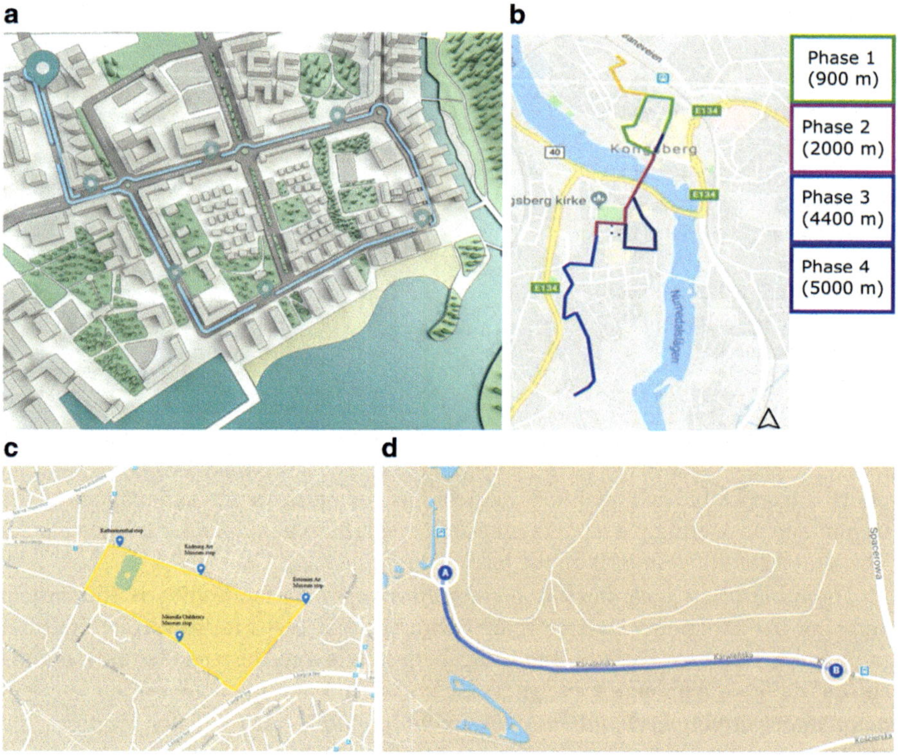

Fig. 4 Depiction of the routes of the buses during the pilots in Helsinki (**a**), Kongsberg (**b**), Tallinn (**c**), and Gdansk (**d**). Path lengths and other parameters are summarized in Table 5

Table 5 Summary of pilots' characteristics in terms of path length, average speed, travel time, and number of stops. For Kongsberg 5000 m refers to phase 4 (see Fig. 4)

Pilot	Path length [m]	Average speed [km/h]	Travel time [min]	Number of stops
Helsinki	2500	7.5	20	7
Tallinn	1700	7	15	4
Kongsberg (p4)	5000	8	37	8
Gdánsk	1800	8.22	14.5	3–5

(eastern terminus) to the Aurinkolahti beach and back. The scheduled duration of the operation was 4 months, from June 2019 to September 2019 (included). During the pilot, the robotic bus drove 2596 km (automatic + manual) with a total of 3932 passengers. In Fig. 5, the driving distance and the number of passengers per day is shown, the chart also shows the rainy days denoted by using blue areas. The pilot was stopped earlier than expected because the shuttle faced a battery-related issue. The vehicle used on the site is manufactured by the French company Navya and was both leased from and operated by the Danish company Holo with a partnership to Metropolia University of Applied Sciences.

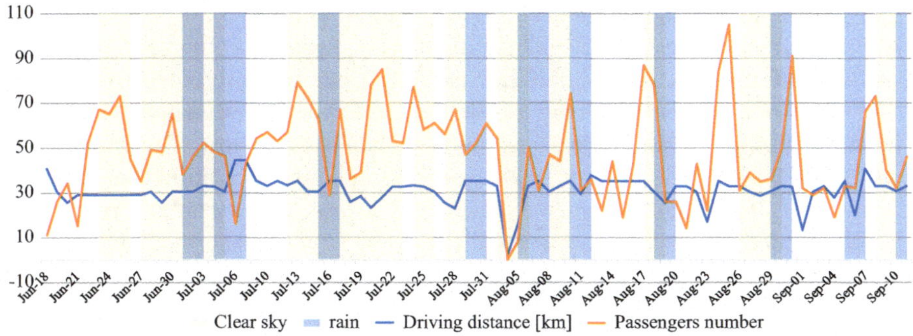

Fig. 5 The chart shows the distance driven by the vehicle per day (blue line) and the number of passengers (orange line). Blue areas denote the rainy days during the pilot, whereas the light-orange areas are the full sunny days

Sometimes the operator was obligated to switch to from automatic mode to manual mode for different reasons, for example, leaves falling from trees have been wrongly detected as obstacles causing many hard brakes that can endanger passengers. The reduction of false object detections (also known as false positive) is an important direction for research as the robustness of the algorithms is not yet sufficient for an effective drive. Frequent false detections oblige the operator to switch to manual mode, but on the other hand, a missing detection would generate a crash, and for this reason, the manufacturer still preserves conservative policies to ensure safety for passengers. Heavy rain and strong wind also affected the operation causing interruptions or switch to manual mode.

During the pilots, a base station to improve GPS localization was used. The base station was installed on the roof of the highest building on the route. The station had to be rebooted manually a few times during the pilot months, this was not always easy or even impossible as a thunderstorm prevented operators to restart the base station.

One of the main issues that unfortunately cannot be truly quantified is related to localization and mapping. In literature, there are no effective and fully working methods to update the map and perform localization in real-time during the bus operation. The result is that the map is firstly recorded before the pilot, the autonomous bus then drives always on the same map. Growing vegetation, snowbanks, and building renovation generates slight changes in the original map, making the localization capability of the vehicle to decrease drastically.

6.2 Experience from Kongsberg

The pilot in Kongsberg took place from October 2018 to June 2019, transporting 2.064 passengers. The vehicle used during the pilot was provided by Easymile,

EZ10 Generation 2, and organized in cooperation with the local private company AppliedAutonomy and the Municipality of Kongsberg. Defining the pilot route was one of the first tasks done and part of a phase 1 preproject in the form of a workshop. Several alternatives were considered and evaluated with regards to safety, usefulness, visibility, feasibility, scalability, and cost. The chosen route was extended in several steps, both to ensure security and fulfill the legislative requirement for gradual implementation and to assess the public's response at the same time. A new risk assessment had to be submitted to the national traffic authority for each stage of the route scaling. See Fig. 4b for the phases of the route scaling.

In October 2018, the phase one started, and a bus was put into operation on a short route (900 m) to be considered as an initial test. At the end of October 2018, the route was extended into phase two, now the path length was 2 km (including the route of phase 1). In April 2019, the route was extended further in phase three, now reaching from one end of town to the other (4.4 km, including the routes of phases one and two). The pilot was in operation until June 2019.

A challenge in the route planning arose when road works had to be executed on part of the pilot route. As a result, the route was adapted (as increased in length from 4.4 km to 5 km) and a new risk analysis had to be conducted.

The vehicle operators were employees of Vy (the national Norwegian railway company). Each operator must have a valid driving license and be trained by the vehicle provider and on the pilot track. A total of six vehicle operators and employees of Vy were trained and certified by Applied Autonomy. The operators had extensive experience with public transport, which proved to be useful as it helped with the evaluation of traffic patterns, quick learning the vehicle behaviors, and good interaction with passengers. The operators also reported any anomalies of the vehicle and the route and made proposals for improvements.

6.3 *Tallinn Pilot Study*

The first long-term open traffic robotic bus pilot in Estonia started its operations on August 28, 2019, in Kadriorg Park, which is located right next to the Presidential Palace and Kadriorg Palace. The operations were paused December 21, 2019, due to winter conditions (overnight charging too complicated when outside weather below 5 degrees Celsius). The bus is manufactured by the French company Navya and was delivered by the Danish company Holo who is the contract partner to Tallinn Transport Department. The operators are students from Tallinn University of Technology who passed the 2-week training organized by Holo. Before the start of operations, the bus had to pass an exam organized by the Estonian Road Administration to ensure the safety of the bus and its capability to drive in open traffic.

The bus drives in a circle around the park and takes the passengers to the Estonian Art Museum. The bus runs regularly from Tuesday to Sunday between 10.00 and

6.00 (till 18.00 on Thursday, Saturday, and Sunday) and carries passengers free of charge. The bus seats eight passengers at a time. The bus drives around the Kadriorg Park and has four stops: Katharinenthal cafeteria located close to the Kadriorg tram stop, Kadriorg Art Museum, Estonian Art Museum, and Miiamilla Children's Museum (Fig. 4c). All passengers can download the Letsholo app from Google Play or Apple App Store to see the real-time location of the bus. Approximately 100 people were using the service during the operational days, although there have been several issues that have influenced the stability of operation.

6.3.1 Weather Conditions and Issues of the Pilot

This section will give an overview of how weather conditions affected the pilot in Tallinn. The section is largely based on the chat log with operators. The chat was used for everyday communications between the operators and responsible personnel from Tallinn Transport Department and Tallinn University of Technology. Atlas.ti software was used to filter out issues related to weather issues. Code words such as "leaves," "rain" ("vihm"), "temperature" ("temperatuur"), "snow" ("lumi"), "battery" ("aku"), "tree branches" ("puuoksad"), "wind" ("tuul"), "snow" ("lumi"), and "slush" ("lörts") were used to find relevant discussions. Relevant parts were later marked as quotes, and the code words were attached to them.

First, operators were surprised that the autonomous bus had so many issues, and other sites of Holo operations in Norway (Oslo) and Finland (Helsinki) had similar issues with constant downtime. The most common issues were related to technology, traffic, and weather.

Technology The main reason for several technology-related errors was that the technical support was managed from the distance (from Denmark and/or France). For example, there was an issue with doors not working properly, and it took one operator 4 days to understand how to open and close these doors. In addition, the distant problem-detection decision tree made an assumption that most challenges are related to issues with the software, although the problem with doors was actually a mechanical one.

Traffic The operation was ceased for weeks due to one traffic accident with a heavy-good-vehicle that ignored the automated bus and hit it at a slow speed. In addition, every-day operations were influenced by cars parked on the road (often illegally), cars driving in the opposite direction (illegally). In addition, the bus does not understand the concept of congestion, it just starts beeping once there are cars waiting at the traffic junction.

Weather All operators agreed that the weather had an impact on the operation. During the operations from the end of August till December 21, the main issues linked to weather conditions were related to precipitation, temperature, and seasonal changes.

All the weather conditions were also related to seasonal changes. Rain, falling, and already fallen leaves were the main weather-related issues in September and October. While falling and already fallen leaves made the bus to have the emergency stop 10–15 times per circle, the combination of leaves with heavy rain made it impossible for a bus to drive smoothly. It must be pointed out that precipitation in autumn 2019 was above the norm of previous years: September 115% of the norm, October 144% of the norm, November 85% of the norm (no operations in November because of technical issues). Such changes in weather need to be taken into consideration in the development of sensor technology.

In December, temperature started to become an issue because of two reasons. When the bus was not operating, it was stored in the outside tent located at the parking lot of the Estonian Art Museum as there was no warm garage in the vicinity that could be used. After each day of operation, the bus was left to the tent with its battery charging for the next morning. Battery charging issues started when the outside temperature fell below +5 degrees Celsius as it did not charge properly. The cold temperature made it also necessary to turn on the air conditioning. This decreased the daily operating hours because of the increased power consumption.

7 Conclusions

This chapter covered several aspects related to autonomous driving in a real-urban environment with a specific focus on weather-related issues. Clearly, going out from the testing scenarios introduces a high number of challenges to overcome and unpredictable effects. Due to the hardware and software limitations, it is very hard to design and implement fault-tolerant systems and to ensure that autonomous driving works in every weather condition and situation. At the current state, the most advanced autonomous driving archived is SAE level 4, which means full autonomy in a limited operational domain. The operational domain defines the terms and conditions when the vehicle is considered reasonably safe and can operate in autonomous mode. An operational domain can be defined according to weather conditions, traffic conditions, geographical area, etc. However, it must be said that the automated shuttles used in this studied have shown autonomous driving capabilities of SAE level 3, too many times the operator had to take over the vehicle in case of emergency.

The experience gathered from real-world piloting of automated buses for public transportation is a valuable achievement, highlighting how many practical issues may occur during a pilot. The main challenge faced during the pilots of Sohjoa Baltic is the technical immaturity of the robotic buses. Their reliability is not yet on the level that operators would like it to be. The pilots have been done with a relatively small budget, which in many cases does not allow operators to have spare vehicles in case of failures. Therefore, the operation has had many cancellations because of technical issues with the only bus available on the site.

Fig. 6 Roadmap of predicted future evolution of the usage of autonomous driving in public areas

From the authors' perspective, all these challenges will be overcome in the next future, but further testing and pilots are still required to identify possible causes of danger and make the vehicles safer. The authors' vision is summarized in the roadmap presented in Fig. 6, foreseeing further pilot projects at low speed for the next 5 years. It will probably be possible to see an autonomous bus in continuous operation, though with many limitations only in 5–10 years from now. This may seem a pessimistic vision, though a realistic one. The full automation in mixed traffic for rural and urban areas will be seen in roughly 15 years from now, this will be the real integration in the mobility as a service (MaaS) solution. Only when the service will be fully integrated with the public service, the real strengths of autonomous public transportation will be fully exploited.

A fundamental aspect to be further investigated in research and product development is the localization and mapping for autonomous vehicles, as there is no well-established method to update the map during the operation. The current methodology is to record one single map at the very beginning of the pilot, and never change it to allow the robot to localize in it. However, the environment is in a continuous change, trees grow, leaves fall in the autumn, snowbanks may be formed, road pavement changes, buildings get older or renovated, resulting in a lack of localization capability. Employing highly skilled engineers to constantly record and update maps is not a cost-effective solution to the problem.

One of the major challenges in the Helsinki pilot was related to incorrectly parked vehicles and drivers parking on bus stops. This reveals two additional major issues a technical issue and a social issue. The technical issue is that the robotic bus cannot adapt to humans' behavior, making the interaction in a shared environment very complex. The social issue is related to the awareness of road users that they share the environment with autonomous machines, even though simple solutions like leaflets informing about the pilot, studies shared on the windscreens of incorrectly parked vehicles reduced the problem. This is assumed to be a change in the behavior of people frequently visiting the area with their vehicles. This raises the most interesting question: Are we, as humans, ready to see robotic vehicles in our cities?

Acknowledgments The financial support of EU-Interreg through the Sohjoa Baltic project, Grant Number R073, is gratefully acknowledged.

References

1. Federal Highway Administration, *Road Weather Connected Vehicle Applications – Interim Report* – January 11, 2013 FHWA-JPO-14-124
2. European Road Safety Observatory "Annual Accident Report 2018"
3. N. Kalra, S.M. Paddock, Driving to safety: How many miles of driving would it take to demonstrate autonomous vehicle reliability? Transp. Res. A Policy Pract. **94**, 182–193 (2016)
4. A. Lobo, S. Ferreira, I. Iglesias, A. Couto, Urban road crashes and weather conditions: Untangling the effects. Sustainability **11**(11), 3176 (2019)
5. Z. Liu, Y. He, C. Wang, R. Song, Analysis of the influence of foggy weather environment on the detection effect of machine vision obstacles. Sensors **20**(2), 349 (2020)
6. S. Zang, M. Ding, D. Smith, P. Tyler, T. Rakotoarivelo, M.A. Kaafar, The impact of adverse weather conditions on autonomous vehicles: How rain, snow, fog, and hail affect the performance of a self-driving car. IEEE Veh. Technol. Mag. **14**(2), 103–111 (2019)
7. J. Kim, J. Canny, Interpretable Learning for Self-Driving Cars by Visualizing Causal Attention, in *2017 IEEE International Conference on Computer Vision (ICCV), Venice*, (2017), pp. 2961–2969. https://doi.org/10.1109/ICCV.2017.320
8. S.1108–116th Congress (2019–2020), Algorithmic Accountability Act of 2019, https://www.congress.gov/bill/116thcongress/senate-bill/1108
9. B. Goodman, S. Flaxman, European union regulations on algorithmic decision-making and a "right to explanation". AI Mag. **38**, 50. arXiv:1606.08813 (2016)
10. J. Ainsalu, V. Arffman, M. Bellone, M. Ellner, Haapamäki, et al., State of the art of automated buses. Sustainability **10**(9), 3118 (2018)
11. S. Roychowdhury, M. Zhao, A. Wallin, N. Ohlsson, M. Jonasson, Machine learning models for road surface and friction estimation using front-camera images, in *2018 International Joint Conference on Neural Networks (IJCNN)*, (IEEE, Piscataway, 2018, July), pp. 1–8
12. SAE Automotive, *Taxonomy and Definitions for Terms Related to Driving Automation Systems for on-Road Motor Vehicles*, Report J3016_201609 (SAE Automotive, Warrendale, 2016)
13. A. Carreras, X. Daura, J. Erhart, S. Ruehrup, Road Infrastructure Support Levels for Automated Driving. In proceedings of the 25th ITS world Congress, Copenhagen, Denmark, International Transport Forum, (2018, September), pp. 17–21
14. A. Rassõlkin, R. Sell, M. Leier, Development case study of the first Estonian self-driving car, ISEAUTO. Electr. Control. Commun. Eng. **14**(1), 81–88 (2018)
15. R. Sell, M. Leier, A. Rassõlkin, J.P. Ernits, Self-driving car ISEAUTO for research and education, in *2018 19th International Conference on Research and Education in Mechatronics (REM)*, (IEEE, Piscataway, 2018, June), pp. 111–116
16. B. Hofmann-Wellenhof, H. Lichtenegger, J. Collins, *Global Positioning System: Theory and Practice* (Springer, Berlin, 1997)
17. B.L. Decker, *World Geodetic System 1984*, Technical Report, DTIC Document (Defense Mapping Agency Aerospace Center, St Louis, 1986)
18. K. Park, H. Chung, J. Lee, Dead Reckoning Navigation for Autonomous Mobile Robots, In proceedings of intelligent autonomous vehicle, Madrid, Spain, March 1998, Elsevier, (1998), pp. 775–781
19. C. Chen, H. Zhu, M. Li, S. You, A review of visual-inertial simultaneous localization and mapping from filtering-based and optimization-based perspectives. Robotics **7**(3), 45 (2018)
20. M. Bijelic, T. Gruber, W. Ritter, Benchmarking image sensors under adverse weather conditions for autonomous driving. IEEE Intelligent Vehicles Symposium (IV), Changshu, China, (2018, June.). pp. 1773–1779. IEEE
21. R. Gallen, A. Cord, N. Hautière, D. Aubert, Towards Night Fog Detection through Use of in-Vehicle Multipurpose Cameras. IEEE intelligent vehicles symposium (IV), Baden-Baden, Germany, (2011, June). pp. 399–404. IEEE

22. M.S. Alluhaidan, I. Abdel-Qader, Visibility enhancement in poor weather-tracking of vehicles, in *Proceedings of the International Conference on Scientific Computing (CSC),* pp. 183–188. The Steering Committee of The World Congress in Computer Science, Computer Engineering and Applied Computing (WorldComp) (2018)
23. M. Hadj-Bachir, De Souza P., LIDAR sensor simulation in adverse weather condition for driving assistance development. HAL Id : hal-01998668. (2019)
24. C. Goodin, D. Carruth, M. Doude, C. Hudson, Predicting the influence of rain on LIDAR in ADAS. Electronics **8**(1), 89 (2019)
25. S. Michaud, J. Lalonde, P. Giguère, in IEEE International Conference on Intelligent Robots and Systems, in *Towards Characterizing the Behavior of LiDARs in Snowy Conditions*, (Hamburg, Germany, 2015)
26. I. Bilik, O. Longman, S. Villeval, J. Tabrikian, The rise of radar for autonomous vehicles: Signal processing solutions and future research directions. IEEE Signal Process. Mag. **36**(5), 20–31 (2019)
27. T. Peynot, J. Underwood, S. Scheding, Towards reliable perception for unmanned ground vehicles in challenging conditions, in *2009 IEEE/RSJ International Conference on Intelligent Robots and Systems*, (2009, October), pp. 1170–1176
28. B. Schoettle, Sensor fusion: A comparison of sensing capabilities of human drivers and highly automated vehicles. Sustain. Worldw. Transp. **12**, 1–42 (2017)
29. P. Radecki, M. Campbell, K. Matzen, All weather perception: Joint data association, tracking, and classification for autonomous ground vehicles. arXiv preprint arXiv:1605.02196 (2016)
30. N. Druml, O. Veledar, G. Macher, G. Stettinger, S. Selim, J. Reckenzaun, S.E. Diaz, M. Marcano, J. Villagra, R. Beekelaar, J. Jany-Luig, PRYSTINE-technical progress after year 1, in *2019 22nd Euromicro Conference on Digital System Design (DSD)*, (IEEE, Piscataway, 2019, August), pp. 389–398
31. C. Sakaridis, D. Dai, L. Van Gool, Semantic foggy scene understanding with synthetic data. Int. J. Comput. Vis. **126**(9), 973–992 (2018)
32. M. Cordts, M. Omran, S. Ramos, T. Rehfeld, M. Enzweiler, R. Benenson, U. Franke, S. Roth, B. Schiele, in Proceedings of the IEEE Conference on Computer Vision and Pattern Recognition (CVPR), in *The Cityscapes Dataset for Semantic Urban Scene Understanding*, (Las Vegas, Nevada, United States, 2016)
33. A. Geiger, P. Lenz, C. Stiller, R. Urtasun, Vision meets robotics: The KITTI dataset. Int. J. Robot. Res. **32**(11), 1231–1237 (2013)
34. M. Zhang, Y. Zhang, L. Zhang, C. Liu, S. Khurshid, DeepRoad: GAN-based metamorphic testing and input validation framework for autonomous driving systems, in *proceedings of the 33rd ACM/IEEE international conference on automated software engineering*, (Montpellier, France, 2018, September), pp. 132–142
35. R. Heinzler, P. Schindler, J. Seekircher, W. Ritter, W. Stork, Weather influence and classification with automotive lidar sensors. arXiv preprint arXiv:1906.07675 (2019)
36. M. Bijelic, P. Kysela, T. Gruber, W. Ritter, K. Dietmayer, Recovering the unseen: Benchmarking the generalization of enhancement methods to real world data in heavy fog, in *Proceedings of the IEEE Conference on Computer Vision and Pattern Recognition Workshops*, (2019), pp. 11–21
37. S. Hasirlioglu, A. Riener, Challenges in object detection under rainy weather conditions, in *First International Conference on Intelligent Transport Systems*, (Springer, Cham, 2018, November), pp. 53–65
38. M. Jokela, M. Kutila, P. Pyykönen, Testing and validation of automotive point-cloud sensors in adverse weather conditions. Appl. Sci. **9**(11), 2341 (2019)
39. M. Bijelic, F. Mannan, T. Gruber, W. Ritter, K. Dietmayer, F. Heide, Seeing through fog without seeing fog: Deep sensor fusion in the absence of labeled training data. arXiv preprint arXiv:1902.08913 (2019)
40. M. Bellone, G. Reina, L. Caltagirone, M. Wahde, Learning traversability from point clouds in challenging scenarios. IEEE Trans. Intell. Transp. Syst. **19**(1), 296–305 (2017)

41. L. Caltagirone, M. Bellone, L. Svensson, M. Wahde, LIDAR–camera fusion for road detection using fully convolutional neural networks. Robot. Auton. Syst. **111**, 125–131 (2019)
42. S. Grigorescu, B. Trasnea, T. Cocias, G. Macesanu, A survey of deep learning techniques for autonomous driving. J. Field Rob. **37**, 362 (2019)
43. S. Segura, D. Towey, Z.Q. Zhou, T.Y. Chen, Metamorphic testing: Testing the untestable. IEEE Softw. **37**, 1 (2018)
44. S.L. Brunton, J.N. Kutz, *Data-Driven Science and Engineering: Machine Learning, Dynamical Systems, and Control* (Cambridge University Press, Cambridge, 2019)
45. I. Goodfellow, Y. Bengio, A. Courville, *Deep Learning* (MIT press, Cambridge, MA, 2016)
46. G.J. Brostow, J. Fauqueur, R. Cipolla, Semantic object classes in video: A high-definition ground truth database. Pattern Recogn. Lett. **30**(2), 88–97 (2009)
47. G. Pandey, J.R. McBride, R.M. Eustice, Ford campus vision and lidar data set. Int. J. Robot. Res. **30**(13), 1543–1552 (2011)
48. J.L. Blanco-Claraco, F.Á. Moreno-Dueñas, J. González-Jiménez, The Málaga urban dataset: High-rate stereo and LiDAR in a realistic urban scenario. Int. J. Robot. Res. **33**(2), 207–214 (2014)
49. M. Rezaei, M. Terauchi, Vehicle Detection Based on Multi-Feature Clues and Dempster-Shafer Fusion Theory, in *Pacific-Rim Symposium on Image and Video Technology*, (Springer, Berlin/Heidelberg, 2013, October), pp. 60–72
50. S. Hwang, J. Park, N. Kim, Y. Choi, I. So Kweon, Multispectral Pedestrian Detection: Benchmark Dataset and Baseline, in *Proceedings of the IEEE Conference on Computer Vision and Pattern Recognition*, (Boston, MA, USA, 2015), pp. 1037–1045
51. G. Neuhold, T. Ollmann, S.R. Bulo, P. Kontschieder, The Mapillary Vistas Dataset for Semantic Understanding of Street Scenes, in *International Conference on Computer Vision (ICCV)*, (Venice, Italy, 2017), pp. 5000–5009
52. W. Maddern, G. Pascoe, C. Linegar, P. Newman, 1 year, 1000 km: The Oxford RobotCar dataset. Int. J. Robot. Res. **36**(1), 3–15 (2017)
53. F. Yu, W. Xian, Y. Chen, F. Liu, M. Liao, V. Madhavan, T. Darrell, BDD100K: A diverse driving video database with scalable annotation tooling. arXiv:1805.04687 (2018)
54. P. Wang, X. Huang, X. Cheng, D. Zhou, Q. Geng, R. Yang, The apolloscape open dataset for autonomous driving and its application. IEEE Trans. Pattern Anal. Mach. Intell. **42**, 2702 (2019)
55. H. Caesar, V. Bankiti, A.H. Lang, S. Vora, V.E. Liong, Q. Xu, A. Krishnan, Y. Pan, G. Baldan, O. Beijbom, nuScenes: A multimodal dataset for autonomous driving. arXiv preprint arXiv:1903.11027 (2019)
56. Q.-H. Pham, P. Sevestre, R.S. Pahwa, H. Zhan, C.H. Pang, Y. Chen, A. Mustafa, V. Chandrasekhar, J. Lin, A*3D dataset: Towards autonomous driving in challenging environments. arXiv:1909.07541 (2019)
57. Z. Che, G. Li, T. Li, B. Jiang, X. Shi, X. Zhang, Y. Lu, G. Wu, Y. Liu, J. Ye, D2-City: A large-scale dashcam video dataset of diverse traffic scenarios. arXiv:1904.01975 (2019)
58. J. Geyer, Y. Kassahun, M. Mahmudi, X. Ricou, R. Durgesh, A.S. Chung, L. Hauswald, V.H. Pham, M. Mhlegg, S. Dorn, T. Fernandez, A2d2: Audi autonomous driving dataset. arXiv preprint arXiv:2004.06320 (2020)
59. J. Jeong, Y. Cho, Y.S. Shin, H. Roh, A. Kim, Complex urban dataset with multi-level sensors from highly diverse urban environments. Int. J. Robot. Res. **38**(6), 642–657 (2019)
60. K. Behrendt, R. Soussan, Unsupervised Labeled Lane Markers Using Maps, in *Proceedings of the IEEE International Conference on Computer Vision Workshops*, (Seoul, Korea, 2019)
61. G. Ros, L. Sellart, J. Materzynska, D. Vazquez, A.M. Lopez, The SYNTHIA Dataset: A Large Collection of Synthetic Images for Semantic Segmentation of Urban Scenes, in *Proceedings of the IEEE Conference on Computer Vision and Pattern Recognition*, (Las Vegas, Nevada, United States, 2016), pp. 3234–3243
62. S.R. Richter, Z. Hayder, V. Koltun, Playing for Benchmarks, in *IEEE International Conference on Computer Vision (ICCV)*, (Seoul, Korea, 2017), p. 3

63. A. Gaidon, Q. Wang, Y. Cabon, E. Vig, Virtual Worlds as Proxy for Multi-Object Tracking Analysis, in *Proceedings of the IEEE Conference on Computer Vision and Pattern Recognition*, (Las Vegas, Nevada, United States, 2016), pp. 4340–4349
64. Y. Cabon, N. Murray, M. Humenberger, Virtual KITTI 2. arXiv preprint arXiv:2001.10773 (2020)
65. T. Peynot, S. Scheding, S. Terho, The Marulan data sets: Multi-sensor perception in a natural environment with challenging conditions. Int. J. Robot. Res. **29**(13), 1602–1607 (2010)
66. T. Gruber, M. Bijelic, F. Heide, W. Ritter, K. Dietmayer, Pixel-accurate depth evaluation in realistic driving scenarios. arXiv preprint arXiv:1906.08953 (2019)
67. A. Dosovitskiy, G. Ros, F. Codevilla, A. Lopez, V. Koltun, CARLA: An open urban driving simulator. arXiv preprint arXiv:1711.03938 (2017)

ITS Performance Evaluation in Direct Short-Range Communication (IEEE 802.11p) and Cellular Network (5G) (TCP vs UDP)

Muhammad Naeem Tahir and Marcos Katz

1 Introduction

In the 21st century century, we have witnessed the development and advancement of new astonishing technologies, which have deeply influenced human life in numerous ways by making it more convenient, comfortable, and easy. Intelligent Transportation System (ITS) have rapidly developed into a highly relevant field of its own, inspired and motivated by the constant development in wireless technologies such as cellular and ad hoc networks. Vehicular networking IEEE 802.11p (Direct Short Range Communication) and 3rd Generation Partnership Project (3GPP) Vehicle-to-Everything (C-V2X) have primarily focused on improving the vehicular and road safety and efficiency and they also support/enable entertainment services in vehicular networks. 3GPP introduced an initial version of C-V2X communication to use cellular communication in Release 14, where intelligent vehicles act as mobile devices. DSRC (Direct Short Range Communication) is normally used in a vehicular ad hoc network (VANET) and is particularly a kind of mobile ad hoc network (MANET). A potential alternative to DSRC could be to use cellular technology for vehicular communication. DSRC and cellular networks can be used together in a hybrid mode, as seen in Fig. 1. Vehicular communication provides an up-to-date road weather information, precollision warning, and improves the traffic management system, as seen in Fig. 1 [1, 2].

M. N. Tahir (✉)
Finnish Meteorological Institute (FMI), Helsinki, Finland
e-mail: naeem.tahir@fmi.fi

M. Katz
Center for Wireless Communication-Networks and Systems, Faculty of Information Technology and Electrical Engineering, Erkki Koiso-Kanttilan katu 3, University of Oulu, Oulu, Finland
e-mail: marcos.katz@oulu.fi

U. Z. A. Hamid, F. Al-Turjman (eds.), *Towards Connected and Autonomous Vehicle Highways*, EAI/Springer Innovations in Communication and Computing,
https://doi.org/10.1007/978-3-030-66042-0_10

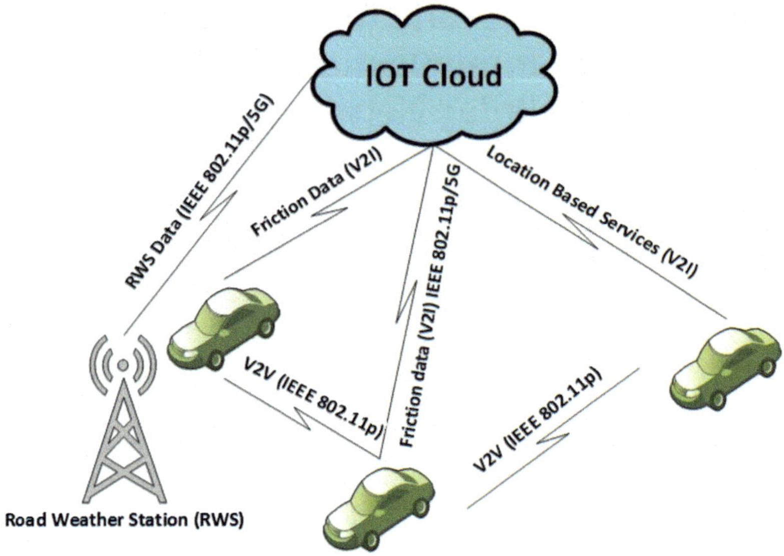

Fig. 1 Vehicle to RWS and vehicle to cloud (DSRC communication)

The growing importance of data in vehicular networks underline the necessity of understanding the impact of mobility protocols and the mobile environment on the performance of ITS applications. Certainly, this requirement is extensively acknowledged by the vehicular networking research community with a plethora of recent research dealing with numerous aspects of measuring, characterizing and improving data performance on VANET and cellular networks (e.g., C-V2X) using Transmission Control Protocol (TCP) or User Datagram Protocol (UDP) [3].

Vehicular Communication can be categorized into three types: Vehicle-to-Infrastructure (V2I), Vehicle-to-Vehicle (V2V) and Vehicle-to-Broadband Cloud (V2B) [4].

V2V: It provides a platform for vehicles to exchange road weather information and warning messages for driver assistance.

V2I: It provides a communication link between vehicle and road weather station considering environmental sensing and monitoring. It provides real time weather information and traffic updates for drivers.

V2B: It provides a communication link between vehicles and management center via road weather stations. In V2B, vehicles may communicate via wireless broadband 4G/5G. The broadband cloud could include the traffic and monitoring data as well as infotainment data for vehicle tracking and real-time driver assistance.

In Fig. 1, we can see the architecture of the vehicular network; V2V and V2I scenarios as VANET and cellular networking. V2I and V2V communications are established by using DSRC (802.11p)/cellular network (5G), and V2B communication uses the cellular wireless broadband that offers high-speed wireless access [4]. This fusion of VANET and cellular networking is called hybrid communication; and in this case, we consider a cellular network (5G test network) and DSRC (IEEE 802.11p). We have implemented this approach in our test network in Sodankylä, Finland. In vehicular communication, the transmission protocols at the transport layer are used for end-to-end data transmission called UDP and TCP. TCP offers a connection-oriented communication with the mechanisms of rate adaptation and assisting end-to-end transfer of application data through numerous IP hops and requires bidirectional communication between hops for acknowledgments. Nevertheless, UDP does not have any error retrieval mechanism [5], and on the contrary the TCP, it uses the retransmission technique to recover the lost data making the communication reliable efficiently. TCP has a drawback compared to UDP that its reliability mechanism increases the transmission delay of the data in case of packet loss, which will be discussed later in this chapter [3, 4].

The Finnish Meteorological Institute (FMI) has devised and implemented a large set of pilot use-cases for this purpose for both IEEE 802.11p and 5G test networks. In this chapter, we have compared the transport layer in vehicular communications (V2V & V2I) by exploiting the road traffic information. Our pilot measurements evaluated and compared UDP and TCP to choose, which could be the most suitable and adequate mode of vehicular communication using IEEE-802.11p and 5G test network [5]. Multiple performance evaluation metrics are used in this work such as throughput, packet loss percentage, average packet/s, average packet size, latency, data rate and jitter. The rest of the chapter is organized as follows. In Sect. 2, we present the ITS-related work with DSRC and cellular network. Section 3 demonstrates an overview of ITS protocol architecture i.e., UDP and TCP. In Sect. 4, we present the pilot scenarios for the IEEE 802.11p and 5G test network considering TCP and UDP at the transport layer in vehicular networking. Section 5 presents the analysis of the results, while Sect. 6 ends up with the conclusion of this chapter.

2 Intelligent Transportation System (ITS) with Cellular (5G) and DSRC (IEEE 802.11p)

The term Intelligent Traffic System (ITS) refers to the idea of incorporating transport infrastructure and vehicles with the latest communications technology. This combination plays an important role in better management of the current transport systems so that they function and operate more efficiently and effectively, as presented in Fig. 2.

Currently, new technological advancement in traffic management systems includes different control instruments and sensors. These instruments, extensively deployed on the roads and highways, provide road weather information, traffic jam alerts, speed limit, traffic signals, and accident warnings. Among other things, ITS aims at providing assistance in the design and development of the latest traffic

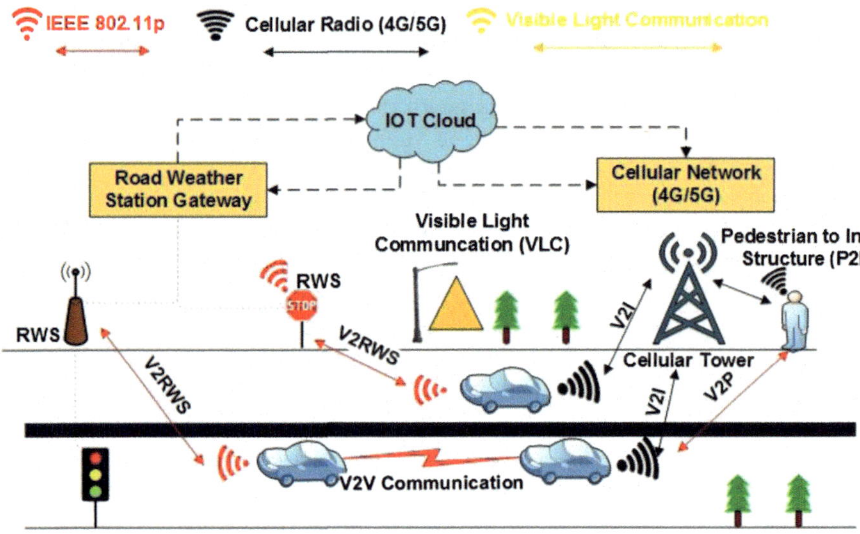

Fig. 2 Future Intelligent Transportation System (ITS)

monitoring framework, being able to make run-time decisions about the traffic modeling structure with best available substitute route, to illustrate the control system [4, 5]. Within the framework, traffic devices and sensors can be demonstrated as intelligent agents connected to centralized data base with communication to the external traffic environment. Developing techniques in machine learning (ML) and artificial intelligence (AI) have assisted the latest advances in an intelligent control by optimal decision-making through intelligent agents that can adapt their behavior conferring to a developed information base.

In ITS, the vehicular communication system delivers cooperative services related to road weather and road safety alerts, etc., as the ITS service alerts illustrated in Fig. 3 [6]. Vehicles equipped with advanced instruments like telematics, sensors, and cameras are typically the evolving smart vehicles. These intelligent vehicles have the capability to sense the current nearby environment and road weather situation, paving the way to the state-of-the-art ITS with great safety and efficiency. Constant research and efforts are being done by numerous governing bodies, which will allow V2V, V2I, and V2B communications in new automobiles as a part of future ITS, as seen in Fig. 2.

It is critical to design and develop cooperative applications making full use of vehicular networking infrastructure, which requires seamless communication in mobile environments, that is, VANET (vehicular ad hoc networking). In VANET, cooperative applications require seamless communication links among V2V, V2I, and V2B. The IEEE 802.11p standard with the licensed ITS band 5.9 GHz (5.85–5.925 GHz) has been developed and launched in 2012 for vehicular communication system. In Europe, 802.11p is used as a source for the ITS-G5 standard, providing support to the Geo-Networking protocol by European Telecommunications Stan-

Fig. 3 Examples of typical ITS service alerts

dards Institute (ETSI) for European frequency bands and channels [6]. In vehicular networking, cellular communication systems offer a far better coverage by default, in contrast to VANET. 5G test network with LTE-A (Long Term Evolution—Advanced) is a foundation for vehicular cloud services in vehicular environments provided by several vehicle manufacturers. The 5G test network does not natively support direct V2V communications and particularly in the high-density vehicular environment, the network can be overloaded by beaconing signals of the vehicles. Additionally, the response time for safety hazards and required instant messaging in V2V is also a crucial issue. The most feasible solution to resolve this issue is to combine cellular network and VANET collectively to make a hybrid communication system [7, 8].

Having a hybrid vehicular network solution with the road weather infrastructure and road safety amenities would allow drivers, road management companies and the automated mechanism of vehicles having better road traffic management and to avoid accidents.

3 ITS Protocol Architecture

Every communication system has some particular considerations during the process of the designing standards. By taking into account these considerations, industry, academia and research institutions can design the standards to study and evaluate their performance for their communication system. Similarly, for ITS, we need to have some specific considerations when designing the standards for the ITS. So, that a network architecture which defines how the vehicles participating in vehicular networking and communication with the other vehicles and infrastructure (RWS).

For vehicular networking architecture, we have available ITS protocol stack structure with ETSI specifications having six layers, shown in Fig. 4. This ITS protocol model is structured by four upright layers with two parallel layers [9].

Based on Fig. 4 above, we present a structure of three protocol architectures: ETSI, OSI, and IEEE, as seen in Fig. 5. The ETSI and IEEE 802.11p protocol architecture is compared in reference to the OSI model. The ETSI protocol model is built on the idea to allow the use of various protocols on intermediary and lower layers. Generally, the structure of the ETSI-ITS protocol stack is introduced by Next Generation Protocol group, where the Physical and Media Access Control (MAC) layers are defined by the ITS-G5 Protocol (see ETSI ES 202 663), which is largely based on IEEE 802.11p [8]. Therefore, several possibilities for the user application requirements might be approachable to the applications of the upper layer. As an individual, the user application may need different requirements for communication, that is, reliability, delay, etc. It is generally a good idea to have a protocol like IEEE 802.11p supporting multiple protocols to satisfy the requirements of different user applications. IEEE 802.11p is built on the base of the OSI model, but IEEE 802.11p is more concentrated on the two lower layers: the network and transport layer as well as the access layer (Physical and MAC). The IEEE 802.11p permits to work outside

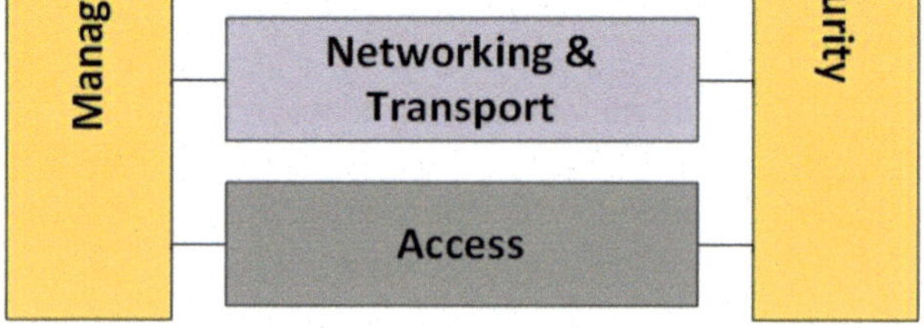

Fig. 4 ITS protocol stack

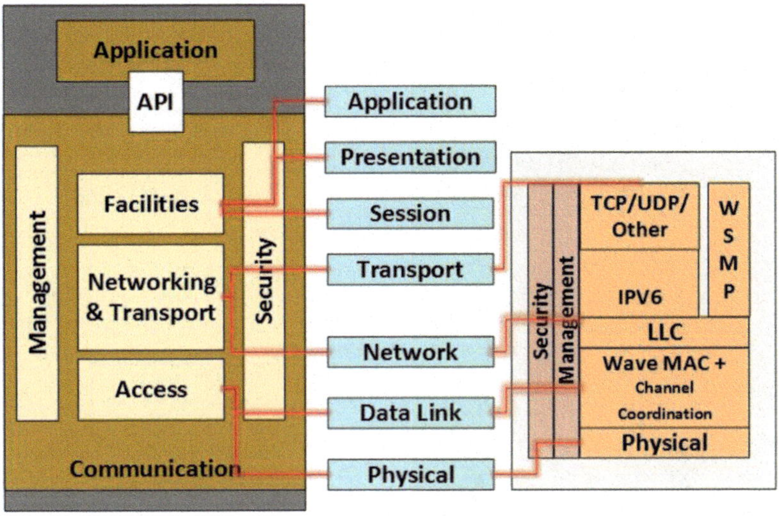

Fig. 5 Comparison between the ETSI/ISO, the OSI model, and the IEEE architectures. (Source: IEEE, ETSI & OSI)

of the Basic Service Set (BSS), providing the opportunity to have a medium access in vehicular networking with fast variability of the network because of high-speed vehicles. The IEEE 802.11p provides the advantage that the ITS node saves time in the search and selection process, but the frequency channel should be predefined in the ITS infrastructure [10].

In 802.11p, the physical layer uses OFDM modulation and has a bandwidth range of 10–20 MHz. The IEEE 802.11p bit rates are available between 27/54 Mbps, but three commonly available bit rates for all the ITS-scenarios are 3, 6, and 12 Mbps. In IEEE 802.11p, the MAC layer uses the Enhanced Distributed Coordination Access (EDCA) technique, which describes four queues depending on the information priority (from high-low), AC_VI (Video), AC_BK (Background), AC_BE (Best Effort), and AC_VO (Voice). Lastly, the Decentralized Congestion Control (DCC) is used in DSRC for channel saturation [9, 10].

The below mentioned Fig. 6 illustrates the 5G test network radio protocol architecture, which is divided into two main parts, namely, the radio access and the core network. The core network, also known as nonaccess stratum (NAS), involves all the functionality required to establish the connection between external IP networks and cell towers. Each element is a distinct server that executes a particular set of functionalities. The 5G test network protocol architecture entities are mentioned below.

PDN-GW: The Packet Data Networks (PDN) Gateway facilitates the communication with external IP networks.

S-GW: The serving gateway is the mobility anchor and performs as a router by transferring data among PDN-GW and base stations.

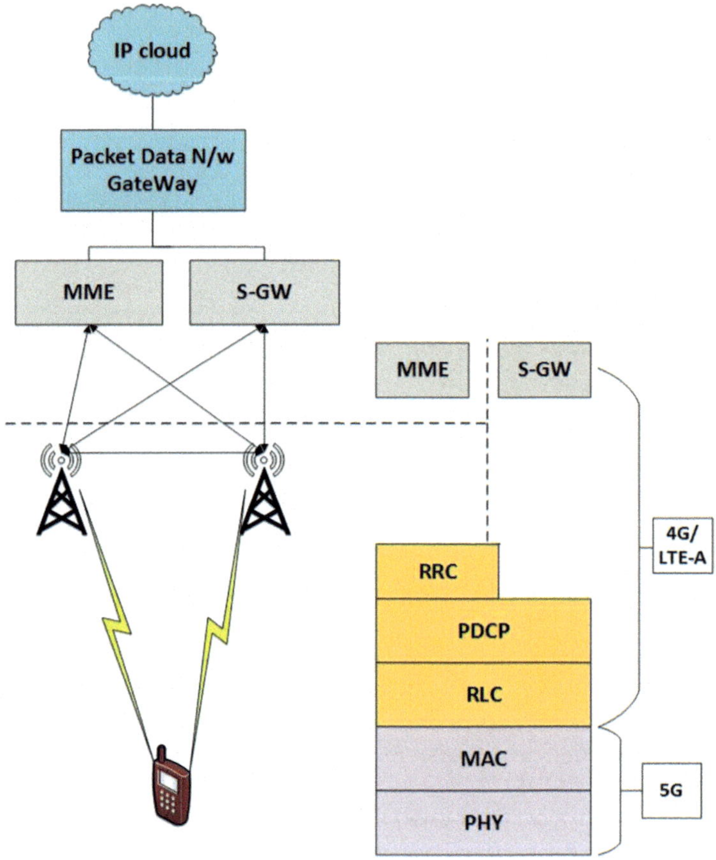

Fig. 6 5G test network framework with protocol architecture

MME: MEE stands for the Mobility Management Entity Controlling high-level mobile functions through handover and signaling commands. The radio access network also stated as access stratum (AS), is the module responsible for the radio connection set-up between a mobile device and Base Station (BS), called User Equipment (UE) [10, 11].

The most significant layers in the 5G test network protocol stack are Packet Data Convergence Protocol (PDCP), Radio Link Control (RLC), Media Access Control (MAC), and Physical (PHY). To maintain a connection between UE and MME using the signaling system, it has an additional layer between UE and eNodeB called Radio Resource Control (RRC). RRC is the layer that exists between eNodeB and UE that exist at the IP level (Network Layer). The relative 5G test network Framework with Protocol Architecture is shown in Fig. 6.

In IEEE 802.11p and 5G, the packets are encapsulated and transferred through the network, using the transport layer for providing a host-to-host application

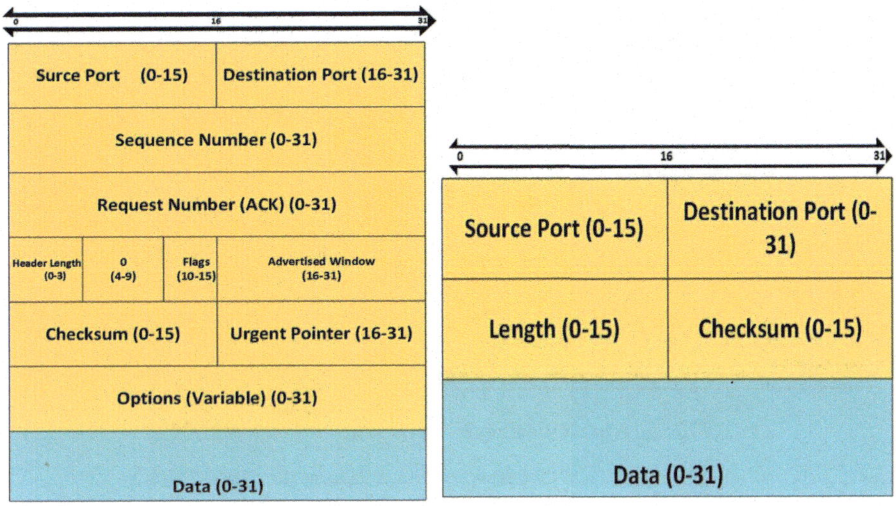

Fig. 7 TCP and UDP segment header

environment. The most common transport layer protocols are UDP and TCP. The major difference between UDP and TCP is the assurance of communication reliability; TCP is a connection-oriented protocol, while UDP is connection-less protocol. TCP is used for connection-oriented communications as it can provide acknowledgment and retransmission capabilities. To establish a connection in TCP, the protocol uses a three-way handshake (SYN, SYN-ACK, and ACK, respectively). Before a client attempts to connect with a server, the server must first bind to and listen at a port to open it up for connections: this is called a passive open. Once the passive open is established, a client may initiate an active open. TCP has three additional packets to establish a connection between peers and then transmits the actual data. This results in a packet header size of 20–60 bytes with TCP segment header size is 4 bytes, as illustrated in Fig. 7. The TCP has a maximum packet size of 1500 bytes and minimum packet size of 20 bytes. Furthermore, TCP also offers the benefit of a systematic transfer of data packets with error correction capability. TCP is typically used for applications that are not time-critical because of its comparative slow speed due to retransmission of lost packets [11, 12]. To terminate a TCP, four-way handshake (FIN, ACK, FIN, and ACK) is used, with each side of the connection terminating independently.

In TCP, the Nagle's algorithm approach is used to enhance the data transfer efficiency by combining numerous small request bytes into a single TCP segment so that the ratio between header data to payload is more proficient. TCP headers take up 40 bytes and there are plenty of applications that can emit a single byte of payload. There is a combination of four algorithms, namely TCP uses to offer congestion control, congestion avoidance, fast recovery, fast retransmit and slow start. In these algorithms, the packet loss indicates the congestion and TCP will send the all change

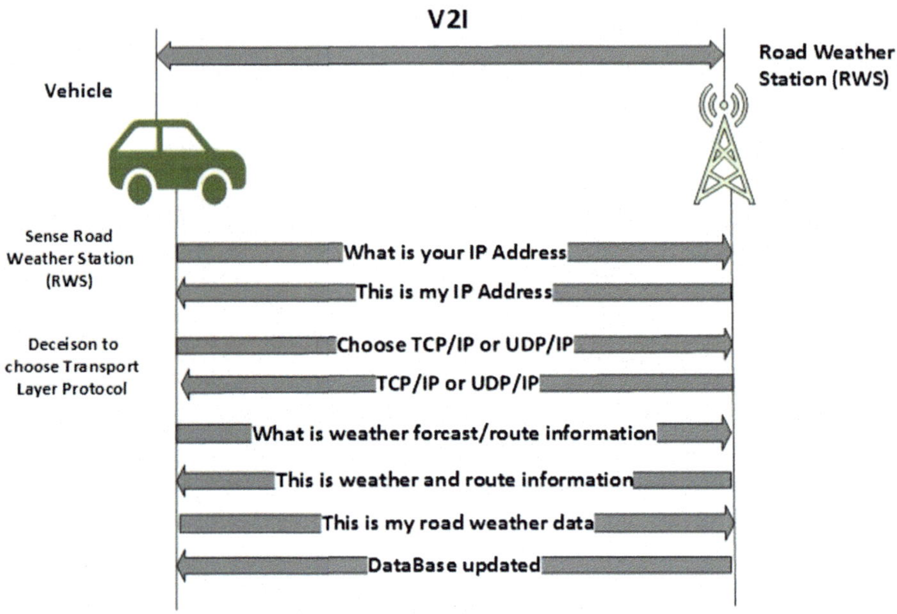

Fig. 8 Measurement scenario V2I (Vehicle-to-RWS) using TCP/UDP

number of packets before waiting for acknowledgments for packets. These changes affect the available bandwidth, as well as alter the delays by providing source of jitter.

With the use of UDP protocol, messages/datagrams are transmitted without any prior communication setup, and UDP data packets are normally broadcast over a network to everyone who is listening on the particular UDP port. On IP-based networks, distinct network addresses are used to support UDP broadcast messages. There is not any inherited acknowledgment functionality in UDP and that is one of the reasons that UDP provides unreliable communication service. In UDP, some datagrams may be lost or arrive in an out-of-order fashion. On the other hand, it decreases the overhead with eight bytes of packet header size and the UDP segment header consists of 32 bits, as seen in the Fig. 8. UDP is comparatively fast, hence used for applications that require fast data transmissions, such as games. The above mentioned Table 1 illustrates the difference between transport layer TCP and UDP [12, 13]. In contrast of UDP, TCP has the Nagle algorithm feature that escalates the network latency, so in TCP Nagle's algorithm approach is not beneficial for real-time applications.

Table 1 Comparison between TCP and UDP

Comparison description	UDP	TCP
Definition	UDP sends the unreliably in un-ordered fashion	TCP establishes a connection between two hosts in-order byte-stream
Acronyms	User Datagram Protocol	Transmission Control Protocol
Connection type	Connectionless	Connection-oriented
IETF RFC	RFC768	RFC793
Speed	Fast	Slow (due to retransmission)
Reliability	Unreliable	Highly reliable
Sequencing	No sequencing	Segment sequencing
Header size	8 bytes	20 bytes
Acknowledgment	No acknowledgment or retransmission of lost packets	Acknowledgment of data and retransmission if the user requests
Connection set-up	Connectionless, data are sent without setup	Connection-oriented, the connection must be established prior to transmission
Data interface for application	Message-based	Stream-based
Lost data	No retransmission of lost data and no windowing	Retransmission of lost data and flow control by windowing

4 Demonstration of Pilot Scenarios for IEEE 802.11p and 5G Test Network Using TCP and UDP

In this section, we demonstrate the operational pilot scenarios for IEEE 802.11p and 5G test network using TCP and UDP in V2V and V2I scenarios. For these pilot measurements, shown in Figs. 8 and 9, we present the two different process structures to demonstrate the detailed operational work for each V2I and V2V scenario. The Fig. 8 illustrates the V2I communication structure.

The information exchange process between RWS and the vehicle system functionalities are presented as separate process structures supplemented with gray arrows representing different V2I communication functionalities. Figure 9 illustrates the V2V communication structure, the information exchange process between vehicular systems functionalities are presented as separate process structures.

Figures 8 and 9 illustrate the information process structure that is comprised of two-way communication, V2V and V2I, with a complete interaction process presented in detail. Our main objective is to fragment the vehicular communication (V2V, V2I) process structures into distinct phases, providing thus an opportunity to analyze and investigate the communication procedure. In the V2V and V2I communication process, the insertion-points enabling the communication transport layer protocols UDP/TCP in VANET or 5G test network were implemented. To

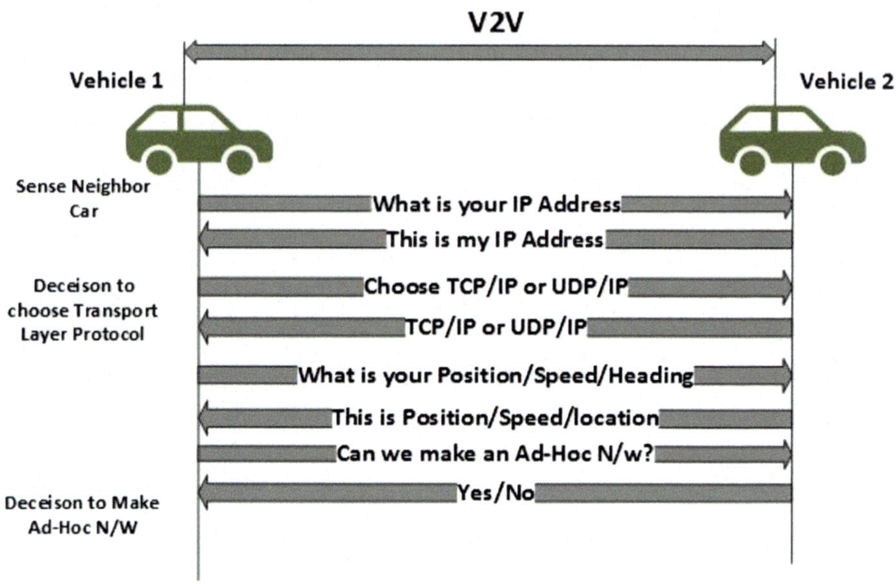

Fig. 9 Measurement scenario V2V (Vehicle-to-Vehicle) using TCP7UDP

get the safe, reliable, and heterogeneous vehicular networking, it is tempting to consider the use of the TCP/IP family of protocols to support ITS applications in pilot scenarios. These protocols are employed in order to select the transmission protocol with better performance at the transport layer to exchange road weather information using IEEE802.11p/5G test network.

IEEE 802.11p TCP and UDP packet captures can be seen in yellow color in Figs. 10 and 11, respectively, indicating captured packet locations from either RWS1 or RWS2. The IEEE 802.11p vehicular networking has been provided by Cohda MK5 radio transceivers. Vehicles used SUNIT-F-series vehicle PC for User Interface (UI) to investigate and evaluate the performance of IEEE 802.11p considering TCP and UDP. For the 5G test network measurements, a Samsung S7 smartphone with the SUNIT F-series vehicle user interface is used for pilot measurements.

Tables 2 and 3 show the used parameter settings for IEEE 802.11p and 5G test network pilot measurements for V2I and V2V scenarios at the transport layer.

For pilot measurements, we used two road weather stations and two vehicles on the FMI test track while driving and collecting data in V2V and V2RWS modes. For all test measurements, we used Python program and Iperf software for sending TCP and UDP packets to RWS and for the analysis of the IEEE 802.11p and 5G test networks. The collected data from pilot measurements were analyzed by using Wireshark and Origin 2019b programs. The road weather data are collected from different road friction instruments like Teconer RCM 411 and WCM 411 installed in vehicles and road weather station sensors on the vicinity area. The road weather

Fig. 10 IEEE 802.11p (V2I) TCP vs UDP packet capture in the Sodankylä test track

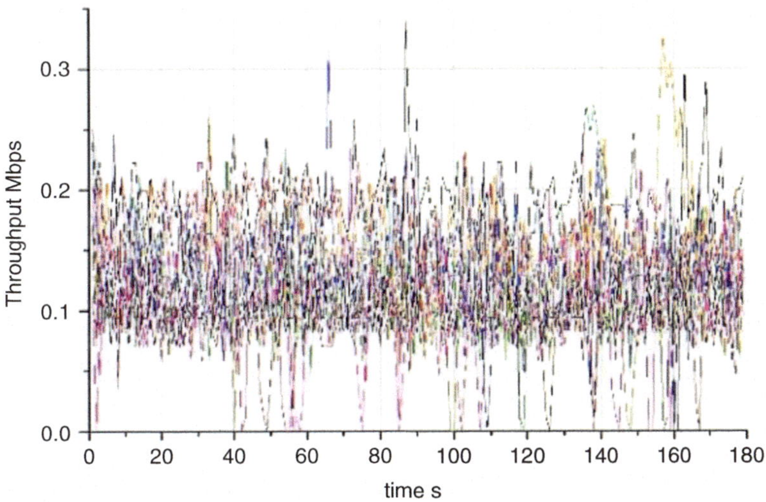

Fig. 11 IEEE 802.11p (V2I) TCP 13 pilot measurements

station sensor data, presented in Table 4, were collected during IEEE 802.11p and 5G test network measurements operating in V2I and V2V modes.

4.1 IEEE 802.11p (V2I) TCP vs UDP Packet Capture

In the first stage of IEEE 802.11p pilot scenario, the vehicles exchanged data with RWS in V2I communication mode while driving on the test track. For V2I (V2RWS) communication, the RWS delivered up-to-date road weather data to the

Table 2 802.11p Parameter settings

Parameters	Settings
Transmission power	−10 to +23 dBm
Frequency band	5.9 GHz
Modulation technique	BPSK, QPSK, 16QAM, 64QAM
Maximum transmission rate	27/54 Mbps
Data traffic	Bidirectional
Symbol duration	16, 8, 4 us
Bandwidth	5, 10, 20 MHz
Supply voltage	12V
Temperature	−40 °C to +85 °C
Maximum range	1000 m

Table 3 5G test network parameter settings

Parameters	Settings
Transmission power	41.8 dBm
Frequency band	2.3 GHz
Modulation technique	QPSK, 16QAM, 64QAM
Data rate (each user)	10 Mbps
Symbol duration	66.66 us
Data traffic	Bi-directional
Bandwidth	40 MHz
Supply voltage	230 volts
Antenna gain (Tx and Rx)	19 dBi
Maximum range	1000–1700 m

Table 4 Road weather station collected data

Parameter	Type of sensor	Measured height/depth (m)
Temperature	PT-100	2
Humidity	HMP-45D	2
Wind direction and speed	Thies Clima-2D Ultrasonic Anemometer	6.5
Air humidity	Vaisala HMP-155	4.5
Current weather and visibility	Vaisala PWD-22	6
Road state (surface)	Vaisala DSC-111	4.5
Road temperature (surface)	Vaisala DST-111	4.5
Infrared camera	Zavio B7210 Full HD	4.5

vehicles while passing the RWS. The resulting connectivity is presented in Fig. 10 by showing the yellow marks and pointing the locations where the packets were received by a vehicle in the V2I scenario. The packet capture in V2I (V2RWS) communication scenario has been done in 13 drives on the 1.7 km test track and the measured throughput is shown in Fig. 11 (TCP) and Fig. 12 (UDP). The pilot measurements fluctuate abruptly due to the vehicle distance from the RWS, the line-

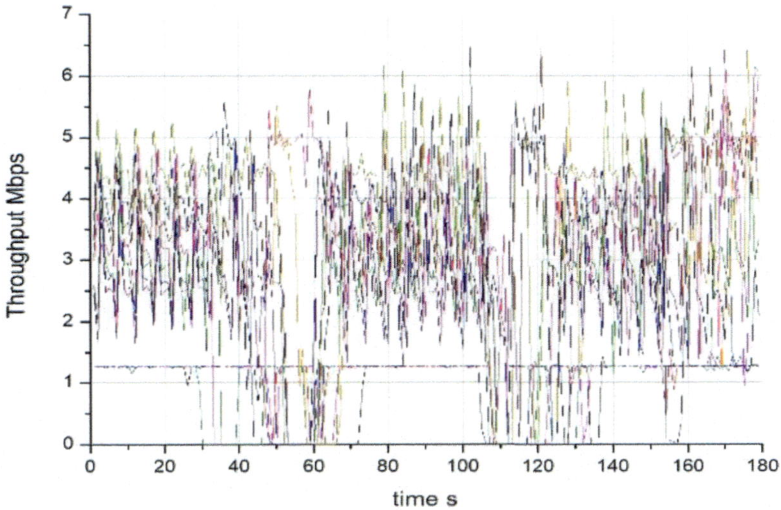

Fig. 12 IEEE 802.11p (V2I) UDP pilot measurements

Fig. 13 IEEE 802.11p ((V2V)) TCP vs UDP packet capture in the Sodankylä test track

of-sight network availability, periodically shadowed by the tall trees on test track, as well as the used program for pilot measurements.

In the second stage of the IEEE 802.11p pilot scenario, the vehicles exchanged the collected data from RWS in V2V communication mode while driving on the test track. The encountering vehicles exchanged the latest road weather data received from RWS through the ad hoc network. The resulting connectivity is presented in Fig. 13 by showing the yellow marks and pointing the locations where the TCP and UDP packets were received in the V2V scenario.

The packets capture in the V2V communication scenario was performed in 13 drives each for TCP and UDP packet capture; Figs. 14 and 15 show throughput measurements for TCP and UDP respectively. Test vehicles were driving in the same

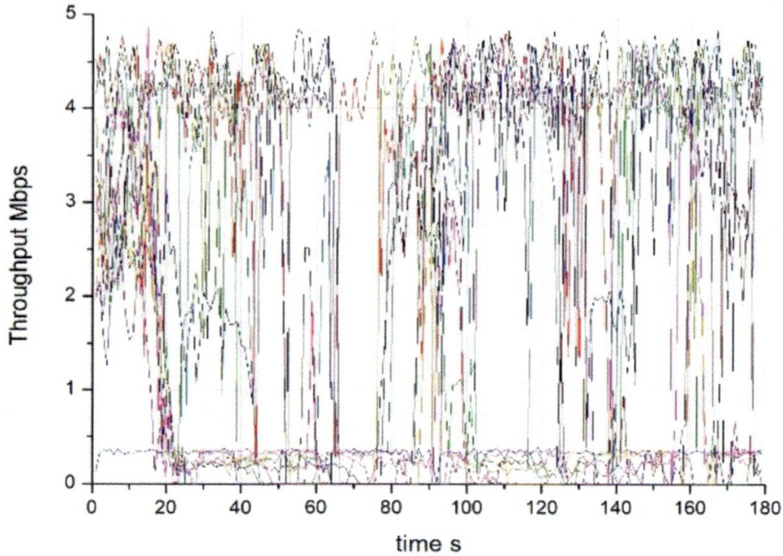

Fig. 14 IEEE 802.11p (V2V) TCP 13 pilot measurements

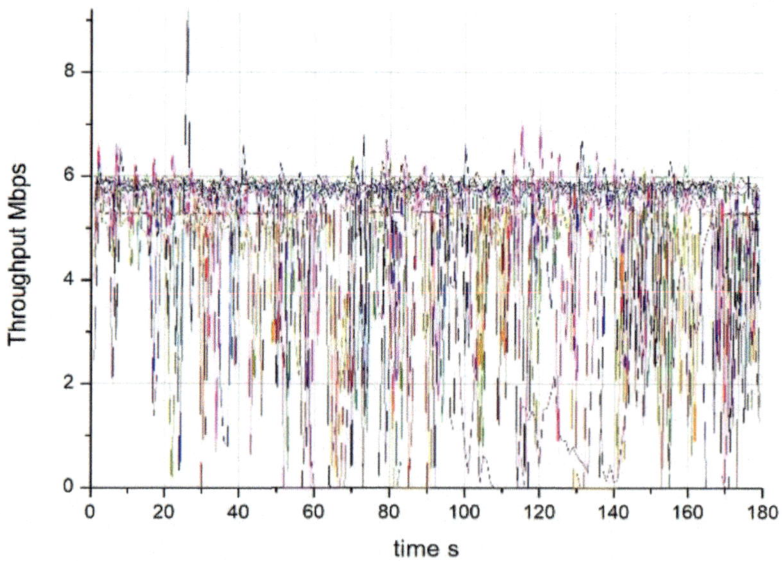

Fig. 15 IEEE 802.11p (V2I) UDP pilot measurements

lane, as well as traveling across each other on the test rack. It can be noticed as in the previous cases that the data throughput in the pilot measurements varies abruptly due to similar reasons as previously explained. The UDP measurements in the Fig. 15 illustrate that the continuous transmission of packets during pilot drives.

Fig. 16 5G test network TCP vs UDP packet capture at Sodankylä, Finland

4.2 5G Test Network Pilot Measurements

At this stage of pilot measurements, we performed V2I 5G measurements considering TCP and UDP. The vehicle encountering another vehicle transfers the RWS information (V2RWS) to other vehicle (V2V). Thus, distributing real-time RWS information, extending RWS range in an ad hoc network and getting updated warnings alerts from RWS and vehicles will help to avoid accidents. TCP and UDP connectivity by using 5G test network is presented in Fig. 16 by showing the yellow marks and pointing the locations of packet capture in the V2I scenario. The packet capture using the 5G test network in V2I communication scenario has been performed in a total of 13 drives for both TCP and UDP cases. The corresponding packet captures showing the throughput are displayed in the Figs. 17 and 18 respectively. The Fig. 17 shows the rapidly fluctuating TCP throughput measurements. As mentioned before, these fluctuations can be attributed to the varying vehicle's distance with RWS, as the line-of-light network availability shadowed by the tall trees on test track. The limitations of the used program also affects the measurement of throughput for the pilot. The UDP pilot measurements in the Fig. 18 suggests more consistent spikes and continuous packet transmission for test drives on a test track.

5 Performance Evaluation of DSRC & Cellular Network Using TCP and UDP

5.1 IEEE 802.11p Performance Analysis Using TCP and UDP

In this section, we analyze the performance of IEEE 802.11p in vehicular ad hoc networks (V2V and V2I). We evaluated the performance of the IEEE 802.11p-

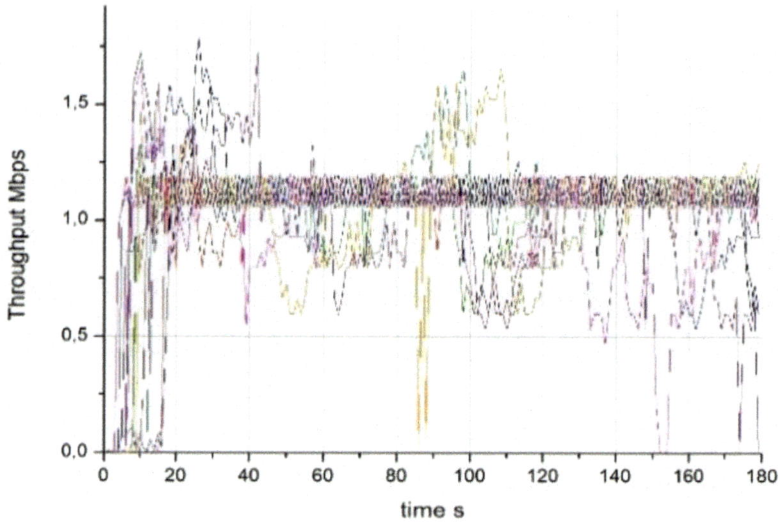

Fig. 17 5G test network (V2I) TCP pilot measurements

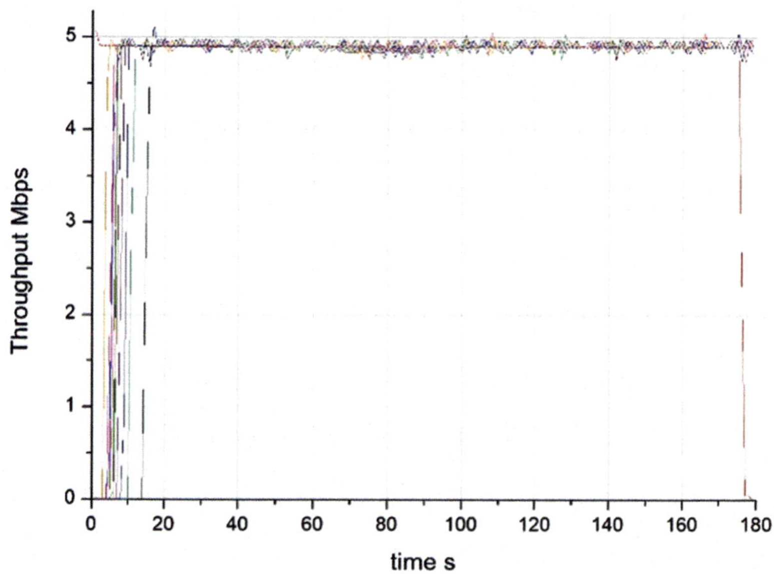

Fig. 18 5G test network (V2I) UDP pilot measurements

based V2V and V2I scenarios on the transport layer (TCP/UDP). Table 5 illustrates the performance evaluation for V2I (V2RWS) while Table 6 shows the V2V performance considering TCP and UDP.

Table 5 IEEE 802.11p: V2I

Analysis parameters	TCP	UDP
Average speed	30 km/h	30 km/h
Time span (s)	180	180
Average packet/s	80	257.59
Average packet size	1431	925
N/w latency (ms)	179.754	110.29
Throughput (Mbps)	0.394	2.21
Data rate (Mbps)	0.220	2.19
Packet loss (%)	22	38

Table 6 IEEE 802.11p: V2V

Analysis parameters	TCP	UDP
Average speed	30 km/h	30 km/h
Time span (s)	180	180
Average packet/s	74	582
Average packet size	1433	925
N/w latency (ms)	102.93	79.39
Throughput (Mbps)	1.79	5.09
Data rate (Mbps)	0.212	2.06
Packet loss (%)	20	35

In vehicular pilot measurements, Nagel's algorithm was turned on in TCP measurements, as TCP is more appropriate for reliable data transmission. Although UDP measurements with IEEE 802.11p architecture could be able to produce fast data transmission with a bit more probability of packet loss. It can be seen in Table 5 (V2I/V2RWS) and Table 6 (V2V) that the average packet size of TCP and UDP differs, and this have an impact on network latency, data rate and packet loss. The large sizes of the TCP packets require longer times to transfer, resulting in more collisions and data loss at the transport layer [14]. For that reason, network latency is quite high in TCP in contrast to UDP, which ultimately effects the data rate and throughput in TCP. We used Tahoe and Reno packet loss and delay strategy in this case, if an ACK times out retransmission time-out (RTO) slow start is used and both algorithms decrease congestion window to one maximum segment size (MSS). TCP waits for 200 ms for a full packet of data to send it again.

It can also be noticed in Tables 5 and 6 that the packet loss probability in UDP is high compared to that of TCP, and this is due to the continuous transmission of packets and lack of acknowledgment feature in UDP. The throughput analysis of IEEE 802.11p is presented in the Fig. 19 considering V2I and V2V scenarios using TCP and UDP. In this performance analysis, the UDP (V2I and V2V) is best in terms of throughput compared to TCP (V2I and V2V). It is important to notice that V2I (TCP) throughput was quite low and did not perform well in our test measurements due to packet loss and network latency (Windowing). The network latency is quite high with V2I (TCP) because TCP is very sensitive to network latency and packet loss depending on congestion control mechanism. Basically, IEEE 802.11p is a

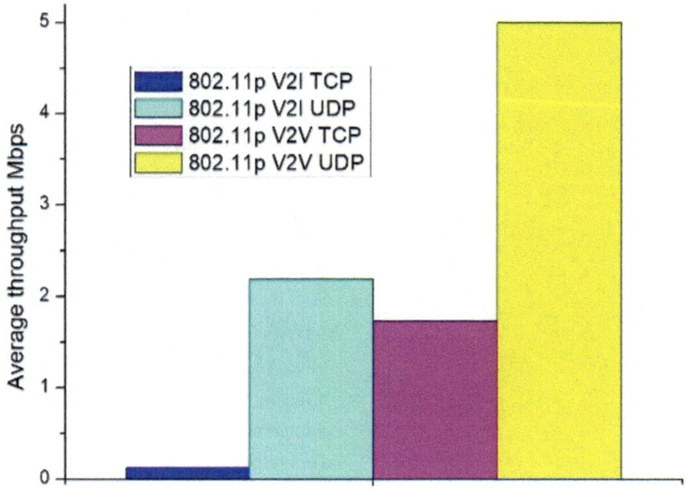

Fig. 19 Throughput analysis of IEEE 802.11p in V2V and V2I

Table 7 Performance analysis of 5G test network

Analysis parameters	TCP	UDP
Time span (s)	180	180
Bandwidth (Mbps)	1.05	4.62
Avg packet/s	113.2	368
Avg packet size (bytes)	1409	958
Throughput (Mbps)	1.12	4.96
Lost packet (%)	28	39
Jitter (ms)	–	5.76

narrowband communication that performs well with UDP continuous transmission of packets; that is why it performs well in our pilot measurements [14, 15].

5.2 5G Test Network Performance Analysis Using TCP and UDP

In this section, we analyze the performance of the 5G test network. We evaluate the performance of the 5G test network on the transport layer (TCP/UDP) in terms of the V2I scenario. Table 7 illustrates the performance evaluation for cellular communications considering TCP and UDP in the 5G test network.

It can be seen in the abovementioned Table 7 that the average packet size of TCP and UDP differs, as it was in IEEE 802.11p case, affecting network latency, throughput and packet loss. The large packet size in TCP means longer transfer times, resulting in more collisions and data loss. This explains the quite high network latency in TCP that ultimately affects throughput in TCP in contrast to UDP.

Fig. 20 Throughput analysis of 5G test network in V2V and V2I

The packet loss probability in UDP is high compared to the TCP case, due to the lack of acknowledgments and continuous transmission of packets. As compared to UDP, the blockage time in TCP (>1 s) causes packet drop (%) and ultimately affects the throughput. Nonetheless, the large buffer size can provide some compensation for packet loss in TCP, but it will increase the communication period (latency) in contrast to UDP [14, 15]. In the 5G test network, TCP offers some features such as retransmission of lost packets and reordering of packets, but generates network latency and jitter. Jitter is the delay between data transmission and receiving, that the reason that jitter is calculated in UDP but not in TCP. Because of TCP features. It has an unacceptable level of jitter for real-time applications.

In Fig. 20, the performance analysis of TCP and UDP is presented. UDP performed well in our pilot measurements with continuous transmission of packets. The delay handling and congestion control mechanism of TCP are very important in the 5G test network because of high frequencies and channel sensitivity to a wide range of factors that affect the throughput, that is, delay, attenuation, outage, etc.

6 Conclusion

In this chapter, we have studied the feasibility of using TCP and UDP with IEEE 802.11p and 5G test network in real vehicular environments (V2V and V2I). Our results indicate that TCP provides quality of service (QoS) in modern vehicular networks by predicting and optimizing their performance in different automotive environments and applications. On the contrary, UDP in IEEE 802.11p performed well for V2V and V2I scenarios with connectionless packet transmission and avoiding the overhead of processing delay in the network. In the 5G test network, the network performance was significantly reduced when using TCP in V2V and V2I scenario instead of UDP. UDP performance was adequate in the 5G test network because packet loss is tolerated in UDP rather than waiting for acknowledgments and retransmitted packets, which may not be the best choice for real-time applications.

Our pilot measurements show some interesting results that offers end-to-end delays less than 200 ms and throughputs of up to 5Mbps, accomplishing the active road safety, as well as cooperative traffic efficiency applications requirements.

Acknowledgment Here, I would like to say a special thanks to Timo Sukuvaara and Kari Mäenpää from Finnish Meteorological Institute (FMI). I am grateful to them that I have had the pleasure to work during the development and execution of measurement scenarios. They also provided technical assistance and guidance during the whole research period.

References

1. C.-C. Chang, et al., "*Cooperative Vehicular Surrounding Sensing System*." 23rd ITS World Congress, Melbourne, Australia, 10–14 October 2016
2. K.C. Dey, A. Rayamajhi, M. Chowdhury, P. Bhavsar, J. Martin, Vehicle-to-vehicle (V2V) and vehicle-to-infrastructure (V2I) communication in a heterogeneous wireless network–Performance evaluation. Transp. Res. C: Emerg. Technol. **68**, 168–184 (2016)
3. M. Eriksson, *A Performance Evaluation on the Use of IEEE 802.11 for Long Range Communication* (2005)
4. K. Kairbek, *TCP Performance in 5G Systems with Dynamic Blockage* (Master's thesis) (2018)
5. A. Kaufmann, T. Stamler, S. Peter, N.K. Sharma, A. Krishnamurthy, T. Anderson, TAS: TCP acceleration as an OS service, in *Proceedings of the Fourteenth EuroSys Conference 2019* (Dresden, 2019, March), pp. 1–16
6. S. Kuhlmorgen, I. Llatser, A. Festag, G. Fettweis, *Performance Evaluation of ETSI GeoNetworking for Vehicular Ad Hoc Networks. IEEE Vehicular Technology Conference*. (2015). https://doi.org/10.1109/VTCSpring.2015.7146003
7. M. Klapez, C.A. Grazia, L. Rold, M. Casoni, IEEE 802.11 p under congestion in an Infrastructure-to-Vehicle communication approach, in *2019 AEIT International Conference of Electrical and Electronic Technologies for Automotive (AEIT AUTOMOTIVE)*, (IEEE, Piscataway, 2019, July), pp. 1–6
8. ETSI, T, Intelligent Transport Systems (ITS); European profile standard for the physical and medium access control layer of Intelligent Transport Systems operating in the 5 GHz frequency band. Draft ES **202**, 663 (2010)
9. Z.H. Mir, F. Filali, LTE and IEEE 802.11 p for vehicular networking: A performance evaluation. EURASIP J. Wirel. Commun. Netw. **2014**(1), 89 (2014)
10. I. Petrov, T. Janevski, *TCP 5G for Next Generation High Speed Mobile Networks* (2017). https://www.researchgate.net/publication/321668068_TCP_5G_for_next_generation_high_speed_mobile_networks
11. F.H. Rahman, A.Y.M. Iqbal, S.S. Newaz, A.T. Wan, M.S. Ahsan, Street Parked Vehicles Based Vehicular Fog Computing: TCP Throughput Evaluation and Future Research Direction, in *2019 21st International Conference on Advanced Communication Technology (ICACT)*, (IEEE, Piscataway, 2019, February), pp. 26–31
12. M.G. Rubinstein, F.B. Abdesslem, M.D. De Amorim, S.R. Cavalcanti, R.D.S. Alves, L.H.M.K. Costa, O.C.M.B. Duarte, M.E.M. Campista, Measuring the capacity of in-car to in-car vehicular networks. IEEE Commun. Mag. **47**(11), 128–136 (2009)

13. T. Sukuvaara, ITS-Enabled advanced road weather services and infrastructures for vehicle winter testing, professional traffic fleets and future automated driving, in *Proceedings of the 2018 ITS World Congress, Copenhagen, Denmark*, (2018, September), pp. 17–21
14. M.N. Tahir, K. Maenpaa, T. Sukuvaara, Evolving Wireless Vehicular Communication System level comparison and analysis of 802, 11 p, 4G 5G, in *2019 2nd International Conference on Communication, Computing and Digital systems (C-CODE)*, (IEEE, Piscataway, 2019), pp. 48–52
15. H. Ullah, N.G. Nair, A. Moore, C. Nugent, P. Muschamp, M. Cuevas, 5G communication: An overview of vehicle-to-everything, drones, and healthcare use-cases. IEEE Access **7**, 37251–37268 (2019)

Autonomous Shuttle Development at Universiti Malaysia Pahang: LiDAR Point Cloud Data Stitching and Mapping Using Iterative Closest Point Cloud Algorithm

Muhammad Aizzat Zakaria, Baarath Kunjunni, Mohamad Heerwan Bin Peeie, and Georgios Papaioannou

1 Introduction

Autonomous vehicle (AV) development has gained popularity nowadays due to the rapid development in sensor technology. The ability of the vehicle to navigate autonomously is a promising factor in reducing the accident rate on the road, and eliminating fatalities caused by human errors by giving more authorities for the intelligent machine to operate autonomously [1].

The autonomous vehicle has seen a recent interest among researchers around the world. The main aim is to reduce the number of accidents [1] and decrease traffic levels [2]. A large sum of money has been invested into the autonomous vehicle development in many countries such as United States, Finland, Japan, China, Singapore and Dubai.

As far as Malaysia is concerned, the development of the autonomous vehicle is not as aggressive as in other developed countries, as Malaysia is a country which focuses on the agricultural industry, such as palm oil, rubber, paddy and

M. A. Zakaria (✉) · B. Kunjunni
Autonomous Vehicle Laboratory, Automotive Engineering Centre, Universiti Malaysia Pahang, Pekan, Malaysia

Innovative Manufacturing, Mechatronic and Sports Laboratory, Faculty of Manufacturing and Mechatronic Engineering Technology, Universiti Malaysia Pahang, Pekan, Malaysia
e-mail: maizzat@ump.edu.my

M. H. B. Peeie
Autonomous Vehicle Laboratory, Automotive Engineering Centre, Universiti Malaysia Pahang, Pekan, Malaysia

G. Papaioannou
KTH Vehicle Dynamics, Department of Engineering Mechanics, KTH Royal Institute of Technology, Stockholm, Sweden

U. Z. A. Hamid, F. Al-Turjman (eds.), *Towards Connected and Autonomous Vehicle Highways*, EAI/Springer Innovations in Communication and Computing,
https://doi.org/10.1007/978-3-030-66042-0_11

coconut, and electronics industry. From the automotive perspective, Malaysia has two national carmakers: Proton and Perodua, which developed the in-house R&D development with some collaboration with global automotive key players. Furthermore, due to the larger Malaysian Automotive Industry, the research works are broader and provide better ground for the development of the autonomous vehicle in Southeast Asia [3].

The autonomous vehicle development in R&D can be traced back to late 2012, when a first prototype of the autonomous vehicle called IREV-01 has been developed by the Universiti Teknologi Malaysia with a basic trajectory tracking control [4]. Then, the IREV-02 has been developed using the Proton Preve platform with the collaboration with Proton. The goal was mainly to study the characteristic of the steering control system. Furthermore, the Vehicle System Engineering (VSE) team has advanced further with the in-wheel Compact Electric Vehicle (CEV) development and the prototype of the autonomous vehicle using the Proton Exora platform. The first vehicle to use is 3D LiDAR in the navigation system [5].

Besides the development of autonomous vehicle in open roads, there are several researches on autonomous vehicle in a specific area such as university campus [6, 7]. Autonomous vehicle in campus areas can provide transport services throughout the campus and reducing the number of vehicles in the campus. Furthermore, people with visual and mobility impairment will benefit from this service [6]. This would encourage people to use the public transportation, thus reducing the number of vehicles on the campus road. This creates a better environment with less traffic and lower pollution. There is also a research development by the Autonomous Vehicle Laboratory team from Universiti Malaysia Pahang regarding the autonomous vehicle development. Recently, research has been published on the autonomous vehicle stability control and navigation [8, 9]. An algorithm for the perception and the mapping solution of the autonomous system is developed by the team to achieve the autonomous navigation of a compact shuttle. The algorithm is embedded on the Proton Persona and currently, the team is developing the compact shuttle to test and study their algorithm. Figure 1 shows the prototype of the compact shuttle that we developed in our laboratory for algorithm implementation and testing, while Fig. 2 illustrates the vehicle being calibrated at our university on a straight line.

The technology of the shuttle is developed in-house with many advanced sensors to achieve the autonomous navigation, for which it is important to use accurate and dense metric maps to measure object locations in three dimension (3D) [10]. There are several sensors such as camera, global positioning system (GPS), inertial measurement unit (IMU), radar and light detection and ranging (LiDAR) used by researchers to obtain real-time information on the surroundings [10–13]. LiDAR is an active remote sensing technique that works by emitting short pulse of laser and measures the range between the sensor and the surroundings [14]. So, the purpose of the sensor is to perceive the environment in three dimension and to provide the map data of the environment for further processing and classification. For example, the LiDAR data can be processed to detect specific shape that are determined by the algorithm to classify objects such as pedestrians, cars, large vehicles, and buildings.

Point cloud data are set of data point in space which are generally produced by 3D scanners such as LiDAR. These scanners measure various points on external

Fig. 1 UMP autonomous shuttle with LiDAR sensor mounted on top of the vehicle

Fig. 2 LiDAR sensor calibration and testing at Universiti Malaysia Pahang

surface of objects or surroundings generating point clouds which are then used to generate 3D CAD models. The measurement is based on the intensity of the light from the laser scanner array inside the LiDAR. The intensity value for each LiDAR point will represent the light energy value that is reflected back from the transmitter to the receiver. In most cases, the sensor will only provide the current frame of the scan and each frame of the scan has not linked each other from the dataset that was received from the LiDAR. In order for the data to be useful for navigation, each frame has to be linked each other to represent a single map of the environment. The process of merging the different frame of the Point Cloud data based on the differential motion of the vehicle is called stitching and mapping.

The need to have many data testing as possible for mapping, classifications and machine learning has motivated us to replicate the experimental testing using a vehicle simulator. Therefore, in this paper, the development of the mapping algorithm for the autonomous shuttle is discussed using iterative closest point (ICP) algorithm since it has been widely used with point cloud data [15–17].

In this paper, the data are simulated as close as possible to our real environment at our university so that we can replicate the same algorithm using experimental set of data that were collected.

2 Methodology

The data are gathered from the open-source vehicle simulator in Linux-based operating system where the data are simulated based on the real navigation trajectory: a roundabout cornering [18]. The simulated data are constructed at roundabout since this is the road that the autonomous shuttle will be passing through inside our university. The roundabout junction is common at the university as seen in Fig. 3, showing the need to replicate the data in simulation so that more analysis can be done. Also, the point cloud data are required to be simulated as close as possible to our current environment, enabling us to do further processing and analysis.

According to Fig. 4, the development initiates by constructing the vehicle simulation environment in the open-source vehicle simulator engine. The vehicle simulator engine acts as server into our system where the algorithms, the data logging and the post-processing of our data act as clients to the system. In this simulator, the environment, including the vehicle and the pedestrians, is spawned randomly to give the dynamic noise into our system as in the real implementation that we will be testing. Examples of the vehicle simulation environment are presented in Figs. 5a, 5b, 5c and 5d. More specifically, Figs. 5a and 5b illustrate the view of the camera sensor prior to entering the roundabout at straight line and during taking the roundabout respectively. Figures 5c and 5d present the third view of the vehicle during the roundabout and after taking the roundabout. The LiDAR is configured to be placed on top of the vehicle, similar with our autonomous shuttle, and the frame reading of the point cloud is shown in Fig. 5d.

Our ego vehicle starts at straight line for few seconds, then drives toward the roundabout, and finally it enters the roundabout; simulating our university common path where we have many similar roundabout configurations. The data are then published via robotic operating system (ROS) bridge whereby the ROS bridge acts as middleman between our simulator server and our ROS system as in Fig. 6 [19].

ROS system will publish the LiDAR data in Pointcloud2 format whereby the data are then recorded in *bag format. The recorded data are then post-processed whereby the poincloud2 format is then converted into point cloud format for compatibility reading in the simulation software. The point cloud data format is

Fig. 3 Roundabout junction at Universiti Malaysia Pahang

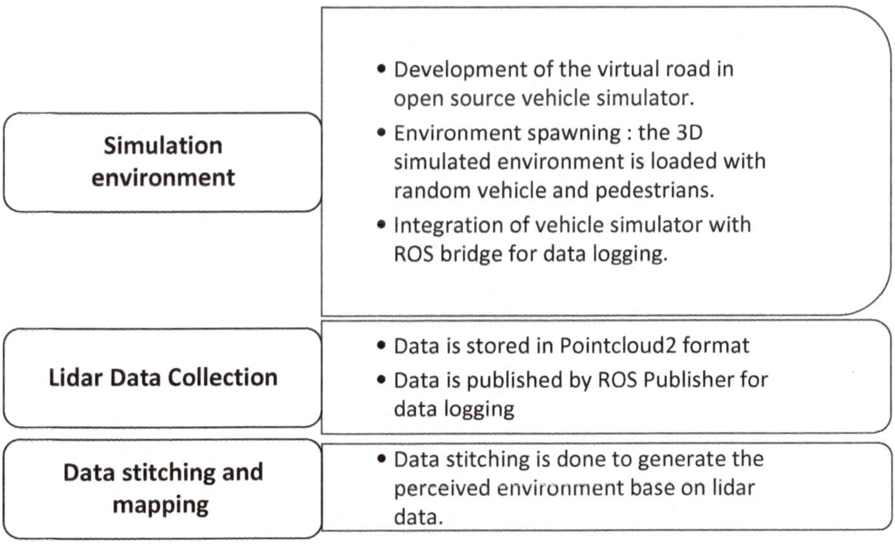

Fig. 4 Process flow of the LiDAR processing

stored as an array of matrix. Thousands of data are stored in each array for each sample time as in Table 1. Each array represents a frame which consists of individual data on the environment. For example, array 1 and array 2 consist of two moving point cloud data. Both data are identical with a slightly different because the vehicle is moving. It can be said that the data have a differential movement value between each frame set.

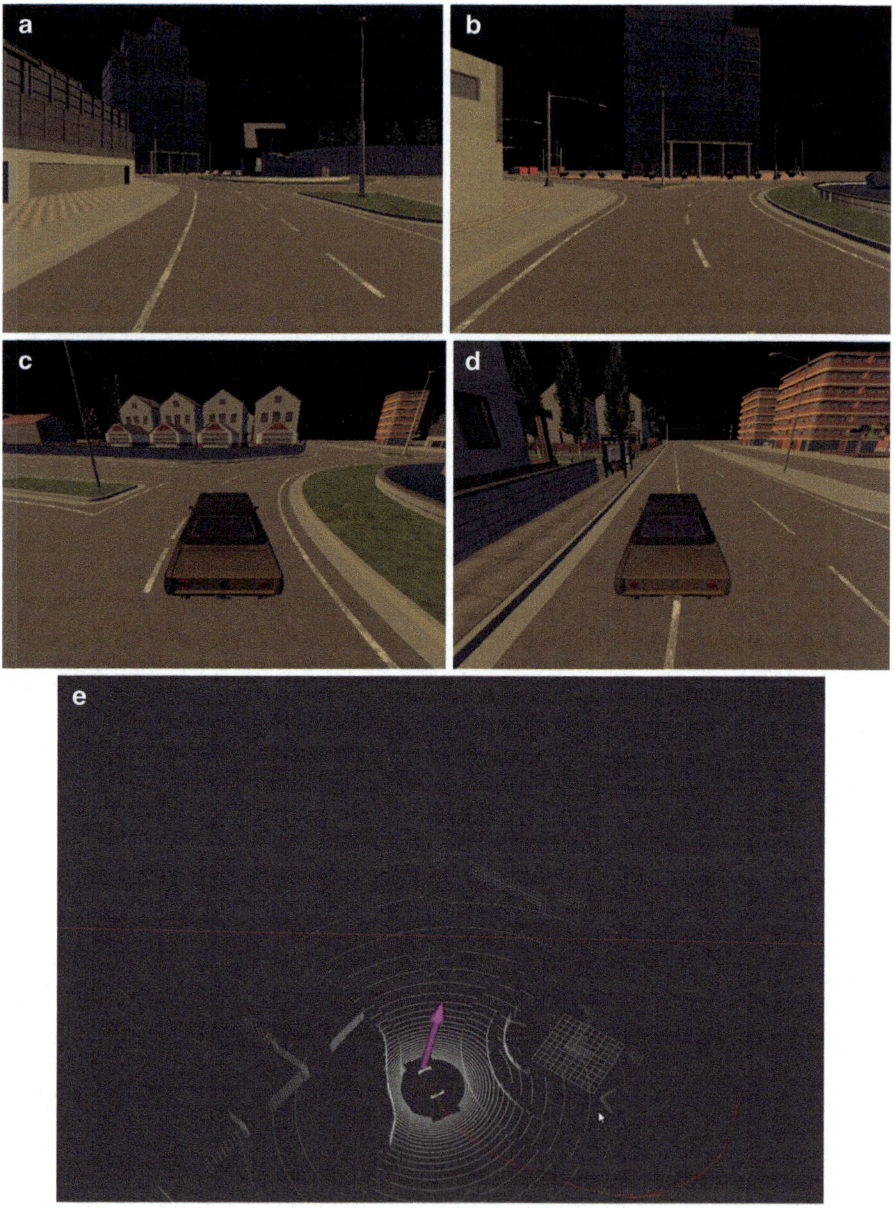

Fig. 5 (**a**) Camera sensor view of the vehicle simulator at straight line. (**b**) Camera sensor view of the vehicle simulator at roundabout cornering. (**c**) Vehicle taking roundabout corner. (**d**) Vehicle finish taking the roundabout. (**e**) Sample of point cloud data at each frame during roundabout manoeuvre. The red line indicates the trajectory motion of the vehicle

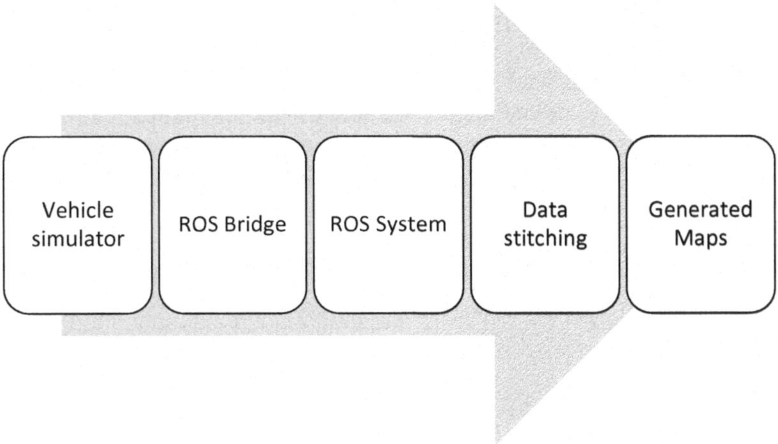

Fig. 6 Flow process of map generation

Table 1 Example of the point cloud array format store in N data

Time step 1	Pointcloud array {1}
Time step 2	Pointcloud array {2}
Time step 3	Pointcloud array {3}
...	...
...	...
...	...
Time step N	Pointcloud array {N}

3 Iterative Closest Point Algorithm for 3D Point Cloud Registration

From the previous section, we have discussed that two closest set of point cloud data are identical and have differential motion between sets of data due to the movement of the vehicle. However, up to this stage, the point cloud data are only recorded as raw format which means the data cannot be interpreted as a meaningful representation. For example, we cannot easily classify types of road that the vehicle was navigating since the point cloud data for each frame are limited to the view of the frame data. Therefore, registration of point cloud is a crucial step to compare the data at different frame and to document changes and geometric deformations of the observed environment [17, 20].

Therefore, we need a post-processing technique to detect the variation and registration process to transform the acquired point cloud to a common global coordinate system. This is because the coordinate system that we obtained from LiDAR data is relative to the sensor position mounted on top of the vehicle. The process is commonly used in the simultaneous localisation and mapping (SLAM) technique in order to get the map data of the environment [21]. One of the most widely used algorithms is the iterative closest point algorithm (ICP). The algorithm works by comparing two clouds of point (reference and current) trying to find

3D correspondences between points. Then, it outputs the translational and rotation matrices which the application will provide the best match with regard to the source in terms of minimum distance [22]. Furthermore, the Euclidean alignment of two overlapping sets of 3D point clouds is solvable using the ICP method [23].

The method is relatively simple to implement with less computation time as no other sensor fusion is required. The data only rely from single source of data, which is LiDAR. However, the drawback of this method is the data may reach a wrong convergence if the sensor is not calibrated properly. The pseudo code of the algorithm is shown in Table 2. In this algorithm, some of the API are from MATLAB point cloud processing API to process and merge the data.

Table 2 Iterative closest point implementation pseudo code

```
Algorithm 1: Iterative closest point
Function rosbagread()

Input: Simulator environment data store in ROS *bag format.

bag ← input.bag

#Getting time data from *bag file
bagtime ← bag.Time

#Select respective pointcloud data based on ROS sensor message type.
selectPC ← bag (/sensor_msgs/poinclouds)

Global PC ← selectPC #read message file

Function ICPAlgorithm()
#Looping from initial data to end of pointcloud data
for i = initial to N pointcloud
PCCurrent ← PC(i)

#Loop from start frame until last frame

        While FrameNotFinished do

                #define fix and moving frame
                Reference = moving

                #apply ICP registration
                Tform = pcregistericp(moving, reference,
                        'Metric','pointToPlane','Extrapolate', true);
                alignPC #align pointcloud

                #update environment
                PCScene ← PCmerge(PCScene, alignPC)
        End while
End function
#display Pointcloud
PCShow
```

4 Results

In this section, the results from the algorithm are shown. The example of raw point cloud data at each frame is displayed in Figs. 7a and 7b. We can see that the frame is identical as they are in the same environment, but the relative location of the sensor is different therefore made the stitching algorithm possible.

The algorithm starts by stitching two iterative set of point cloud first as in Fig. 8a. At this point, the height of the building can be seen from the stitching. As the vehicle move forward to the roundabout, the LiDAR gather more point cloud data to be merged (Fig. 8b and Fig. 8c). More detailed information has been registered to the map especially on the right side of the data where the new data were recorded. It can be seen that more detailed points were added such as the updated structure of the building.

As the vehicle is driving around the roundabout, the contour of the road also can be seen clearly such as the road divider, junctions and roundabout shape. There is slight tone variation (in purple colour), which indicates the slight contour of the road. The junctions, dividers and roundabout have slight tone variations to indicate that this is part of the road surface type. For large buildings and vehicles, the yellow and green colour are observed in a single map. The final stitch data using ICP is shown in Fig. 8d using 300 sets of point cloud data. We can further expand the map by having more LiDAR data with the movement is relative to its previous position of the maps.

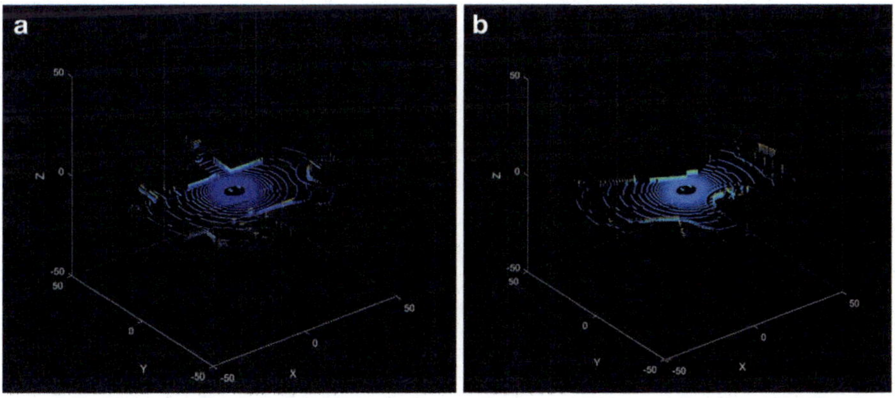

Fig. 7 (**a**) Raw point cloud data at frame initial frame. (**b**) Raw point cloud data at consequence frame

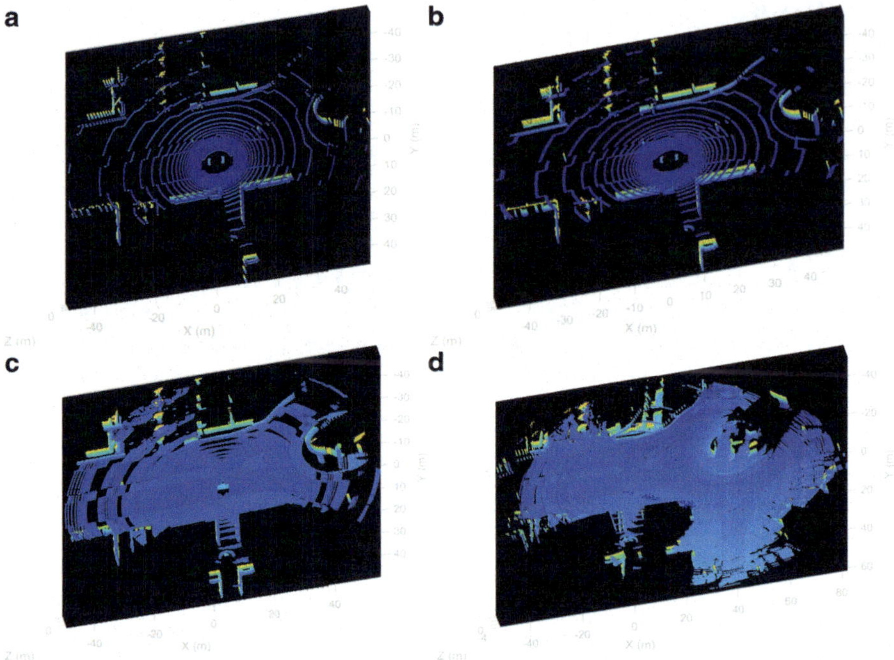

Fig. 8 (**a**) Two initial stitching point cloud array. (**b**) 10-point cloud stitching array. (**c**) 50-point cloud stitching array. (**d**) Final 300-point cloud stitching array

5 Conclusion and Future Works

From the results, the maps data can be generated using the ICP algorithm. The mapped environment is useful for further processing algorithm such as classification and object detection algorithm as all the point cloud frame has been combined into one single useful map. However, the mapping can be further improved by using additional sensor fusion technique with another sensor inside the vehicle such as IMU, odometer and GPS. It can be seen from the ICP algorithm that the larger dataset may have overlapping error if such data are not properly calibrated and tuned.

Acknowledgements The author would like to thank Ministry of Higher Education Malaysia (KPT) and Universiti Malaysia Pahang (www.ump.edu.my) for financial supports given under FRGS/ 1/2018/TK08/UMP/02/1, RDU1903139. Not forgotten to Universiti Malaysia Pahang for providing a flagship grant with vote number RDU192201 for the development of the autonomous shuttle prototype inside the campus. Special thanks to all Autonomous Vehicle Laboratory members, research assistants for the data logging and fabrication process.

References

1. K. Bimbraw, Autonomous cars: Past, present and future. 12th Int. Conf. Informatics Control. Autom. Robot. **01**, 191–198 (2015)
2. M. Salazar, F. Rossi, M. Schiffer, C.H. Onder, M. Pavone, On the interaction between autonomous mobility-on-demand and public transportation systems, in *2018 21st International Conference on Intelligent Transportation Systems (ITSC)*, (2018), pp. 2262–2269. https://doi.org/10.1109/ITSC.2018.8569381.
3. U.Z.A. Hamid, S.Z. Ishak, F. Imaduddin, Current landscape of the automotive field in the ASEAN region: Case study of Singapore, Malaysia and Indonesia – a brief overview. Asean J. Automot. Technol. **1**(1), 21–28 (2019)
4. M.A. Zakaria, H. Zamzuri, R. Mamat, S.A. Mazlan, A path tracking algorithm using future prediction control with spike detection for an autonomous vehicle robot. *Int. J. Adv. Robot. Syst.***10** (2013). https://doi.org/10.5772/56658
5. K.A. Zulkepli, H. Zamzuri, M.A.A. Rahman, W.J. Yahya, M. Aizzat, M.Z.A. Zakaria, F.R.A Zakuan, N.H. Faezaa, I-drive: Modular system architecture and hardware configuration for an intelligent vehicle research platform. ARPN J Eng App Sci **12**, 4259–4264 (2017)
6. L.C. Básaca-Preciado et al., Intelligent transportation scheme for autonomous vehicle in smart campus. *Proc. IECON 2018 – 44th Annu. Conf. IEEE Ind. Electron. Soc.*, 3193–3199 (2018). https://doi.org/10.1109/IECON.2018.8592824
7. P. Marin-Plaza, A. Hussein, D. Martin, A. de la Escalera, iCab use case for ROS-based architecture. Robot. Auton. Syst. **118**, 251–262 (2019). https://doi.org/10.1016/j.robot.2019.04.008.
8. K. Baarath, M.A. Zakaria, M.H. Bin Peeie, U.Z.A. Hamid, A.F.A. Nasir, An investigation on the effect of lateral motion on normal forces acting on each tires for nonholonomic electric vehicle: Experimental results validation, in *The IAVSD International Symposium on Dynamics of Vehicles on Roads and Tracks*, (2019), pp. 1643–1650
9. W.K. Chew, M.A. Zakaria, Rover Car Outdoor Localization for Navigation Tracking Using Differential Global Positioning System Estimation. *Symp Intell Manuf Mechatron*, 535–556 (2019)
10. R. Miyamoto et al., Vision-based road-following using results of semantic segmentation for autonomous navigation. *IEEE Int. Conf. Consum. Electron. – Berlin, ICCE-Berlin***2019**, 174–179 (2019). https://doi.org/10.1109/ICCE-Berlin47944.2019.8966198
11. D. Droeschel, M. Schwarz, S. Behnke, Continuous mapping and localization for autonomous navigation in rough terrain using a 3D laser scanner. Robot. Auton. Syst. **88**, 104–115 (2017). https://doi.org/10.1016/j.robot.2016.10.017
12. T. Wan et al., RGB-D point cloud registration via infrared and color camera. Multimed. Tools Appl. **78**(23), 33223–33246 (2019). https://doi.org/10.1007/s11042-019-7159-6
13. S.A.S. Mohamed, M.H. Haghbayan, T. Westerlund, J. Heikkonen, H. Tenhunen, J. Plosila, A survey on Odometry for autonomous navigation systems. IEEE Access **7**, 97466–97486 (2019). https://doi.org/10.1109/ACCESS.2019.2929133
14. X. Zhang, C. Glennie, A. Kusari, LiDAR using a weighted anisotropic iterative closest point algorithm. IEEE J Sel Top Appl Earth Obs Remote Sens **8**(7), 3338–3346 (2015)
15. F.A. Donoso, K.J. Austin, P.R. McAree, How do ICP variants perform when used for scan matching terrain point clouds? Robot. Auton. Syst. **87**, 147–161 (2017). https://doi.org/10.1016/j.robot.2016.10.011
16. H. Liu, T. Liu, Y. Li, M. Xi, T. Li, Y. Wang, Point cloud registration based on MCMC-SA ICP algorithm. IEEE Access **7**, 73637–73648 (2019). https://doi.org/10.1109/ACCESS.2019.2919989
17. R. Marani, V. Renò, M. Nitti, T. D'Orazio, E. Stella, A modified iterative closest point algorithm for 3D point cloud registration. Comput. Civ. Infrastruct. Eng. **31**(7), 515–534 (2016). https://doi.org/10.1111/mice.12184
18. A. Dosovitskiy, G. Ros, F. Codevilla, A. Lopez, V. Koltun, CARLA: An open urban driving simulator. *Proc 1st Annu Conf Robot Learn*, 1–16 (2017)

19. C. Crick, G. Jay, S. Osentoski, B. Pitzer, O.C. Jenkins, Rosbridge: Ros for non-ros users, in *Robotics Research*, (Springer, 2017), pp. 493–504
20. D. Chetverikov, D. Svirko, D. Stepanov, P. Krsek, The trimmed iterative closest point algorithm. *Proc. – Int. Conf. Pattern Recognit.* **16**(3), 545–548 (2002). https://doi.org/10.1109/icpr.2002.1047997.
21. Q. Tian, Y.F. Gao, G.L. Li, J.X. Song, A novel global relocalization method based on hierarchical registration of 3D point cloud map for mobile robot. *2019 5th Int. Conf. Control. Autom. Robot. ICCAR 2019*, 68–73 (2019). https://doi.org/10.1109/ICCAR.2019.8813720
22. H. Sobreira et al., Map-matching algorithms for robot self-localization: A comparison between perfect match, iterative closest point and normal distributions transform. J. Intell. Robot. Syst. Theory Appl. **93**(3–4), 533–546 (2019). https://doi.org/10.1007/s10846-017-0765-5.
23. P. Kamencay, M. Sinko, R. Hudec, M. Benco, R. Radil, Improved feature point algorithm for 3D point cloud registration. *2019 42nd Int. Conf. Telecommun. Signal Process. TSP 2019*, 517–520 (2019). https://doi.org/10.1109/TSP.2019.8769057

The Deep Learning Method for Image Segmentation to Improve the Efficiency of Data Processing Without Compromising the Accuracy of an Autonomous Driving Country-Road Pilot System After Image Classification

Kathrin Kind-Trueller, Maria Psarrou, and John Sapsford

1 Introduction

1.1 Statement of the Problem and Its Context

Machine learning (ML) is likely the most promising subfield of research on artificial intelligence [1]. During data analysis, ML algorithms learn from the input data and make predictions with a large data size [2]. In terms of image size processing for driving, current driving assistance systems require substantial throughput and storage in electronic control units 'application-specific integrated circuits (ASIC), image segmentation with high-resolution quality and accuracy can resolve the cost-issue currently making ML expensive in the automotive industry among other subjects. Furthermore, ML methods for image classification and image size processing for these systems utilise currently high-resolution and large images. The necessary pre-developed intelligent algorithms that can reduce the data size to be processed without jeopardising the quality, and that are needed as inputs for processing a task on the system, may not have been thoroughly validated. These results, therefore, need to be interpreted with caution.

K. Kind-Trueller (✉)
University of Hertfordshire, Hatfield, UK

M. Psarrou
Computational Neuroscience, University of Hertfordshire, Hatfield, UK

J. Sapsford
Computational Science, University of Hertfordshire, Hatfield, UK

U. Z. A. Hamid, F. Al-Turjman (eds.), *Towards Connected and Autonomous Vehicle Highways*, EAI/Springer Innovations in Communication and Computing,
https://doi.org/10.1007/978-3-030-66042-0_12

1.2 Aim and Objectives

1. To investigate the effects of individual vehicle movements (left/right) and make the classification algorithm intuitively understandable through visualisation; the first step was to use different machine learning algorithms [3] to train and analyse the weights of the learned models, followed by seeking a way to prepare this information so that the relation between the model of the ML algorithm (from data and pre-processed image segmentation) and the original image could be determined via a visualisation that ensured that the features in the original image were coherent and in accordance with the visualisation while showing a direct relationship among the mathematical attributes of the respective features of a correlating image.
2. To identify the most powerful areas of the image for the classification result, even though their resolution has been reduced and can still be used for training (road and road signs are detected in different light conditions); to achieve this goal, a suitable visualisation form had to be found; as a representation, a heatmap was made, placed over the original image around the main image areas to identify how much discrepancy exists compared to the original image, so that the learning algorithm could still be used for object classification.

2 Background

Some research results referenced in this work are based substantially on the results of the research on biological nervous systems and our nervous systems [4]. The biological nervous system supplies neurons and synapses as well as neural processes for our image classification process and object recognition, forming structural and functional templates for the construction of artificial neural networks. One of the areas of application particularly influenced by the innovations of the development of artificial neural networks (ANN) is image recognition [5]. It was first introduced in this section as the basics of image processing from the evaluation of individual pixels to simple and subsequently more complex structures. It depicts an example of the operation of filters in image processing using so-called convolutions [6, 7], leading to the introduction of the construction of neural networks, the use of different learning methods for these networks as well as the elements of deep learning in ANN. At the end of this chapter, a simple validation concept illustrates the performance of image processing in neural networks.

2.1 Image Processing

Image processing involves an array of research and engineering areas in which image processing techniques are used. In particular, it has a rapid development in image sensor and computer technology with continuously increasing storage capacities and computing power; miniaturisation as well as providing cost-effective computer services on the internet and the widespread use of camera systems, for example, in advanced driving assistance systems, are commonly used.

Even today, computers are widely used by other machines for the evaluation and display of images and image content, such as electronic control units in passenger cars [8]. Methods such as pattern recognition enable these machines—in addition to exact inputs—to be dynamic and use less concrete signal processing to segment and classify the information acquired on making a choice regarding how to further process and react to a driving activity to a certain extent. In order to continue the performance of the machines in image processing, improving and developing new applications can tirelessly attempt to bring out the outstanding orientation skills, people and skills to reach, simulate and reproduce objects by recording and processing visual signals to simulate human cognition. They can be utilised for vehicle positioning and navigation in a complex vehicle environment.

In addition to other senses, humans and animals mainly receive signals in the form of light stimuli via photoreceptors in the eye as individual pixels, which are processed in the visual cortex [9]. Correspondingly, the pixels of seeing humans and animals are used in computer science raster graphics, which are supplied to the image processing units. They provide information on colour space and colour depth and are arranged in the form of pixels grid [10].

This pixel information can then be used as input for the light stimuli processed in the visual cortex of the human brain ANN, to imitate the functions of the brain. Studies in computer science, particularly artificial intelligence, include not only the recognition of predetermined patterns but that computers can feed on the information, independently learn patterns and apply recognition in the object [9]. The basics of acquiring these new skills, such as the convolution, are discussed in detail below.

2.2 Convolution

A convolution is an arrangement of pixels such as an edge, so the change of the intensities of neighbouring pixels according to a pattern is one of the most widely used methods for image processing [11] (Fig. 1).

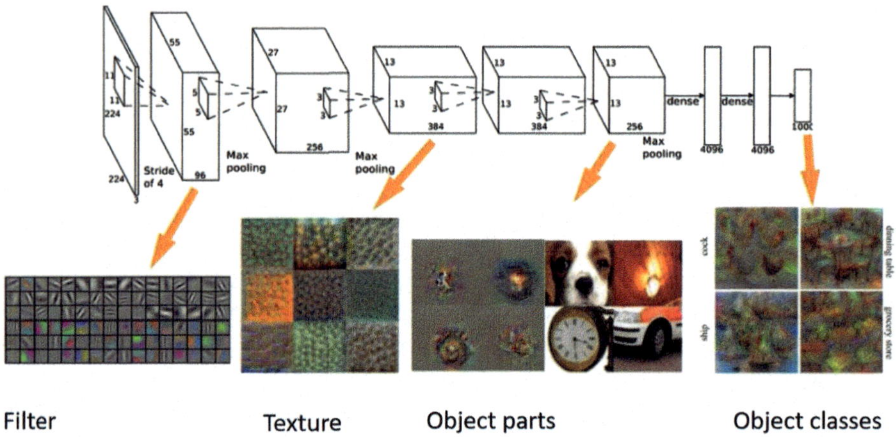

Fig. 1 A typical CNN architecture source: (https://www.mathworks.com/content/mathworks/www/en/discovery/convolutional-neural-)

2.3 *Neural Networks and Learning Algorithms*

The goal of machine learning is to estimate the functional relationship between input and output data [1]. Since our brain is the most efficient machine learning and cognition organ, it is obvious that the structures of a neural network aim to replicate the brain. The different approaches to make and learn ANN are presented below.

2.4 *Artificial Neural Networks*

Artificial neural networks consist of artificial neurons, which, in keeping with the biological model of the nerve cell, that is, weight inputs over one activation function to generate an output [4], simulate the natural behaviour of natural neural networks. An ANN includes an input layer, possibly further intermediate layers and an output layer. In addition, ANN has different structures: the 'single-layer feed-forward networks', 'multilayer feed-forward networks' and 'recurrent networks' [2].

The single-layer networks consist of an input layer and an output layer, passing the outputs only in the processing direction ('feed-forward'). Furthermore, multilayer networks and their 'multilayer perceptrons' (MLP) consist of additional layers along with input and output layers, the 'hidden layers' and feed-forward-controlled as well as recurrent networks. Recurrent networks possess, in contrast to the feed-forward-controlled networks, back-aligned edges, so the outputs of one step are again in front of inputs for neuron layers or within the individual layers to serve and thus increase the dynamics of the network [2].

2.5 Deep Learning

There are several methods for learning in neural networks, for example, supervised, empowering (reinforcement) and unsupervised [12]. The basis for the classification of this work is supervised learning.

2.6 Deep Convolutional Networks

Deep learning is a machine learning approach to gain knowledge from experience and solve problems with a hierarchy of understanding solution concepts, which are defined by more straightforward partial solutions [13]. Through this approach, the need for a definition of any knowledge is necessary for the computer to solve a problem the programmer may avoid. The hierarchy of solution concepts makes it possible for the computer to present complex solutions from simpler ones to form approaches. A graph that maps these concepts together consists of many layers and is, therefore, 'deep'. Therefore, we speak of this artificial intelligence approach to deep learning. The following sections explain the techniques of deep learning and the procedure of image recognition in deep networks. Possible weaknesses, as well as their potential solutions, are shown, and a proof of concept for image recognition is carried out [14].

2.7 Image Recognition with Convolutional Neural Networks

The filters for convolutions are applied to the input, and the output is computed to create local links in which each detected input area is networked to a neuron in the output. These connections and the process of image recognition with CNN are shown below.

2.8 Overfitting

When training, neural networks with a vast number of parameters and associated high complexity of the networks can quickly result in the problem of over- or underfitting, which can arise by both a high bias and a high variance [15, 16]. A high bias solves underfitting and a high variance focuses on low bias overfitting.

In the regression, z, for example, given data pairs (x1, y1), (x2, y2), ..., (xn, yn), a function f (x) can be found, which is well suited for mapping to the underlying data, that is, the Error | f (xi) -yi | for a pair of examples (xi, yi) is very small despite the regression of the classification.

If a graph was to be plotted, the graph could perfectly be mapped to the values since every polynomial with a degree of at least n – 1 on a record of size n can be mapped, however, is very erratic and does not follow the course of the pattern of values on the left side. This can be called overfitting since the net has trained too much for any value to hold true. This feature has a low bias but a high variance [7]. Therefore, in the future, there could be a prediction regarding the relationship between x and y, and it is unlikely that this function will yield an accurate result, especially if $x < 1$, the first degree polynomial on the left. The page has a high bias because the model is inflexible. Moreover, independent of how strongly the pattern of the data is curved can always only form straight lines that could be assumed underfitting. The middle polynomial represents a good fit on the data, and no over- or underfitting is visible. Overfitting is usually a significant problem in neural network training and is much more common than underfitting. When the model is responded by overfitting, it is too strong for noise, such as other objects that are next to each other along with the object recognised in the image instead of the provided pattern to train [16]. The more complex the chosen model, the lower is error rate it gets of the training set, and it reaches zero with high complexity [10].

In order to prevent overfitting, different methods were used. The simplest method was to provide additional available training data. If no other data were available, the complexity of the nets was reduced, for example, by reducing the number of hidden layers or using Dropout [6].

2.9 Proof of the Feasibility of Artificial Neural Networks

Before the practical implementation of a solution, the strategies described above are intended with the help of selected, current frameworks for software development in KNN and based on a simple prototype to create a proof of concept. This prototype uses the MNIST dataset as the basic data, consisting of a collection of 60,000 grayscale images with handwritten numerals used by the United States Federal Statistical Office staff and comes from American high school students [17]. They were brought to a uniform size of 28×28 pixels from larger black and white images and centred. This dataset was used to evaluate many new models for the training used by ANN, as due to its size, normalisation of the images is ideally suited to the correctness of a model to check and fine-tune for better recognition performance and lower error rate before training the actual image data of interest with new models.

The development of the handwritten digit recognition prototype used here takes place in the Python programming language, alongside C++ TensorFlow could also be used. TensorFlow is an open-source library for numerical calculations using data flow graphs [1].

In order to implement efficient calculations, usually—as in the present work—Python was applied; other libraries such as NumPy were also used, which require computationally intensive operations such as matrix multiplication in order to reduce the algorithm size by using other, more powerful programming languages such as C++ while frequently changing back to Python, which, unfortunately, may have created a lot of overhead. Usually, code size is especially noticeable on computing GPUs (graphics processing units) and it may be a source causing data transfer speed issues. In order to fix this problem, TensorFlow offers the ability to graph with interacting operations executed outside of Python, rather than calculating single compute-intensive operations in Python.

3 Discussion of the Methods and Methodology Used to Explore or Address the Issue, Selection of Approach, Selection of Techniques and Selection of Evaluation Approach

3.1 Methodologies Introduction

There are several frameworks for neural learning from early program libraries such as OpenCV since 2000 to further frameworks such as Torch, Theano, Caffe, Neon and TensorFlow. While Theano, Neon and TensorFlow are based on Python, Caffe is for C ++ and Torch the lesser-used scripting language is employed in Lua. Regarding Theano, all frameworks are prepared to use several GPUs for the calculation. The thematic introduction took place via a Udacity course on deep learning with TensorFlow. In this course, the author got into the subject of the programming language Python, with which the mode of operation of deep learning could be represented memorably. Due to the excellent documentation and provided source code, models with TensorFlow and Python for the practical implementation of the concepts of deep learning were selected in this work.

Additional frameworks for neural learning include new program libraries such as OpenCV from 2000 to Torch, Theano, Caffe, Neon and TensorFlow. On Theano, all frameworks are prepared to use several GPUs for the calculation. In addition to the listed contexts, the IBM Machine Learning Stack is still under development (Fig. 2).

3.2 Deep Learning Methods for Image Classification

In order to extend this project, the target is to investigate the optimisation methods to improve the efficiency of data processing without compromising the accuracy of the driving assistance system input—even though the algorithm reduces the size of data—while still delivering qualitative images.

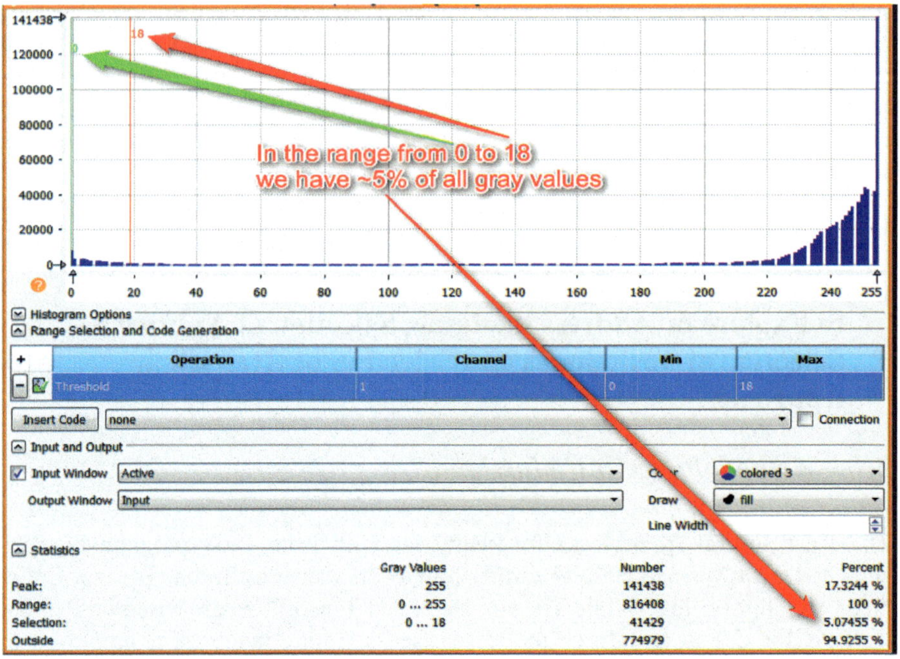

Fig. 2 Min max gray principle

3.3 Image Segmentation

An algorithm finds a (real) object in the image. In image processing terms, the intelligent algorithm determines a region (i.e. a collection of pixels) within an image that looks somehow special or different, for example, causing:

(a) Darkness or brightness
(b) Enclosed by edges
(c) Texture

These characteristics warrant a classification of the results based on local grey value statistics.

In most cases, the algorithm starts the segmentation based on darkness or brightness.

- Simple operator algorithm
- threshold.hdev
- The algorithm was careful about fixed thresholds
- threshold2.hdev

3.4 Application of the Methods

In the case of visual concept detection, the raw data are digital images, basically colour information (mostly in the RGB colour model) for each pixel of the picture. In order to sum up this information, many papers use the *Bag of Words* model (BoW model) borrowed from the text classification [18] training methodology: analogous to visual concept detection. The goal of text classification is to create a classificatory using sorting given text into one or more categories, which are often the content or used to characterise the theme of the text. The BoW model has a methodology for text representation [15] for this purpose; a text is represented as a frequency distribution over a given vocabulary. This vocabulary can, for example, be extracted from a large body of text by collecting all the words appearing in the texts, with the terminating words usually being eliminated. The stop words are very common in texts and usually have no relevance to the topic of the text. Additionally, articles, conjunctions and prepositions are typical examples of stop words. Moreover, representing a single text now counts for each word from the vocabulary, how often in the text it is examined and creates a histogram containing a class for each word.

This histogram represents the frequency distribution of each vocabulary for a single text and is called 'Bag of words', also referred to as 'there is no information containing more about the text structure and the positions of the words'. The metaphor in this name is that of a bag into which all the words of the text are individually thrown. Subsequently, all words can be taken out of the bag and counted, but all information about their original arrangement is lost. The result of the feature extraction is an n-element, in this case, vector, where n is the size of the vocabulary. In this work, BoW was attempted to be applied for *image classification* by treating *image* attributes as *words* [15].

The methodology for developing and evaluating an algorithm for classification (Adaboost) [19] and unsupervised (Kerner) (UKR) was followed. The scope and preparation of the sample for semi-supervised learning [1, 12] need to be further investigated. However, one limitation was observed. That is there is a limitation in the division of the sample data packages for cross-validation as well as training data and algorithms, in that they are publicly available along with the data, which means the data have not been quality assured, that is, quality criteria such as reliability and maintainability could not be guaranteed.

The focus in this work lies in the analysis of the usability of the training algorithms to improve the success rate of a classifier by adding it to the training

Fig. 3 System description and system architecture. (Source: Author)

data. The synthetic trajectories are derived from the training trials, underlying a sample using the unsupervised kernel presented in generated regression (UKR).

There are different methods of construction as well as views on the training data to be discussed. As such, with different views on the training data, statistical independence is to be attempted to maximise the base classifiers used in an ensemble that could have a direct influence on the success rates of an ensemble [20]. The parameterisation was also discussed regarding the deterministic classifier architectures used in this work (see Figs. 3 and 4), subsequently, on the training data and the classifier architectures which were designed.

As mentioned in the objectives and background, basic classifiers were evaluated for their suitability for semi-supervised learning in the ensemble of image segmentation. Later in a phase, the base classifiers were reduced to a manageable number. The formation of an ensemble with the basic classifiers was analysed, and different methods for the formation of the overall decision in classifying still have to be further explained in more detail in addition to its suitability for methodology and therefore the results should be interpreted with caution. In order to solve for the issue of improving data quality, some algorithms used for semi-supervised learning got basic quality assurance based on a validation of the training results (behaviour) as defined in software-based requirements and software architecture and design. First, some approaches to mitigating the uncertainty of a classification decision were presented, followed by two procedures for determining the criteria [4] for adopting new ones; if the segmented image data were of adequate quality, they could be used for training [5].

In order to calculate the probabilities for the numbers to be evaluated, among others the 'multinomial logistic regression' (softmax) was used to estimate group membership or a corresponding probability [11]. First, the class membership factors of the inputs (in classes 0–9 for each digit) were summed up. Second, these

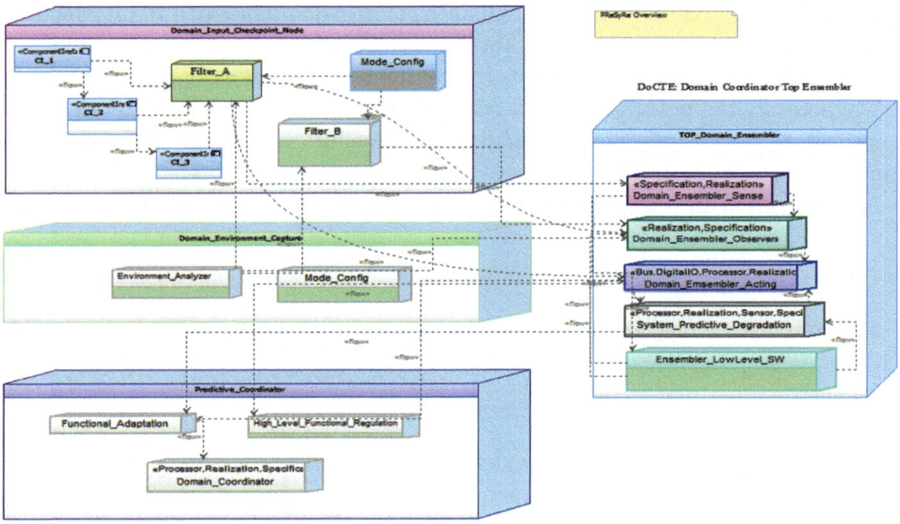

Fig. 4 Software system created for simulation environment

membership factors were converted into probabilities. To assign the membership of a picture to a class, the weighted sum of the pixel intensities was formed; if the weighting at a high pixel intensity was negative, it caused the image to not be divided into the appropriate class [17]. A positive weight indicates that the image belongs to the corresponding class.

In order to control influences more independent of the input, a bias should also be introduced. For the class membership θ of class with an input x, the following equation result was arrived at:

$$\theta i = Xj\ Wi, jxj + bi \tag{1}$$

This is Wi for the weights, bi for the bias of class i and j for the index of the sum of the pixels of the input image x. Using Softmax, therefore, the class affiliations θ in changed probabilities y are as follows:

$$y = \mathrm{softmax}\ (\theta) \tag{2}$$

Softmax provides an activation function that shapes the output into a probability distribution over ten classes. The Softmax function was as follows:

$$\mathrm{softmax}\ (x) = \mathrm{normalize}\ (\exp\ (x)) \tag{3}$$

In detail, one gets the following function:

$$\mathrm{softmax}\ (x)\ i = \exp\ (xi)\ P\, j\ \exp\ (xj) \tag{4}$$

The exponentiation in this context means that an additional unit of classification increases the weighting of a hypothesis multiplicatively. On the other hand, less classification results in a hypothesis recovering a portion of its original weight [5]. A hypothesis can never get zero or negative weight. Subsequently, the weights are normalised by Softmax, so they add up to one and thus form a valid probability distribution.

The result is thus a compact of the following equation:

$$y = \text{soft max } (W\,x + b) \tag{5}$$

The model for classifying the MNIST record can now be trained using Softmax. For evaluating the model, common cross-entropy was applied as a cost function as follows:

$$Hy\ 0\ (y) = \text{-}Xiy0\ ilog\ (yi) \tag{6}$$

In this function, y is the predicted probability distribution, and y0 is the actual distribution, according to the digit on the respective image. The cross-entropy then measures how inefficient the predictions of the model are. The evaluation of the learning outcomes is in addition to the textual edition of the accuracy and cost ('cost' or 'loss') after every training step, and the TensorBoard is included in TensorFlow. This visualises data previously recorded during execution and the graph of the neural network beyond that [14]. The implementation of the training and the evaluation of the results are described in a further chapter.

3.5 Selection of Convolutional Neural Networks

The proof of concept is intended to introduce deep convolutional neural networks involving Softmax and cross-entropy without the use of exclusive models. This chapter looks at well-known models and their specifics and those concerning their suitability for use in image recognition in dynamic systems. All models have the goal of recognising objects, for example, using larger datasets, methods to avoid overfitting and more efficient algorithms to improve. A vital approach that can significantly improve recognition is the use of CNN [10].

There are currently several particularly successful models of CNN [10] achieving outstanding results in the various international image recognition and classification competitions. Accordingly, two of the best-placed architectures of recent times have been evaluated for this work and are described below.

3.5.1 SuperVision AlexNet

The Krizhevsky et al. Featured Architecture AlexNet was the most successful in the visual recognition format 'ImageNet Large-Scale' during 2010 and achieved a top 5 error rate of 17.0% and a top 1 error rate of 37.5%. Even 2 years later, an improved architecture by Krizhevsky et al. was compared to a top 5 error rate of 15.3% with the second-placed model of 26.2%, occupying the first place [10]. Due to the favourable results and free availability of documentation and source code, this model was used to verify suitability for the image recognition and navigation of the robot vehicle selected. The authors did not follow the usual course of many earlier models, which have the functions for the output of the neurons. Instead, regarding the training time, the much faster f (x) = max (0, x) was selected. This approach is by the authors [10], previously by Nair and Hinton [21], called rectified linear units (ReLUs), which accelerates CNNs by many loops compared to the use of tanh.

3.5.2 GoogLeNet Inception

Another model that achieves outstanding results in the ILSVRC14 is Inception, presented by Szegedy et al. [22]. The architecture is characterised by the fact that with a significantly increased depth and width of the neural network, the computational effort remains constant [11].

The architectural decisions are based on the optimisation of the network on the Hebbian learning rule, which states that the more common neuron A is active simultaneously with neuron B, the sooner these two neurons interact with each other in the network [23].

In the meantime, the model has been further refined and available in version 3 as Inception V3 [22]. The authors Szegedy et al. in further adjustments promise a top 5 error rate of 3.5%, which is an improvement of 25% compared to the best-published results and as the ILSVRC2014 achieved error rate is almost halved.

3.5.3 Procurement and Preparation of Training Data

For the training of the proof of concept as well as the models to be tested, different datasets were loaded from different sources. The proof of concept is, as described above, to obtain the MNIST dataset from Yann LeCun's website. This dataset does not have to be cleared before the training runs and is immediately used. The MNIST [22] dataset is very often used to test new models. This record is cited by the University of Oxford as '17 Category Flower Dataset'. As the name suggests, it has pictures of 17 different flower types; each sample source code is also available on the internet for the models AlexNet and Inception to load and train this default record. The University of Oxford poses alongside the flower record with 102 classes; there are 28 more labelled records available.

The most comprehensive source for labelled records, however, is ImageNet6, whose datasets are also delivered annually in the annual report; ImageNet Large-Scale Visual Recognition Challenge serves as a database. As part of this work, a classic landscape was chosen as an object to be evaluated to test a record loaded by ImageNet [22].

3.5.4 Position Determination by Segmentation

Based on the results from the training runs on the use of the trained model through Inception for the image evaluation, it had to be determined how the evaluation for the position in the image can be used. Accordingly, to determine the position of the vehicle and to control the navigation of the robot, the picture was divided into segments. Each of these segments was then used for evaluation and contained the highest probability of detection. The segments with relatively high detection probability were grouped into middle, left and right. A good recognition rate was used in the tests with segmentation in 12 image parts. These image parts were correspondingly 160 × 160 pixels in size at a camera resolution of 640 × 480.

3.5.5 The Country Road Autonomous Driving System

System Description and System Architecture

System of systems control

System-based cognitive deep learning predictive software-system framework with automotive applications is based on a system of systems structure, in which a robot vehicle depends not only on hardware mechanic, hardware electronics and software within its frame for positioning, location and motion, but required external aid for plausibilisation of a driving path or driving situation from an external source, be high-definition map data or precise satellite navigation. Figure 3 shows the virtual system architecture applied in as part of the quality assurance plausibilisation of the algorithm. If the vehicle left the intended lane centre unintended for $t > 1S$ with a lane drift $d > 35$ cm, it was inferred, within the simulation environment, that the image input was not adequate as a source of vehicle control. These data must be interpreted with caution because the simulation environment did not consider neither the possible friction within the vehicle against the road nor the temperature model of the interactive cross-domain systems. A brief description of the main system components can be found below.

3.6 Sensors

Traditionally, a RADAR is used for long-range detections, while cameras are used for rich sensory inputs. When it comes to sensor configurations for Software Domain Controllers, which is usually a type of GPU, it may be noted that in these processes for value plausibilisation with more than one source of environment recognition form, that is, LiDAR (Light Detection and Ranging), the RADAR's data, could be given what may be known as an *active flag* in which the sensors' data might be marked as having higher reliability; that is, the RADARs may sense the environment in real time as active sensors with a possible correct format for the true driving situation based on their measure principle of transmissions of energy.

Cameras, on the other hand, are *passive sensors*; they can only sense the environment based on energy (in this case, photons) already in the scene.

These sensor details have serious repercussions concerning the types of algorithms we end up using to analyse these data. Computer vision has many powerful tools, but part of good design may have to do not only with knowing what to do but also what not to do when we use sensor data from multiple sources.

Now, computer vision is a very powerful tool, but it should not necessarily associate with *only* camera images. It is possible to also construct LiDAR images with a LiDAR sensor, thereby giving one measured depth alongside classified pixels.

3.7 LiDAR Sensor Data

LiDAR stands for Light Detection and Ranging, a type of sensor that uses light (a laser) to measure the distance between itself and objects that reflect light by sending a series of laser pulses out and measuring the time it takes for an object to reflect that light to the sensor; the longer the reflection takes, the farther an object is from the sensor. In this sense, LiDAR provides *spatially coherent* data and can be used to create a visual world representation.

Below, one can see the output of a LiDAR that Voyage uses; it sends out pulses of light and detects the car's surroundings.

3.8 Point Clouds

Since LiDAR uses laser light, which sends out a thin light, the data it collect end up being many single points, also known as *point clouds*. These point clouds can tell us a lot about an object, such as its shape and surface texture. By clustering points and analysing them, these data provide enough information to classify an object or track a cluster of points over time.

Below, one can see the result of a classification algorithm performed on these points. The red are pedestrians, and green indicates other cars.

3.9 Visual World

LiDAR data provide enough spatial information to create a visual world representation, and industry computer vision techniques can be used to classify objects using not only camera images but also point clouds and other types of spatially coherent data.

3.9.1 Computer Vision in Industry

To help us learn about computer vision techniques and applications, we have with us an industry expert, Tarin Ziyaee, the co-founder and Chief Technology Officer of Voyage, a self-driving car manufacturing company.

At Voyage, computer vision is used in a myriad of applications, for example, for the recognition of what state a traffic light is in, the detection of lanes, and so on.

One peculiar characteristic feature of computer vision is that the techniques that one was learning about cannot only be used with camera images but also images created with *other sensors*. The techniques that one was learning were useful for any data that have *spatial coherency*, and spatially coherent data can think of as any data that *predictably vary over space*, for example, sound. If one hears the sound from a speaker close up, it means the sound was very loud, but the farther away one gets, the softer the sound becomes. So, the volume of sound can give one the spatial information.

In addition to cameras, self-driving cars use sensors like RADAR and LiDAR, which use sound waves and lasers to gather data about a car's surroundings.

3.9.2 Training the Convolutional Neural Networks

Once the models and datasets for the training were selected and prepared, these models could work on the hardware described trained, and the training results were evaluated. The training of networks took place in the order of the scale of the models. At first, the model of the proof of concept was performed and evaluated. This was followed by training and result evaluation of the models by Alex Krizhevsky [22] and GoogLeNet's Inception. These models initially used the previously mentioned standard datasets and then trained with their own pictures.

3.9.3 Training and Optimisation

After choosing the preferred CNN model for training, the following chapter deals with the learning process, which to ensure efficient performance cannot be undertaken on the CPU, but on a more powerful platform such as a GPU. Before the actual training, the training environment, as well as the training and preparation of validation data, was configured. After the end of each test run, the results were analysed and documented.

3.9.4 Why Do we Normalise the Pixel Values of an Image?

Image normalisation is used as a pre-processing step. This is necessary because originally pixel values lie in the range [0,255] which are multiplied and added with weights and biases multiple times; this might lead to very large computations (also referred to as gradient explosion). The images are normalised by dividing the pixel values by their maximum value to convert pixel values to the range [0,1] [17].

3.9.5 How Do we Read Multiple Images and Store them?

During training, generally, multiple images are read into the program and stored in the form of an image array. We read an image and store it in an array, and then read another image and pile it over the first image in the same array. In this way, multiple images can be read and stored them in the form of an array, one over the other [17]. We wanted our model to be generalised for the unseen data. For doing so, we created a validation set using train-test_split to create the validation set.

3.9.6 What Is Train-Test Split?

When training a model, it is important to monitor its performance as it is learning; for this, it may be excluding some part of the dataset to be used for generating predictions at every training iteration. This is known as the train-test split, where the train split is used for training, and the test set is used for performance evaluation [17].

Batch Size

The number of samples used at a training step is known as the batch size. For example, if we have training data that consist of 10,000 samples, but all of it cannot fit into the memory at once, we use a batch size of approximately 100.

So, for every epoch, there were 10,000/100 = 100 steps, and the model was trained in batches of 100 samples each instead of all the 10,000 samples at once. An epoch was complete only once the model had seen all the 10,000 samples.

3.9.7 Image Classification Pipeline

An image classifier is an algorithm that takes in an image as input and produces an output a label "class" that identifies the image. For example, a traffic sign classifier looks at different roads and can identify whether that road contains humans, cars, bikes and so on. It is the ability to distinguish and classify each image based on its contents.

Many types of classifiers are used to recognise specific objects or even behaviours, for instance, whether a person is walking or running. Notably, they all involve a similar series of steps:

1. First, the computer receives a visual *input* from an imaging device such as a camera. A common way of this being captured is in the form of an image or a sequence of images.
2. Each image is then sent through some *pre-processing* steps for standardisation. Common pre-processing steps include resizing an image, or rotating it, to change its shape or to transform the image from one colour to another, for instance, from colour to grayscale. Only after standardisation can each image be compared with the others to further analyse them in the same way.
3. Next, we *extract features*. Features help us define certain objects; they are usually information about object shape or colour. For example, some features that distinguish a car from a bicycle are that a car is usually a much larger and has four wheels instead of two. The shape and the number of wheels would be the distinguishing features of a car. We will discuss more features later in this lesson.
4. Finally, these features are fed into a *classification model*. This step looks at any features from the previous step and predicts whether, for instance, this image is that of a car or a pedestrian or a bike, and so on.

One was programming each of these classification steps manually so that one *really* understands each step.

3.9.8 Classification Techniques

One can see a complete classification pipeline from Fig. 5. For an image classifier, starting with some input images, computer vision techniques are used to process those images and extract features such as distinguishing colours or shapes in that image.

Then, a classifier looks at these features and outputs a class, which is a label that describes the image.

3.9.9 A Classifier Should Predict if Images with Similar Shapes or Colours Would Have the Same Class

We usually tell the classification model what to look for. For example, we are looking at a bunch of images, and we want to classify them into two classes: car

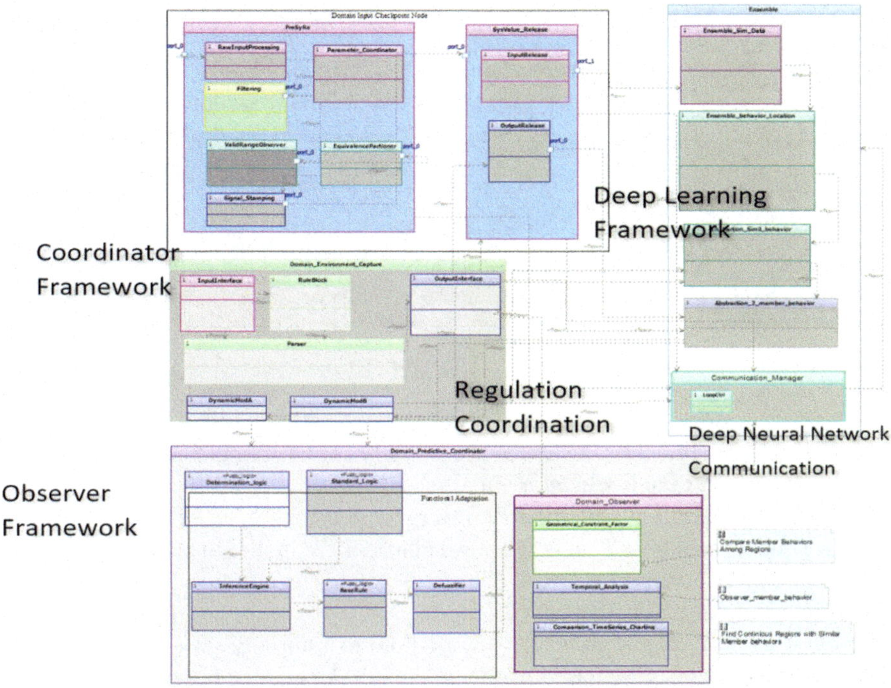

Fig. 5 Image classification pipeline and specific pipeline example applied to classify an image of a car

and not-car. To classify a car, we wrote a program that looks for the different parts of a car, such as wheels, lights and windows if those components were found, we classified an image as that of a car's. One must decide what traits are important to look for, so these results do not rule out the influence of other factors in further data interpretation methods.

3.9.10 Training a Model

Earlier, a simple example of a classification model was mentioned, that classified all images as either *car* or *not-car*. The mentioning in this study that using machine learning techniques, in a model, several examples of cars and not-cars, can be given as input until it learns to recognise them. But how exactly does this work? Like how humans learn, a model must learn from its success attempts and mistakes, and we often call this the training phase. At the beginning of a training phase, a classification model typically performs in an undesirable manner.

The model looks at an image, tries to classify it as a car or not-car and monitors the errors that it makes. For example, if a model mistakenly classifies a car as not-

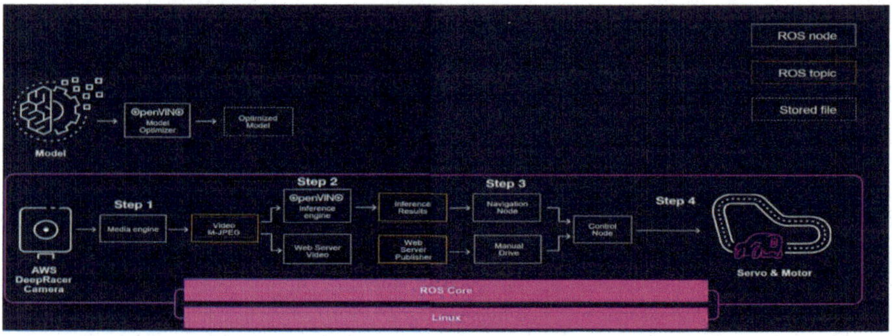

Fig. 6 Training model sample

car, it learns from this mistake, tweaks its classification parameters and improves its performance each time it sees more images.

After many iterations, the model converges on the right set of parameters, and the error-rate comes low. That is when we consider the model to be trained (Fig. 6).

A convolutional neural network adjusts its pattern recognition algorithm until it learns to accurately classify a set of images.

Now, this is a very high-level view of how to train any classification model. Moreover, the details vary based on the type of model and the training algorithm.

3.9.11 Colour Images

Colour images are interpreted as 3D cubes of values with width, height and depth.

The depth is the number of colours. Most colour images can be represented by combinations of only three colour values: red, green and blue. These are known as RGB images; for RGB images, the depth is 3.

It could be helpful to think of the depth as three, stacked 2D colour layers – one layer Red, one Green and one Blue. Together they create a complete colour image (Fig. 7).

Importance of Colour

In general, when one thinks of a classification challenge, such as identifying lane lines or cars or people, one can decide whether colour information and colour images are useful by thinking about one's own vision.

If the identification problem becomes easier for humans due to colour, it is highly possible that it becomes easier for an algorithm to identify colour images, too.

Colour Masking

We know about how colour images are represented numerically. So, I want to show how colour can be used in image analysis and transformation. We started by learning

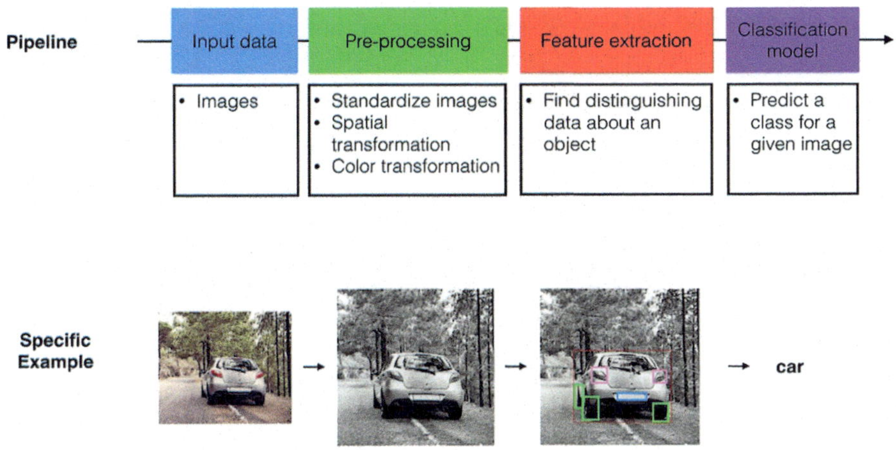

Fig. 7 RGB layers of a car image

Fig. 8 Car on a green screen background

how to use information about the colours in an image to isolate an area in an image. This is easiest to understand with the help of an example.

It was selected as an area of interest using a colour threshold. Colour thresholds are used in several applications, including extensively in computer graphics and video. A *green screen* is commonly used to layer two images or video streams based on identifying and replacing a large green area (Fig. 8).

So, how does it work?

Well, the first step was to isolate the green background, and then replace the green area with another image.

The code should read cv2.inRange().

Green Screen
A new threshold was created to define a colour mask. It was defined based on the lower and upper thresholds (bounds) of the colour that had to be isolated – in this case, green.

```
# Define our colour selection boundaries in RGB values.
```

```
lower_green = np.array([0,180,0]).
```

```
upper_green = np.array([100,255,100]).
```

The colour thresholds created an *image mask*.

```
# Define the masked area.
```

```
mask = cv2.inRange(image, lower_green, upper_green).
```

Then, to mask the image, I made an image called masked_image, a copy of our original image, to manipulate without changing the original image. B asking the part of the image overlapping with the part of the mask that is white or *not* black (= 0) one can select the green screen area.

```
# Mask the image to let the car show through.
```

```
masked_image = np.copy(image).
```

Fig. 9 Masked background

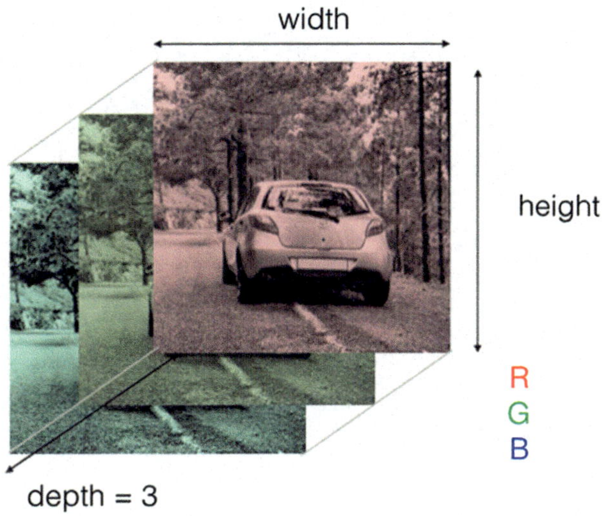

Fig. 10 Green screen replaced with an image of the sky

```
masked_image[mask. = 0] = [0, 0, 0].
```

Then, when we displayed our masked_image, we could see that the image of the car is the only area that was, and the green screen background was gone (Fig. 9).

One can practice masking and add a background. One should get an image that looks similar to the flying car shown in Fig. 10.

Fig. 11 (Top image) A low-frequency sound wave; (bottom) a high-frequency sound wave

3.9.12 Convolution in Self-Driving Cars

Frequency in Images

We have an intuition of what frequency means when it comes to sound. High frequency is a high-pitched noise like a bird chirp or violin; low-frequency sounds are low pitched, like a deep voice or a bass drum. When talking about sound, frequency refers to how fast a sound wave oscillates; oscillations are usually measured in cycles/s Hz. High pitches are made by high-frequency waves. Examples of low and high-frequency sound waves are shown in Fig. 11. On the y-axis is the amplitude, which is a measure of sound pressure that corresponds to the perceived loudness of a sound, and on the x-axis is the time.

High and Low Frequency
Similarly, frequency in images is a *rate of change*. But, what does it mean for an image to change? Well, images change in space, and a high-frequency image is one in which the intensity changes a lot, and the level of brightness changes quickly from one pixel to the next. An image that is relatively uniform in brightness or changes very slowly can be said to be that of a low frequency. This is easiest to understand with an example.

In Fig. 10, we see parts of the sky and background that change gradually, which is considered a smooth, low-frequency pattern.

High-frequency components also correspond to the edges of objects in images, which can help us to classify those objects.

4 Evaluation of the Results

Accuracy
The accuracy of a classification model is found by comparing predicted and true labels. For any given image, if the predicted label matches the true_label, then this is a correctly classified image; if not, it is misclassified.

The accuracy is determined by the number of correctly classified images divided by the total number of images. We tested this classification model on new images; this is called a test set of data. The classifier failed, as the accuracy rating was set too low.

Test Data
The data seen, including the one used to help build a classifier, are called training data. The idea behind creating these two sets is to have one set that one can analyse and learn from (training), and one from which one can get a sense of how a classifier might work in real-time. One could imagine going through each image in the training set and creating a classifier that can classify all these training images correctly, but one would want to build a classifier that *recognises general patterns in data* that work in real-time.

So, one could use a new test set of data to see how a classification model might work in the real world and to determine the accuracy of the model.

Misclassified Images
In this, and most classification examples, there are a few misclassified images in the test set. To see how to improve, it is useful to look at these misclassified images; look at what they were mistakenly labelled as and where the model failed.

Problems Faced
A region-based segmentation methodology based on the pixel values within an image was used. First, the threshold segmentation was applied to an image. The threshold value was set to ensure that the pixel values were different from the objects

and the image's background versus in the upfront area. The objects in the foreground or objects from the background were classified by observing if the values (from pixels) were higher or lower than the given threshold. A division of the regions was achieved via a global threshold; then, multiple thresholds were defined via the form of a local threshold.

Challenges that Remain to Be Resolved
The region-based segmentation achieved so far might not be adequate to train an algorithm after the image has been resized. It is conjectured that cluster-based division allows, using data points, the classification of objects in an image into several groups. The data points in the same groups could be similar to other data points within relevant data group building clusters for areas within an image. Another thesis states that the k-means algorithm would be able to transform an image's objects as point clusters that could easily be understood by the given training algorithm, that is, during the second step of training the machine learning algorithm for vehicle recognition.

The actual algorithm [10] consists of three simple steps:

1. Initialisation: At the beginning, the clusters or their mean values were initialised. In the classical k-means algorithm, the initialisation is done with the random selection of K instances. Different initialisations can lead to different results.
2. Assigning the instances to the clusters: each cluster, CK, has an average, μK. In all instances, the distance between the respective instance and the K clusters, given by their mean value μK, could then be determined by means of a distance function. The instances are assigned to the cluster they resemble the most, that to which they have the least distance.
3. Recalculation of the parameters: after assigning the instances, the mean values of the clusters were recalculated. Subsequently, the assignment described in step 2 was repeated. This is done until no more changes occur in the assignments.

To prevent overfitting, different methods [6] were used. The simplest method was to provide additional training data. If no other data are available, the complexity of the nets then could reduce, such as by reducing the number of hidden layers or by the application of Dropout.

Clustering algorithms and other point mining methods cannot match the pure, unstructured point clusters. Therefore, during the pre-processing phase, another representation of a pure cluster could be chosen. The cluster points are converted into feature vectors. The individual features can then be weighted. A common form of vector space rendering is the bag-of-visual words presentation. This simply uses all the point clusters in a picture as characteristics, such as text. Thus, the size of the vector space is the totality of all different words from all texts. The binary weighting is probably the simplest weighting. For this reason, afterwards, a bag-of-visual-words clustering algorithm was used to process the images for recognition [24].

Using the Matplotlib library, diagrams were added while the cluster analysers were created. An orientation to the implementation could provide the examples from scikit-learn [25]. For the representation of clusters, the size of the matrix was minimised. The individual visualised entries were given a colour according to their cluster membership. If the number of entries was not too large, a diagram was also used to represent the silhouette score values. Silhouette score was already implemented in scikit-learn in the CoLab environment. The average value for each of the entries, as well as the value of each individual entry, was calculated and visualised. For these visualisations, the library Matplotlib was used.

At the same time, all these tasks were tracked daily since August 2nd via Trello for project management and the code was checked daily into GitHub. The SW architecture was drawn using an open-source UML software.

5 Discussion

5.1 Colour Spaces

A *colour space* is a specific organisation of colours; colour spaces provide a way to categorise colours and represent them in the form of digital images.

RGB is red-green-blue colour space. One can think of this as a 3D space, and in this case, a cube, where any colour can be represented by a 3D coordinate of R, G and B values. For example, the colour white's coordinates, 255, 255, 255, give the maximum values for red, green and blue (Fig. 12).

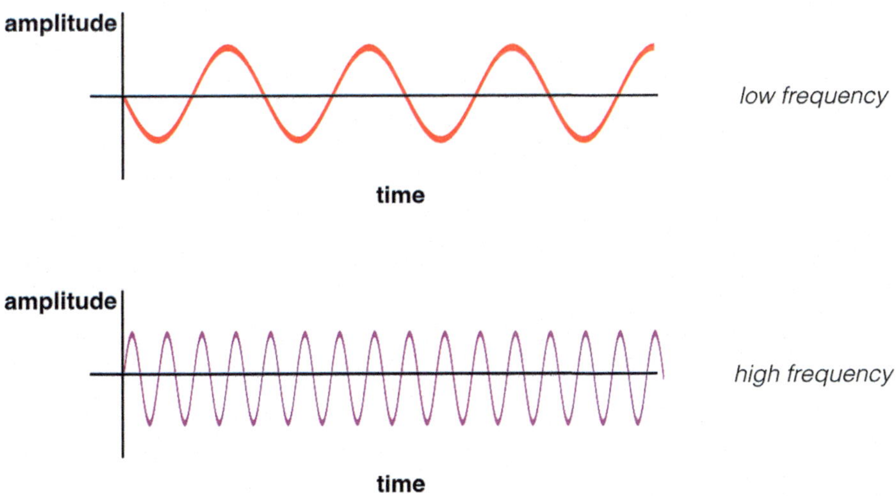

Fig. 12 RGB colour space

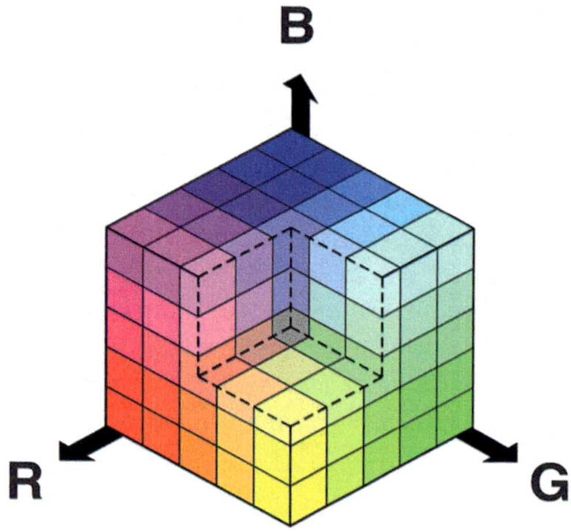

Fig. 13 The derivation of HLS and HSV colour spaces

There are many other ways to represent the colours in images just composed of red, green and blue values.

There is also the *HSV* colour space (hue, saturation and value) and *HLS* space (hue, lightness and saturation). To get some intuition about these colour spaces, one can generally think of *hue* as the value that represents a colour independent of any change in brightness. So, if one imagines a basic red paint colour, then add some white to it or some black to make that colour lighter or darker, but the underlying colour remains the same, so the hue for all these colours remains the same.

HSV and HLS

Lightness and *value* represent different ways to measure the relative lightness or darkness of colour: dark red has a similar hue but much lower value of lightness than light red. *Saturation* also plays a part in this; saturation is a measurement of colourfulness. So, as colours get lighter and closer to white, they have a lower saturation value, whereas colours that are most intense, like a bright primary colour (imagine a bright red, blue or yellow), have a high saturation value. One can get a better idea of these values by looking at the 3D colour spaces shown below.

Most of these different colour spaces were either inspired by the human vision system and/or developed for efficient use in television screen displays and computer graphics (Fig. 13).

6 Image Segmentation Results

6.1 *HSV Conversion*

In the code example, I used HSV space to help detect a green screen background under different lighting conditions. OpenCV provides a function hsv = cv2.cvt colour(im, cv2. COLOUR_RGB2HSV) that converts images from one colour space to another.

After this conversion, I plotted the individual colour channels; it was easy to see that the Hue channel remained constant under different lighting conditions (Fig. 14).

Next, it used the same masking code as before, though only a *Hue channel* colour mask.

6.2 *Image Classification Results*

6.2.1 Day and Night Classification

One might be aware of how to build a classifier. One knows how to analyse colour and brightness in a given image, and that skill alone can help one distinguish between different traffic signs.

We encountered a possible classification challenge. If in a driving situation the system of systems has to classify two types of images, one taken during the day and one at night when the sun has set, the system has to learn when to categorise these images into two classes: day or night (Fig. 15).

This is an important classification challenge for self-driving cars. These cars need to know what kind of conditions they are driving in so that they can safely navigate a road at any time of the day in order to recognise other vehicles and surrounding objects and determine whether it is dark or light outside.

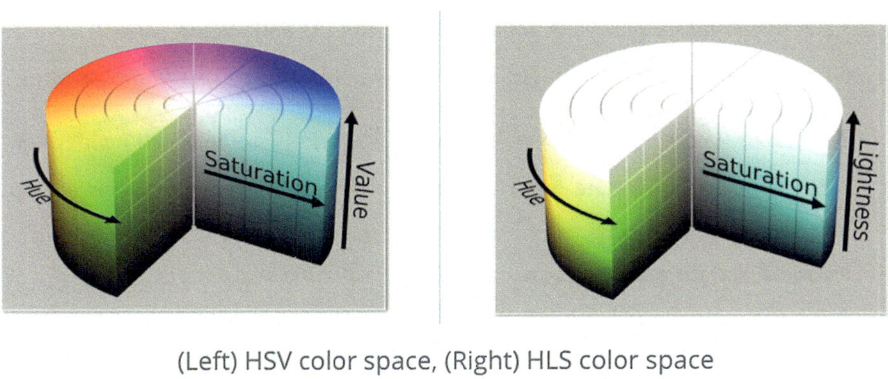

Fig. 14 H, S and V channels for the green screen car image

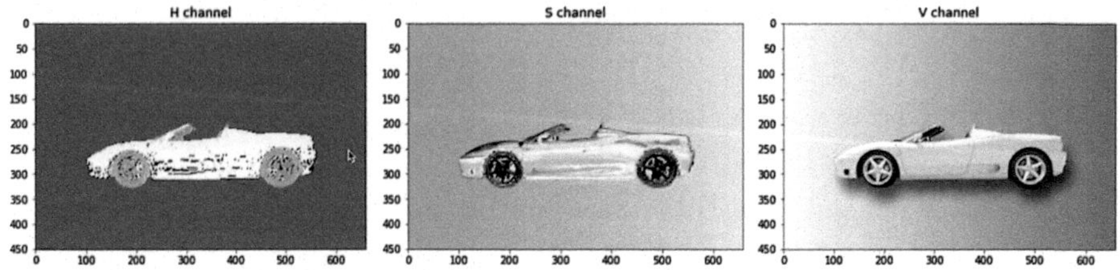

Fig. 15 Examples of day and night images of the same scene

We can walk through each classification step. But what does one think would be the first step in creating a classification model for day and night images? Before one can classify any set of images, one must look at them. *Visualising the image data that one is working with is the first step in identifying any patterns in image data and predictions about it.*

So, we first loaded data in this image and learned a bit about the images we were working with.

6.2.2 Day and Night Image Classifier

The classification dataset consists of 200 RGB colour images in two categories: day and night. For this experiment, the same number of day and night images, 100 images of each, were loaded, that is—a total of 200 images.

It was decided that a classifier with a target to measure accuracy for daily or nightly images would be built. Another target was that the classifier must rely on finding distinguishing features between the two types of images.

The AMOS dataset *(Archive of Many Outdoor Scenes) was used for this purpose.*

6.2.3 Training and Testing Data

The 200 day/night images were separated into training and testing datasets.

- Of these images, 60% were training images, for one to use while creating a classifier.
- The test images were 40%, which were used to test the accuracy of a classifier.

First, some variables were set to keep track of where our images are being stored:

image_dir_training: the directory where our training image data is stored

image_dir_test: the directory where our test image data is stored

Fig. 16 A misclassified image examples. The true_label is "day" and the predicted_label is "night"

6.2.4 Loading the Datasets

The first few lines of code loaded the training day/night images and stored all of them in a variable, IMAGE_LIST. This list contains the images and their associated label ("day" or "night").

For example, the first image-label pair in IMAGE_LIST can be accessed by index IMAGE_LIST[0][:].

6.2.5 Why Do we Need Labels?

One can tell if an image was clicked during night or day, but a computer cannot unless we tell it explicitly with a label, a situation that is particularly important when we are testing the accuracy of a classification model.

A classifier takes in an image as input and generates an output predicted_label that tells us the predicted class of that image. Now, when we load in data, we load in what is called the true_labels, which are the *correct* labels for the image.

In order to test the accuracy of a classification model, a comparison was made between the predicted and true labels. If the true and predicted labels matched, then it means I classified the image correctly. Sometimes the labels do not match, which means I misclassified an image (Fig. 16).

6.2.6 Accuracy

After looking at many images, the accuracy of a classifier is defined as the number of correctly classified images (for which the predicted_label matches the true label) divided by the total number of images. So, if we tried to classify 100 images total, and we correctly classified 81 of them, I would have 0.81 or 81% accuracy.

We can tell a computer to check the accuracy of a classifier only when we have these predicted and true labels to compare. One can also learn from any mistake the classifier makes.

6.2.7 Numerical Labels

It is good practice to use numerical labels instead of strings or categorical labels. They are easier to track and compare. So, for our day and night, binary class example, instead of "day" and "night" labels, we use the numerical labels: 0 for night and 1 for day.

Now that one is familiar with the day and night image data, and one knows what a label is and why we use them, one is ready for the next steps. We built a classification pipeline from start to end.

6.2.8 Average Brightness

Here are the steps we took to extract the average brightness of an image:

1. Convert the image to HSV colour space (the Value channel is an approximation for brightness).
2. Sum up all the values of the pixels in the Value channel.
3. Divide the brightness sum by the area of the image, which is just the width times the height.

This gave us one value: the average brightness or the average value of that image.

Next, one should make sure to look at a variety of day and night images and see if one could think of an average brightness value that could put the images into their respective classes.

The next step was to feed this data into a classifier. A classifier might be as simple as a conditional statement that checks if the average brightness is above some threshold, then this image is labelled as 1 (day), and if not, it is labelled as 0 (night) (Fig. 17).

One can choose to create more features that help distinguish these images from one another; we will soon learn about testing the accuracy of a model like this (Fig. 18).

The pre-processing required in a ConvNet is much lower as compared to other classification algorithms. While in primitive methods, filters are hand-engineered, and with enough training, ConvNets have the ability to learn these filters/characteristics.

The objective of the Convolution Operation is to *extract the high-level features* such as edges from the input image. ConvNets need not be limited to only one Convolutional Layer.

The fully connected layer learns a possibly non-linear function in that space.

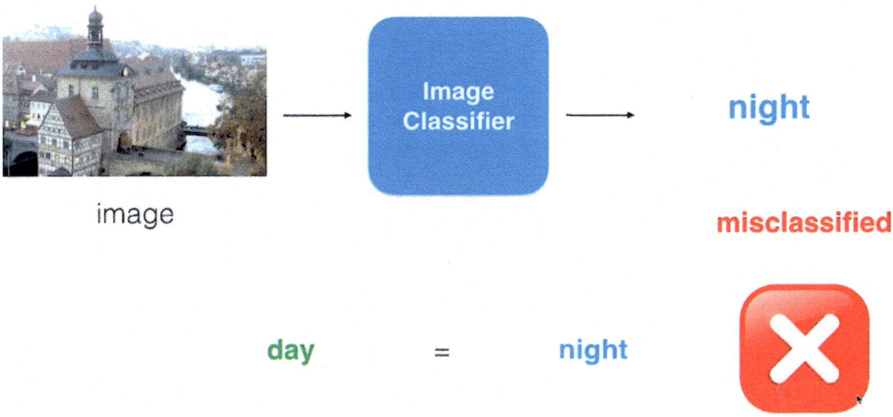

Fig. 17 Hazy output (blurry image as output)

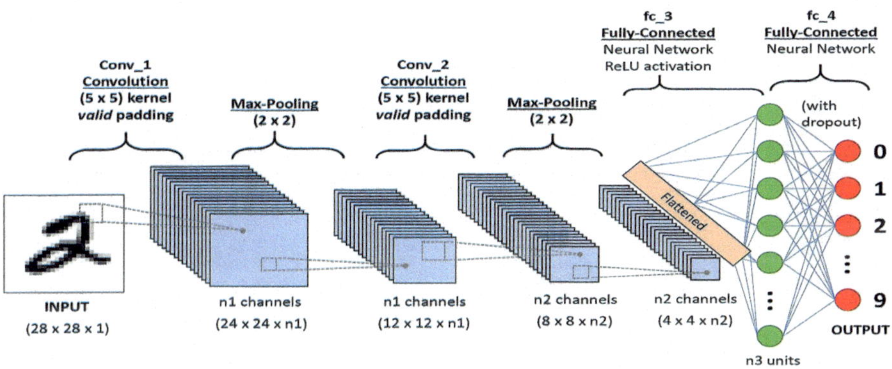

Fig. 18 Clear output (sharp image as output)

Now that we converted our input image into a suitable form for our Multi-Level Perceptron, we shall flatten the image into a column vector. The flattened output is fed to a feed-forward neural network and backpropagation applied to every iteration of training. Over a series of epochs, the model may be able to distinguish between dominating and certain low-level features in images and classify them using the *Softmax Classification* techniquc.

There are various architectures of CNNs available which have been key in building algorithms, which shall power A.I. as a whole in the future. Some of them have been listed below:

1. LeNet
2. AlexNet
3. VGGNet
4. GoogLeNet

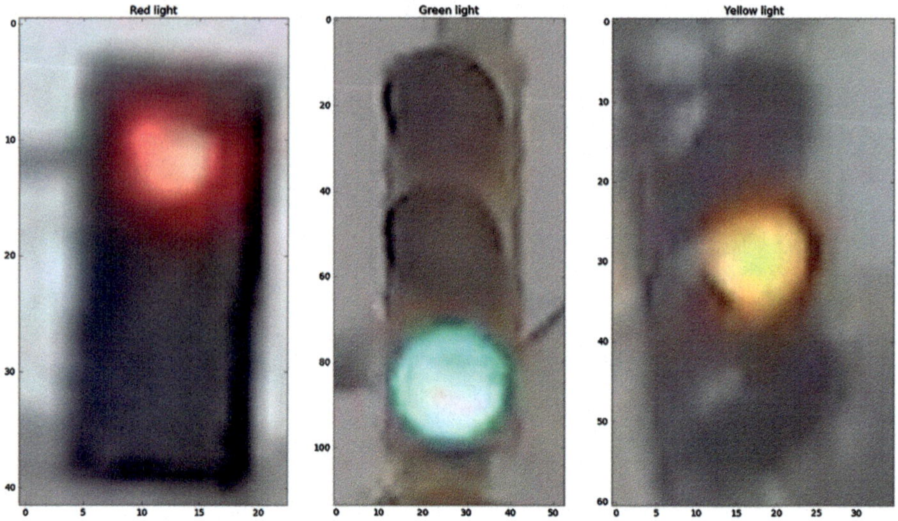

Fig. 19 Images from the dataset. Left to right: red, green and yellow traffic lights

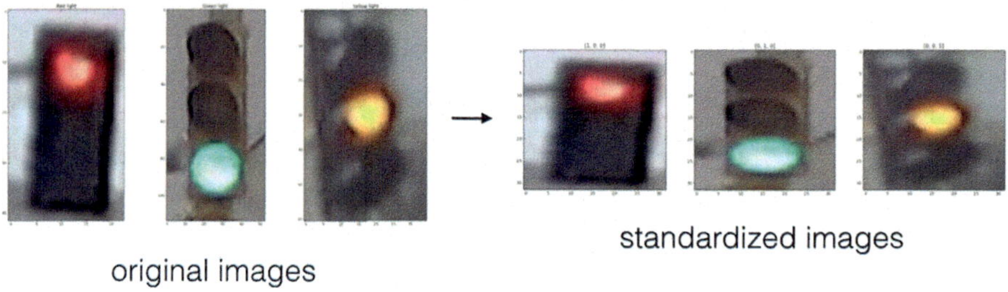

Fig. 20 Pre-processed, standardised images

5. ResNet
6. ZFNet

6.2.9 Traffic Light Classifier

For this task, one was to use knowledge of computer vision techniques to build a classifier for images of traffic lights. One was given a dataset of traffic light images in which one of three lights red, yellow or green – is illuminated (Figs. 19, 20 and 21).

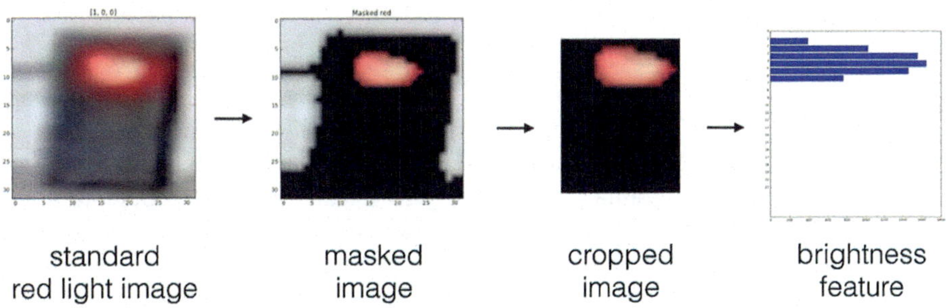

Fig. 21 An example of feature extraction steps

6.2.10 Classification Steps

In the Jupyter notebook, one was to pre-process these images, extract features that helped distinguish the different types of images and use those features to classify the traffic light images into three categories: red, yellow or green. The tasks were broken down into a few sections:

1. *Loading and visualising the data*: The first step in any classification task is to familiarise data; one needed to load in the images of traffic lights and visualise them.
2. *Pre-processing:* The input images and output labels need to standardise, that is, all the input should be of the same type of data and of the same size, and the output should be a numerical label. This way, one can analyse all the input images using the same procedures; anyone would know what output to expect when one eventually classifies a new image.
3. *Feature extraction*: Next, one extracted some features from each image that were used to distinguish and classify these images. This is where one has a lot of creativity; features should include 1D vectors or even single values that provide some information about an image that can help classify it as a red, yellow or green traffic light.
4. *Classification and visualising error*: Finally, one was to write a function that uses features to classify *any* traffic light image. This function was taken applied to an image and output a label was generated. One was also given the code to classify a test set of data, compare a predicted label with the true label and determine the accuracy of the classification model.
5. *Evaluate the model.* The classifier must be >90% accurate and never classify any red lights as green; it is likely that one needed to improve the accuracy of the classifier by changing existing features or adding new features. I would also encourage one to try to get as close to 100% accuracy as possible.

7 Conclusions

The aim of the work was to show that the methods of deep learning in Deep Convolutional Neural Networks used in image recognition can be used to give a robotic vehicle the ability, by means of independently learned patterns of objects, to create objects in its environment analogously to visually perceive people, to orient oneself around objects in space and to navigate autonomously. This goal could not be achieved despite the existing resources being successfully utilised. A simple, practical application cannot hide the unimagined possibilities of using new self-learning methods in artificial neural networks for new applications. In addition, all the models of the Deep Convolutional Neural in the study were examined. Even if, as shown in this work, the speed of learning in artificial neural networks gets affected due to the very large data needed to be processed, image datasets are still strongly correlated with the computing power and storage capacity of the graphics cards used in the computer, the undisputed effectiveness of the concepts and methods and their actual implementation is demonstrated.

8 Suggestions for Future Work

Due to the interdependence of computational power, memory capacity and time required for the training of neural networks, the difficulty was that the computer system used for the study often got overwhelmed due to low computing power and capacity to handle the huge data volume or run times. Nevertheless, it has succeeded by appropriate measures in helping accomplish the set goal in the time available.

The examination of the concepts underlying this work and models of deep learning in image recognition, whose success is essentially based on the use of complex mathematical concepts of linear algebra and statistical methods, were a challenge and impressive because of the convincing ideas. The resultant application with the robotic vehicle can be used for future applications certainly, and it must continue to expand, which continues to be of great interest.

References

1. T. Renelle, *Machine Learning Guide*. URL: http://ocdevel.com/podcasts/machine-learning/2 (2017)
2. M. Copeland, *Thes Difference Between A.I., Machine Learning, and Deep Learning? | NVIDIA Blog*. URL: https://blogs.nvidia.com/blog/2016/07/29/whats-difference-artificial- intelligence-machinelearning-deep-learning-ai/, online, seen on July 7th (2016)
3. A. Udacity, et al, *Udacity on Apollo project*, https://github.com/udacity/Udacity-Apollo, online viewed September 10th (2019)
4. S. Haykin, Network, neural: A comprehensive foundation, in *Neural Networks 2 (2004), Nr*, (2004)

5. R. Cipolla, S. Battiato, G.M. Farinella, *Computer Vision - Detection, Recognition and Reconstruction, Band 285 der Reihe Studies in Computational Intelligence* (Springer-Verlag, Berlin\Heidelberg, 2010)
6. N. Srivastava, G. Hinton, A. Krizhevsky, I. Sutskever, R. Salakhutdinov, *Dropout: A simple way to prevent neural networks from overfitting* (JMLR. org, 2014), pp. S1929–S1958
7. Wan, J., Wang, D., Hoi, S.C.H., Wu, P., Zhu, J., Zhang, Y. And Li, J., 2014. "Deep learning for content-based image retrieval: A comprehensive study", In Proceedings of the 22nd ACM International Conference on Multimedia (pp. 157–166).. ACM. Schmidhu r, J., 2015. Deep learning in neural networks: An overview. Neural networks, 61, pp. 85–117
8. Á. Arcos-García, J.A. Álvarez-García, L.M. Soria-Morillo, Evaluation of deep neural networks for traffic sign detection systems. Neurocomputing **316** (2018)
9. R. Szeliski, *Computer Vision – Algorithms and Applications* (London. Springer Verlag London Limited, 2011)
10. W. Burger, M. Burge, *Principles of Digital Image Processing – Fundamental Techniques* (Springer, London, 2009)
11. D. Liu et al., A review of computer vision segmentation algorithms. Course on computer vision, in *Spring 2012*, (University of Washington, Seattle, 2012)
12. Y. LeCun, Y. Bengio, G. Hinton, *Deep Learning* (Nature Publishing Group, 2015), pp. S436–S444
13. J. Wan, D. Wang, S.C.H. Hoi, P. Wu, J. Zhu, Y. Zhang, J. Li, 2014, Novem r. Deep learning for content-based image retrieval: A comprehensive study
14. In Proceedings of the 22nd ACM international conference on Multimedia (pp. 157–166). ACM. Schmidhuber, J., 2015. Deep learning in neural networks: An overview. Neural Netw., 61, pp. 85–117
15. Y. Zhang, X. Sun, H. Wang, K. Fu, High-resolution remote-sensing image classification via an approximate earth Mover's distance-based bag-of-features model. IEEE Geosci. Remote Sens. Lett. **10**(5), 1055–1059 (2013)
16. S. Haykin, Network, Neural: A comprehensive foundation, in *Neural Networks 2 (2004), Nr*, (2004)
17. Y. LeCun, Y. Bengio, G. Hinton, *Deep learning* (Nature Publishing Group, 2015), pp. S436–S444
18. Y. Zhang, R. Jin, Z.H. Zhou, Int. J. Mach. Learn. Cy r. **1**, 43 (2010). https://doi.org/10.1007/s13042-010-0001-0
19. João Gama at al., Machine Learning, in *ECML 2005: 16th European Conference on Machine Learning, Porto, Portugal, Octo r 3–7, 2005, Proceedings*
20. L. Rokach, *Pattern Classification Using Ensemble Methods. Series in Machine Perception and Artificial Intelligence* (World Scientific, Singapore, 2010)
21. V. Nair, G.E. Hinton, *Rectified Linear Units Improve Restricted Boltzmann Machines* (Department of Computer Science, University of Toronto, Toronto, 2010)
22. C. Szegedy, et al, *Going Deeper with Convolutions*. https://www.cs.unc.edu/~wliu/papers/GoogLeNet.pdf, online, viewed September 10th, 2019
23. Hsu, Chih-Wei; Chang, Chih-Chung; Lin, Chih-Jen U. A.: A Practical Guide to Support Vector Classification. 2003
24. R. Cipolla, S. Battiato, *Giovanni Maria Farinella: Computer Vision – Detection, Recognition and Reconstruction, Band 285 der Reihe Studies in Computational Intelligence* (Springer-Verlag, Berlin\Heidelberg, 2010)
25. T. Renelle, *Machine Learning Guide*. URL: http://ocdevel.com/podcasts/machine-learning/2 (2017)

Index

U. Z. A. Hamid, F. Al-Turjman (eds.), *Towards Connected and Autonomous Vehicle Highways*, EAI/Springer Innovations in Communication and Computing,
https://doi.org/10.1007/978-3-030-66042-0

B

Printed by Printforce, the Netherlands